Catering and Hospitality

NVQ Level 2

Food Preparation and Cooking

Student Guide
Cookery Units

Danny Stevenson, Rowland Foote, Malcolm Ware, Philip Coulthard, Tony Groves, Sarah Brazil, Bob Kenyon, David Klaasen, Harry Tallon

First published in 1993 by:
Stanley Thornes (Publishers) Ltd
Ellenborough House
Wellington Street
CHELTENHAM
Gloucestershire GL50 1YD
United Kingdom

The catalogue record for this book
is available from The British Library.
ISBN 0–7487–1604–1

Reprinted 1993 (twice)
Reprinted 1994 (twice)

The authors and publishers would like to thank Jim Mair, Bill Moorcroft, Joachim Schafheitle and Tony Groves for their help and advice. They would also like to thank the following for permission to reproduce photographs: Rowland Foote (Preparing and cooking dough products; Preparing, cooking and decorating cakes and biscuits; Preparing and cooking fresh pasta), Danny Stevenson (p. 76), Rank Holidays & Hotels Limited (pp. 133, 134, 136), Burton District Hospital (pp. 281) and Regethermic UK Ltd (pp. 272, 282, 284).

The complete material for Food Preparation and Cooking
consists of three books:

Student Guide: *Core Units*
Student Guide: *Cookery Units*
Tutor Resource Pack

Typeset by The New Leaf Book Company, Oxford.
Printed and bound in Great Britain at Scotprint, Musselburgh.

Contents

Planning your time

Introduction

While working in the kitchen, you need to use your time and energy in the most effective manner. In order to achieve this, you will need to plan every activity in advance, making the most of the limited time available to you. Once you have gained experience in food preparation you will be able to plan effectively automatically, but while training you will need to adopt a more strategic approach towards time planning.

The process

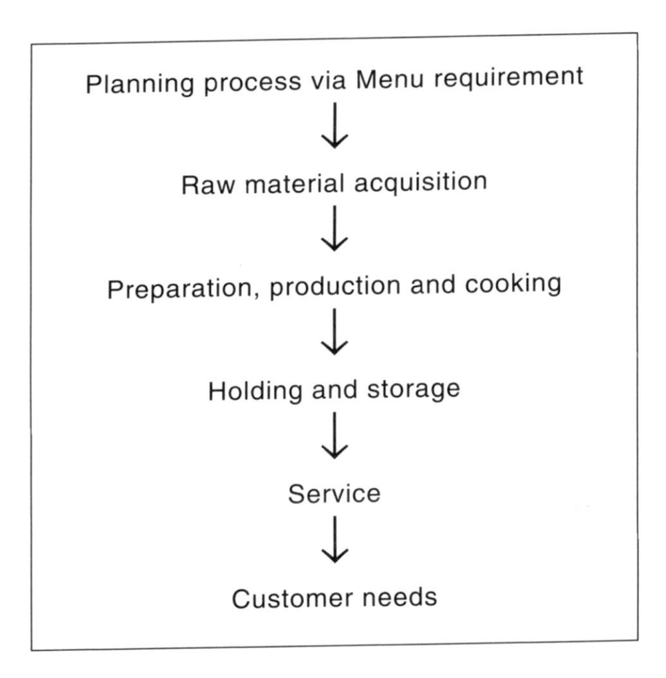

A simple analysis of the task or menu should provide you with the information you need to work out what is required from you, although you may need to consult your supervisor or chef de partie if you have any queries.

Before undertaking any task, ask yourself the following questions:

- What is to be done?
- Why is it to be done?
- Where is it to be done?
- When is it to be done?
- How is it to be done?

In order to use time effectively, you will need information about the tasks you are to perform. Some of this information will be obtained directly from your immediate supervisor, but some may have to be confirmed through checking the recipe or reading your study books.

Basic information may be categorised under the following headings:
- name or menu description of foodstuffs
- preparation prior to cooking
- size of the portion and amounts required
- weight and measure of ingredients
- cooking methods and processes
- oven temperatures and cooking times
- holding times of foodstuffs
- style of service
- time required.

You are aiming to improve your performance in the kitchen and maximise use of energy. When considering time planning, you will also need to consider the following factors:
- arrange items requiring preparation so that the sequence of work requires the least movement
- work methodically, to a set pattern, usually from your left to right
- maintain separate containers for all trimmings
- keep your working area clean at all times, cleaning and washing as you go
- when preparing large quantities of food, organise the food so that lifting and moving items is reduced to a minimum
- have all equipment ready to hand
- arrange all food in containers, trays or bowls at all times. This will help to keep your work area clear and reduce the risks of cross-contamination
- conserve fuel. When not required, gas or electricity should be turned down or off as required
- develop a positive mental attitude to do the job to the best of your ability, improving performance by taking less time to do better work
- plan the best sequence to adopt when several things need to be done.

There are distinct advantages to be gained by planning your time rather than reacting to a potential crisis caused by a lack of careful consideration. You will become a more efficient food handler by taking time to plan, developing skills and knowledge in the process.

Why plan?
- To be more efficient in production.
- To maximise work output.
- To meet production targets and customer needs.

How?
- Knowledge of tasks required.
- Knowledge of techniques.
- Knowledge of machinery and equipment.
- Production of an efficient mise-en-place.

What do you need to analyse?
- Jobs, tasks and menus.
- Check recipes or supervisor's directives.
- Work in a logical and sequential fashion, i.e. longest preparation and cooking times are approached first.

What do you need to be aware of?

- Complete procedures. This means being aware of any necessity to prepare one item in order to produce another, and any collation of items necessary to complete the task.

What do you need to organise?

- Yourself. Self-management is important to ensure constant improvement.
- Materials and equipment.
- Your workplace.

Knowing when advance preparation should stop without sacrificing food quality is really the essence of preparation and planning.

Drawing up a production plan

At this stage you can begin to allocate the time needed for each stage of preparation and handling. Many factors influence a time plan:

- protection of quality. Time planning prevents a crisis from arising because too many items demand attention at the same time
- efficient use of equipment and resources. You will need to plan when to use equipment in order to avoid conflict or over-utilisation of specialised machinery
- protection of quantity. Make sure you measure ingredients accurately
- cooking methods. Always follow technically correct procedures
- using the correct tools for the task. Always use appliances designed for a specific purpose; remember the rule: *use the correct tools for the job.*

Why should materials and equipment be assembled before starting preparation?
In order to ensure that all items are available so work can progress without interruption and with less chance of error.

On what principles should work plans be based?
Proper planning and sequencing to achieve optimum quality with minimum expenditure of unnecessary effort and time.

How can work be done quickly and easily?
By having supplies and equipment arranged for a smooth work flow, so that each motion can be made in the easiest way with the fewest steps.

The basic time plan: an example

Task:	*Piece de boeuf braisé bourguignonne*
Number of portions:	20 covers
Time required:	12.30 p.m. (luncheon service)
Ingredients:	As per recipe
Cooking method:	Braising
Service style:	Table d'hôte, full silver service

Step 1 *2 days prior to requirement:*

- acquire relevant raw materials and check purchasing availability.

Step 2 *1 day prior to requirement:*

- prepare, cook, chill and store brown beef stock
- prepare, trim and lard (with fat) the selected joints
- marinade for 24 hours in a red wine marinade as per selected recipe
- prepare, cook and chill sauce demi-glace or jus lié as required.

Step 3 *Day of requirement:*

0900 hours	Ensure workplace is ready and assemble raw ingredients to prepare and cook.
0930 hours	Seal and prepare joint, commence the braising process at approximately Gas Mark 6 with a lid.
1000 hours	Baste joint with cooking liquor through cooking process.
1030 hours	Prepare appropriate garnish and retain for further cooking.
1100 hours	Remove lid and continue to baste the joint with cooking liquor to achieve desired glaze.
1130 hours	Cook and hold appropriate garnish of turned mushrooms, bacon lardons and glazed brown button onions.
1200 hours	Remove cooked joint and cover, reduce sauce until desired consistency is obtained, adjust seasoning and strain. Immerse joint in prepared sauce to keep moist.
1215 hours	Hold joint and garnish in readiness for service.
1230 hours	Service commence, slice joint to order, sauce and garnish as appropriate.

Identifying time gaps

The gaps in the time plan allow other items of mise-en-place to be undertaken, or other tasks as denoted by your job requirement or supervisor's directive.

Remember:
Advance planning is essential: failing to plan is planning to fail.

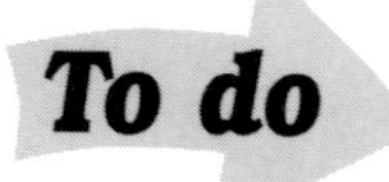

- Plan a comprehensive time plan for a complete menu. Indicate purchase requirements and recipes using the format and information in this chapter.
- List three ways you could save time doing your job without sacrificing quality.
- List the relevant information you require to do your job effectively.

The cookery processes

Introduction

The cookery processes are the *methods* of cooking which are common to all dishes, whether regional or international. Each process requires knowledge of basic skills and techniques which can be transferred throughout different types of cookery.

This chapter outlines each process, giving: a definition of the process, the reasons for using it, the methods used, the food items suitable, the equipment needed and key points to remember for each process.

Boiling

Definition

Boiling is a moist method of cooking where prepared food is cooked in a liquid which contains water (water, aromatic cooking liquor, stock, milk). The boiling action may be quick and rapid (as when cooking green leaf vegetables); or slow, with a gentle surface movement known as *simmering* (used when boiling most foods).

Reasons for boiling foods

Food is boiled in order to:

1 make foods tender, by breaking down and softening starch, cellulose, protein and fibrous material
2 make foods more palatable and digestible
3 make foods safer to eat, by destroying bacteria which can cause food poisoning
4 produce a particular quality in food, of colour, flavour and texture (e.g. boiled cabbage).

Methods of boiling foods

These are divided into two groups:

1 *Food is placed into cold liquid*, brought to the boil and cooked. You would use this procedure in order to:
  a) achieve a clear cooking liquid. Scum and impurities rise to the surface as the liquid comes to the boil and can be skimmed off. This is important when preparing stocks and clarified liquids such as consommes and jellies
  b) work as safely as possible. It is safe and easy to cover food with cold liquid then bring to the boil (see also *Cookery myths*, page 6).

2 *Food is placed into boiling liquid* and cooked. You would use this procedure in order to:
   a) keep cooking times as short as possible
   b) retain as much nutritional value and colour as possible, by keeping the cooking time as short as possible
   c) reduce vitamin loss when cooking vegetables by destroying oxidative enzymes
   d) reduce the risk of burning cereals and starch mixtures such as rice and pastas.

Cookery myths
It is often stated that placing food into very hot or boiling liquid seals in the juices and therefore retains goodness and reduces weight loss. This is a myth which should be ignored.

Food items suitable for boiling

A wide range of foods can be boiled:
1 *meat and poultry.* Boiling is used for the tougher joints and birds, e.g. silverside of beef, rolled brisket and hens. It is a suitable method for producing plain dishes, e.g. boiled gammon with cabbage and parsley sauce
2 *fish and shellfish.* Although it is more desirable to poach most fish, there are some classic dishes which are boiled. Lobsters are usually cooked by boiling
3 eggs and pastas. Examples are boiled eggs and fresh and dried pastas, e.g. noodles, spaghetti and macaroni
4 *fresh and frozen vegetables.* Examples are cabbage, cauliflower, turnips, peas, green beans and potatoes
5 *dried cereals and pulses.* Examples are rice, barley, oats, marrowfat peas, lentils and various dried beans.

Equipment used when boiling foods

Types of equipment include saucepans, stockpots, fast boiling pans, bratt pans and boilers.

Boiling: key points

- Many boiled dishes have long cooking times (e.g. stocks, boiled meats and pulses); you will need to allow for this in your time plan.
- Arrange saucepans of boiling food on the stove so that the correct cooking speed is maintained, i.e. rapid boiling or simmering. Also check saucepans regularly to ensure the correct speed of cooking and to determine degree of cooking.
- Be careful when draining foods with boiling liquid. Stand back from the saucepan, avoiding splashes of hot liquid.
- To avoid food poisoning, cool liquids and foods quickly then store chilled until required for use. Never leave boiled foods and liquor sitting in a warm kitchen.
- Store boiled foods at temperatures below 5 °C/40 °F, for as short a time as possible.
- Thoroughly cook dried beans as under-cooked beans contain a poison which can cause sickness. Dried beans should therefore be boiled rapidly for a minimum of 10 minutes and then simmered for the remaining cooking time.
- Always keep liquid content to a minimum when boiling food to ensure

that valuable nutrients are not lost. Also serve the cooking liquor with the food whenever possible.

- Avoid soaking or storing vegetables in water (except pulses). Also start the cooking of vegetables in boiling liquid whenever possible.
- Cook vegetables in batches and as near to service as possible. Avoid cooking and reheating vegetables.
- Remove fat from the surface of stocks and sauces as it forms.
- Soak out as much salt as possible from salted joints prior to cooking, e.g. hams and gammons.

Poaching

Definition

Poaching is a moist method of cooking where prepared food is cooked in a liquid containing water (water, milk, stock, wine or court-bouillon). The food is cooked at temperatures below boiling point (75–93 °C/167–200 °F) with little or no liquid movement.

Reasons for poaching foods

Food is poached in order to:

1. make foods tender, by breaking down starch, cellulose and protein
2. set or *coagulate* (gel) protein when poaching eggs
3. make foods more palatable and digestible
4. make foods safer, by destroying bacteria which can cause food poisoning
5. produce a particular quality in food of colour, flavour and texture.

Poaching is a gentle method of cooking which is used to cook food items which would break up or lose shape if boiled, e.g. poached eggs, poached fish, delicate fruits.

Methods of poaching foods

Methods of poaching are divided into two groups.

Deep-poaching

Food is covered with the minimum quantity of liquid then gently cooked. In most cases the food is placed into very hot liquid. Large whole fish are an exception: e.g. salmon, which is covered with cold liquid then brought to poaching temperature (to reduce the distortion of the fish when applying heat). Deep-poaching is a plain method of cooking, producing dishes such as poached fruits.

Shallow-poaching

Food is partly covered with the poaching liquor (to two-thirds the height of the food item) and then cooked gently under cover in an oven. This is a more complex method of cooking as the cooking liquor is reduced down and forms the base of the accompanying sauce. Many classic fish dishes are produced in this way.

Food items suitable for poaching

Deep-poaching

Whole fish, portioned fish, shellfish, whole chickens, eggs, fresh and dried fruits.

Shallow-poaching
Small whole fish, e.g. trout, sole, plaice; cuts of fish, e.g. fillets and fish steaks; chicken supremes.

Equipment used when poaching foods

1 *When deep-poaching:* saucepans, shallow-sided pots, fish kettles.
2 *When shallow-poaching:* use plat à sauters and shallow-sided cooking dishes. For fish, oval fish-cooking dishes can also be used.

Poaching: key points

- Allow sufficient time to prepare poached foods which are to be cooked and then served cold. For example, a whole salmon for a cold buffet has to be cooked then thoroughly cooled before it can be moved from its poaching liquor. After this it must be skinned, decorated and garnished.
- Fish may be prepared 1–2 hours prior to cooking. For example, fish for fillets of sole bonne-femme may be kept raw in a chill, trayed up and covered with the cooking paper without any recipe liquid. This reduces the preparation time for the dish, especially during service time.
- Poached eggs are often cooked and kept chilled in ice water. When required for service, the eggs are reheated and then finished as required.
- Check regularly to see that the food is being poached and not boiled; this is important to maintain good quality.
- Always keep liquid content to a minimum when deep-poaching food to ensure that valuable nutrients are not lost.
- Remove any fat from the surface of food as it forms.
- Use reduced sugar syrups when poaching fruits.

Steaming

Definition

Steaming is a moist method of cooking where prepared food is cooked in steam (water vapour) at varying degrees of pressure.

Reasons for steaming foods

Food is steamed in order to:
1 make foods tender, by breaking down and softening starch, cellulose and protein
2 make foods more palatable and digestible
3 make foods safer to eat, by destroying bacteria which can cause food poisoning
4 produce a particular quality in food of colour, flavour and texture (e.g. steamed sponge pudding)
5 keep the loss of soluble nutrients to a minimum, e.g. in vegetables.

Methods of steaming foods

This is divided into two groups: low-pressure and high-pressure steaming.

Atmospheric or low-pressure steaming
Food is cooked at atmospheric pressure or under low-pressure moist steam, usually of 0–17 kN m^{-2}/ 0–2 lb in^{-2}. This is the traditional method

of steaming food: eggs, root vegetables, shellfish and steamed puddings are examples of food items cooked by low-pressure steaming.

High-pressure steaming
Food is cooked at high pressures, usually 70–105 kN m^{-2}/10–15 lb in^{-2}. High-pressure steaming is fast and ideally suitable for most food except steamed puddings and sponge puddings. This type of steaming is a good method of cooking vegetables because there is very little water and air in the cooking chamber and this helps to retain valuable nutrients.

Food items suitable for steaming

- *Eggs.*
- *Fish and shellfish.*
- *Vegetables* (including potatoes).
- *Savoury and sweet puddings* (low-pressure steaming only).

Equipment used when steaming foods

Types of equipment include atmospheric and low-pressure steamers, high-pressure steamers and jet steamers.

Steaming: key points

- Switch on the machine in good time if preheating is required.
- Always turn off the steam before opening the door unless using a low-pressure steaming cabinet (which is not switched off during cooking).
- Always be careful when opening the door after use. Stand behind the door and use it as a shield against escaping steam.
- Cook vegetables as near to service time as required. A high-speed steamer is ideally designed for this purpose.
- Avoid over-cooking food, especially with high-speed steaming where even short periods of over-cooking will destroy nutrients.

Stewing

Definition

Stewing is a moist method of cooking where prepared food (cut into pieces) is cooked in a minimum quantity of liquid. Both the food and the liquid form the stew, so they are always served together. Stewing is an ideal method of cooking for the tougher cuts of meat, poultry and game; and since you are cooking and serving them in their cooking juices, the process also saves valuable nutrients. Stewing is also a term used when slowly cooking fruits to a pulp (e.g. stewed apples).

Reasons for stewing foods

Foods are stewed in order to:
1. make foods tender, by breaking down and softening starch, cellulose, protein and fibrous material
2. make foods more palatable and digestible
3. make foods safer to eat, by destroying bacteria which can cause food poisoning
4. produce a particular quality in food, of colour, flavour and texture (e.g. stewed beef).

Methods of stewing foods

Methods of stewing are grouped according to the following factors:

1 *type of commodity;* e.g. fish, meat, vegetable stews.
2 *colour of stew;* e.g. white and brown stews.
3 *method of preparation:*
   - stews cooked in a prepared sauce (e.g. fricassées)
   - stews where the liquid is thickened at the end of the cooking process (e.g. blanquettes).

Food items suitable for stewing

1 *Fish and shellfish.*
2 *Red and white meats.* Examples are beef, mutton, lamb, veal and pork. The tougher cuts of meat are used for stewing.
3 *Poultry and feathered game.* Examples are chicken, duckling, partridge and pheasant.
4 *Vegetables.* Several vegetables are usually cooked together to form the stew, e.g. onions, garlic, courgettes, aubergines and tomatoes in a ratatouille.
5 *Fruits.* The cooking of apples, pears and rhubarb to form a coarse pulp.

Equipment used when stewing foods

Types of equipment include saucepans, boilers and bratt pans.

Basic techniques of stewing

Blanching

This is done to remove impurities from meat when preparing blanquettes. It is done as follows:

1 Cover the prepared meat with cold water and bring it to the boil.
2 Remove from the stove and place under cold running water to rinse off all the scum which has formed.
3 Drain, then prepare the stew.

Liaising

This is a method of finishing a white stew, using a mixture of egg yolks and cream (*à liaison*). This increases the fat content.

Searing

This is the initial shallow-frying of flesh when preparing brown stews. It is carried out to develop colour and flavour. It is often stated that this procedure seals in the juices and therefore retains goodness and reduces weight loss. This is a myth which should be ignored.

Setting

This is the method of lightly cooking or *stiffening* flesh in fat without developing colour. It is used when preparing fricassées.

Stewing: key points

- Most stews have long cooking times, so make sure that you allow for this in your time plan. For example, beef stew made with shoulder steak may require 2–3 hours cooking before the meat is tender.
- Arrange saucepans on the stove so that the food is only simmering. Stir the stew regularly to prevent burning, and skim off surface fat and impurities.

- Check liquid content of stews during cooking and top up with additional stock as required.
- Remember that stews may be cooked in the oven under cover and this can provide much needed stove space for other items.
- Pay special attention to portion control. Estimating the number of portions from a large quantity of stew is difficult. Use ladles or spoons of standard sizes which will provide the correct portion size and number of portions expected.
- Avoid reheating stews but if they must be reheated, cool quickly and store in a chill below 5 °C (40 °F) for as short a time as possible. When reheating stews, bring to the boil and simmer for 15–20 minutes to avoid food poisoning.
- Trim off as much visible fat from meat, poultry and game as possible before cooking. Also skim the stew regularly during cooking to remove surface fat.
- Where possible reduce the red meat content of stews and increase the quantity of foods which provide fibre, e.g. use vegetable garnishes of beans, brown rice, sweetcorn kernels and low-fat dumplings made with wholemeal flour.

Braising

Definition

Braising is a moist method of cooking, where prepared food is cooked in a covered container with a quantity of stock or sauce in an oven. The food to be braised is usually placed on a vegetable base (*mirepoix*) and the liquid or sauce added to approximately two-thirds the height of the food item. This rule does not apply when braising small cuts of meat and offal such as chops, rump steaks and sliced ox liver, where the food is completely covered with the cooking liquor or sauce to maintain even cooking. When the food is cooked it is portioned and served with the finished sauce or cooking liquor.

Reasons for braising foods

Foods are braised in order to:

1. make foods tender, by breaking down and softening starch, cellulose, protein and fibrous material
2. make foods more palatable and digestible
3. cook and serve foods in their own juices, thus conserving valuable nutrients
4. make foods safer to eat, by destroying bacteria which can cause food poisoning
5. produce a particular quality in food, of colour, flavour and texture (e.g. braised celery).

Methods of braising foods

Methods of braising are grouped according to the colour of the finished dish and the foods to be braised:

1. brown braising of meat, poultry, game, offal and vegetables
2. white braising of sweetbreads
3. braising rice.

Food items suitable for braising

1 *Fresh butcher meats:* including beef, veal and venison.
2 *Fresh offal:* including ox liver and sweetbreads.
3 *Pickled meats and offal:* including ham and pickled tongue.
4 *Poultry and feathered game:* including duck, duckling, pheasant and partridge.
5 *Vegetables:* including cabbage, celery, leek and onion.
6 *Rice.*

Equipment used when braising foods

Types of equipment include braising pans, casseroles, lidded cooking vessels, plat à sauters with lids and bratt pans.

Basic techniques of braising

Basting

This is the process of coating the cooking item with the cooking liquor or sauce during cooking. It ensures that the food item cooks evenly and it also keeps the surface of the item moist during cooking. A glazed shiny appearance on the surface of meat, poultry or game can also be produced by basting occasionally near the end of the cooking period while uncovered.

Blanching

This applies to braising pickled meats, ox liver, sweetbreads and vegetables and means something different in each case.

1 *Blanching pickled meats* (ham or tongue): cover with cold water, bring to the boil and simmer for 20 minutes approximately. This is done to remove excess salt from the pickled meat.
2 *Blanching ox liver:* cover with cold water, bring to the boil, then refresh under cold water. This removes impurities and scum from the liver which would otherwise be present in the sauce.
3 *Blanching sweetbreads:* cover with cold water, bring to the boil and simmer for 10–15 minutes. Refresh the sweetbreads under cold running water and drain. When blanched, the tough membranes and tissue which surround the sweetbreads are trimmed off with a small knife.
4 *Blanching vegetables:* place the prepared vegetables into boiling water, simmer for a specified period of time then refresh under cold running water. Vegetables are blanched for the following reasons:
   - crisp vegetables become limp and easy to shape
   - the process helps to retain colour in vegetables
   - bitterness is reduced in certain vegetables (e.g. mature celery)
   - the process reduces cooking time.

Larding

This process consists of inserting strips of bacon or pork fat through flesh with special needles. It helps to produce a moist, rich eating quality but does increase the fat content.

Marinading

This process consists of soaking meat, poultry or game in a liquid (e.g. wine) with herbs and vegetables. The reason for marinading is to add flavour and increase tenderness. However, increasing tenderness by marinading may be less effective than was previously thought. Marinading times vary depending on type of flesh, size of joint and taste; varying from 2–18 hours.

Searing
This is the initial shallow-frying of flesh when preparing brown braisings. It is carried out to develop colour and flavour. It is often stated that this procedure seals in the juices and therefore retains goodness and reduces weight loss. This is a myth which should be ignored.

Braising: key points

- Special care is required when removing braising pans from the oven, especially when they contain large joints. The pan should be lifted carefully (with correct body posture and movement) using a thick, dry, folded oven cloth on each handle, remembering that the joint may move about as the pan is lifted.
- See also *Stewing: key points* (page 10).

Roasting

Definition

Roasting is a dry heat method of cooking, where prepared food is cooked with the presence of fat in an oven or on a spit.

Reasons for roasting foods

Foods are roasted in order to:

1. make foods tender, by breaking down and softening mainly protein, but also starch, cellulose and fibre
2. make foods more palatable and digestible
3. make foods safer to eat, by destroying bacteria which can cause food poisoning
4. produce a particular quality in food, of colour, flavour and texture (e.g. roast venison).

Methods of roasting foods

1. *Oven roasting.* This is the cooking of food in an oven, mainly by convected heat or forced air convected heat. However, other forms of heat application may also play an important function when roasting: e.g. conducted heat from a roasting tray when roasting potatoes, and radiated heat from the sides of an oven; both of which help to develop colour on the surface of the food. In addition, combination ovens which combine microwave energy or steam with forced air convected heat are also used when roasting.
2. *Spit roasting.* This is the original form of roasting which involves cooking the food by dry heat on a spit which is slowly turned over a heat source such as a charcoal fire, electric elements or gas flames. The main form of heat application is direct radiated heat, but convected heat (hot air) is also present. Conducted heat from metal spit bars may also aid cooking in some instances.
3. *Pot roasting* (poêler). This is included in this section although it is not strictly a form of roasting, being closer to a form of casserole cooking. The food is cooked under cover in an oven with (traditionally) butter as

the cooking fat. An important procedure with this method of cooking is the removal of the lid during cooking to allow the food to develop colour. After cooking, the vegetable base together with the cooking juices provide the basis of the accompanying sauce.

Food items suitable for roasting

Good quality joints must be used when roasting meat, poultry and game.

1 *Butcher meats and furred game:* e.g. beef, veal, lamb, mutton, pork and venison.
2 *Poultry and feathered game:* e.g. chicken, turkey, duckling, grouse and pheasant.
3 *Potatoes and parsnips.*

Pot roasting
Good quality butcher meats, poultry and game (as above) are required for this type of cooking.

Equipment used when roasting foods

1 Types of ovens: general purpose oven, forced air convection oven, combination ovens (e.g. microwave and convection ovens), steam and convection ovens.
2 Spit and rotisserie racks.
3 Small equipment: roasting trays, trivets, temperature probes.

Pot roasting
Various types of casserole dishes or similar cooking utensils can be used.

Basic techniques of roasting

Barding
This involves covering the surface of the roast with slices of pork or bacon fat. This is to prevent the flesh drying out during cooking; but because it increases fat content, it should only be used where necessary, normally with feathered game (grouse, partridge and pheasant).

Brushing with oil and basting
Both of these processes involve lightly brushing the joint with fat before and during cooking. This is done to prevent the surface of a joint drying out and becoming hard (especially lean joints). Basting is the traditional practice of coating the item with the fat. In order to keep the fat content to a minimum, brush with fat rather than baste.

Carry-over cooking
This is the further cooking which takes place after the joint has been removed from the oven.

Larding
See page 12.

Placing in the roasting tray
Butcher meats and furred game: Joints of meat should be placed onto a roasting tray with the fat top upwards. Never place a joint directly onto a roasting tray: always place in onto a bed of bones or vegetables, or onto a trivet.

Poultry and feathered game: Birds should be placed on their sides with the breast downwards then turned during roasting to ensure even cooking.

Searing
This involves starting the cooking of the roast in a hot oven, or shallow frying the item prior to roasting. It is carried out to develop colour and flavour, especially with meat roasts. It is often stated that this procedure seals in the juices and therefore retains goodness and reduces weight loss. This is a myth which should be ignored.

Speed of cooking
The temperature at which an item should be roasted is related to the size of the food item. The larger the item, the lower the cooking temperature. High temperature roasting should be avoided as it increases shrinkage and weight loss. The temperature range when cooking roasts of average size is usually175–200 °C (350–400 °F).

Resting, standing or settling a roast
This refers to removing a roast from the oven after cooking and leaving it in a warm place for a short period (5–15 minutes depending on size). This is to reduce the risk of someone being burned when portioning or carving the joint. The food is also easier to carve or portion after resting.

Roasting: key points

- Remember that roasts are served at a particular degree of cooking, i.e. underdone, medium, well done. This must be carefully considered when preparing your time plan.
- Always keep your hands well protected and wear your sleeves long to avoid burns from spurting hot fat. Take care when removing roasts from the oven. Special care should be taken with large roasts which may move when the tray is lifted.
- Where possible, use lean joints such as rump and good quality topside, and trim off surface fat before serving.

Grilling

Definition

Grilling is a dry heat method of cooking where prepared food is cooked mainly by radiated heat in the form of infra-red waves.

Reasons for grilling foods

Foods are grilled in order to:

1. make foods tender, by breaking down and softening mainly protein, but also starch, cellulose and fibre
2. make foods more palatable and digestible
3. make foods safer to eat, by destroying bacteria which can cause food poisoning
4. produce a particular quality in food, of colour, flavour and texture (e.g. grilled lamb cutlets).

Methods of grilling foods

1. Grilling foods *over* a heat source which may be fired by charcoal, electricity or gas, e.g. steak grills and barbecue type grills.

2 Grilling foods *under* a heat source fired by gas or electricity, e.g. salamander type grills.
3 Grilling foods *between* electrically heated grill bars.

In Methods 1 and 2 above, most of the cooking is done through radiated heat, although some cooking occurs by convection from hot air currents and conduction (when the food is touching hot grill bars). In Method 3, most items of equipment cook the food between very hot ridged metal plates with conduction being the main cooking method.

Food items suitable for grilling

Good quality cuts must be used when grilling meat, poultry and game.

1 *Butcher meats and furred game:* various types of steaks, chops, and cutlets.
2 *Offal and bacon:* e.g. sliced liver, kidneys and gammon steaks.
3 *Poultry and feathered game:* various small birds prepared ready for grilling; e.g. spring chicken, grouse and partridge.
4 *Fish and shellfish:* various small whole fish (sole, plaice, trout); cuts of fish (fillets and steaks); and shellfish such as lobster, large prawns and scampi.
5 *Vegetables:* mainly mushrooms and tomatoes.
6 *Made-up items and convenience foods:* e.g. burgers, bitokes, sausages and sliced meat puddings.

Equipment used when grilling foods

1 Steak grills and barbecue units fired by charcoal, gas, or electricity.
2 Salamanders fired by gas or electricity.
3 Contact grills, infra-red units and toasters.

Basic techniques of grilling

Brushing with oil

This involves lightly brushing the item with fat before and during cooking. It is done to prevent the surface of the item drying out and becoming hard. Basting with fat (coating the item) should be avoided as this increases the fat content.

Flouring items to be grilled

Coating foods with flour prior to grilling only applies to items which do not develop a good colour when cooking. Whole fish, cuts of fish and liver are usually lightly coated with flour when they are to be cooked under a salamander.

Searing

This involves starting the cooking of the item (plain flesh only: not sausages, puddings or breaded items) on a hot part of the grill to develop colour and flavour. It is often stated that this procedure seals in the juices and therefore retains goodness and reduces weight loss. This is a myth which should be ignored (see *Speed of cooking* below).

Speed of cooking

High temperature grilling produces the most suitable infra-red waves to cook food. However, the heat exposure and speed of cooking should cook the food to the correct degree without burning the outer surface. When cooking thick items, such as large steaks and chops, the item may be seared to develop some colour, then the speed of cooking reduced while the item finishes cooking. With thin items such as flattened steaks, small chops and cutlets, the item is usually cooked so that colour and the appropriate degree of cooking are reached at the same time.

Turning an item

Foods being grilled should be turned with tongs or a palette knife. Never stab or pierce foods with a fork *at any stage* of preparation or cooking. This applies to all foods including sausages.

Grilling: key points

- Remember that grilled foods are served at a particular degree of cooking depending on the type of food item and customer choice; i.e. rare, underdone, medium and well done. The time at which you should begin to cook these items is therefore dictated by the service requirements.
- Ensure that foods which can cause food poisoning (e.g. chicken, pork and made-up items such as sausages) are thoroughly cooked. Knowing when food has reached a specific degree of cooking is an important skill which must be learned. One way of quickly determining the degree of cooking is to use a temperature probe. The internal temperature ranges which indicate the various degrees of cooking are as follows:

Red meats:	underdone:	55–60 °C (130–140 °F)
	just done:	66–71 °C (150–160 °F)
	well done:	75–77 °C (167–172 °F)
Chicken, turkey:	cooked through:	77 °C (170 °F)

- Use salt sparingly on grilled foods. This is necessary not only to reduce salt in the diet, but also because by adding salt to an item being grilled you will slow down colour development.
- Where possible, use lean cuts of meat and trim off excess fat before cooking. Also drain the food to remove as much surface fat as possible prior to service.

Shallow-frying

Definition

Shallow-frying is a dry heat method of cooking, where prepared food is cooked in a pre-heated pan or metal surface with a small quantity of fat or oil. Shallow-frying is a fast method of cooking because heat is conducted from the hot surface of the cooking pan directly to the food.

Reasons for shallow-frying foods

Foods are shallow-fried in order to:

1. make foods tender, by breaking down and softening protein, fat, starch, cellulose and fibre
2. make foods more palatable and digestible
3. make foods safer to eat, by destroying bacteria which can cause food poisoning
4. produce a particular quality in food, of colour, flavour and texture (e.g. sauté potatoes).

Methods of shallow-frying foods

Meunière

This method is commonly used for shallow-frying fish and shellfish. The fish is lightly coated with flour before frying and is served with lemon slices and chopped parsley. Nut-brown butter or margarine is poured over the fish when serving; remember that the quantity used should be kept to a minimum to reduce the fat content.

Sauter

This term has three meanings:

1. it is often used as an alternative term for *shallow-frying,* especially when referring to the shallow-frying of small cuts of butcher meats, poultry or game
2. it can refer to a particular type of shallow-frying where you use a tossing action to turn the food while frying, e.g. when cooking sliced potatoes or mushrooms.
3. it can be used when preparing a high quality meat, poultry or game dish served with a sauce. Here a good quality item is shallow-fried to a specific degree of cooking, then removed from the pan while the pan is swilled with stock, wine or sauce. This procedure (known as *déglacer*) uses the sediment lost from the item being cooked, thereby increasing the flavour and aroma of the sauce.

Griddle

This involves cooking items on a lightly oiled metal plate (*griddle plate*). A *ridged surface* is used for cooking small cuts of meat, game and poultry to allow the fat to drain from the meat; while a *flat plate* is used for bakery items such as griddle scones.

Stir-fry

This is the quick-frying of pieces of fish, meat, poultry and vegetables with fat or oil in a wok.

Sweat

This involves slow frying items in a little fat, using a lid, and without allowing colour to develop. Sweating is usually a preliminary procedure used when making certain soups.

Food items suitable for shallow-frying

Good quality cuts must be used when shallow-frying meat, poultry and game.

1. *Butcher meats and furred game:* e.g. various types of steaks, chops, cutlets, escalopes and medallions.
2. *Offal and bacon:* e.g. sliced liver, kidneys and gammon steaks.
3. *Poultry and feathered game:* cuts for sauter and supremes.
4. *Fish:* various small whole fish (sole, plaice, trout) and cuts of fish (fillets and steaks).
5. *Made-up items and convenience foods:* e.g. burgers, bitoks, sausages and sliced meat puddings.
6. *Eggs:* mainly scrambled eggs and omelettes.
7. *Vegetables:* sliced potatoes, mushrooms, onions, tomatoes and courgettes.
8. *Fruits:* e.g. bananas, peaches, apple and pineapple slices.
9. *Batters and doughs:* e.g. crêpes, scones and pancakes.

Equipment used when shallow-frying foods

Equipment used includes frying pans, omelette pans, crêpe pans, plat à sauter pans, sauteuses, bratt pans, griddle plates and woks.

Basic techniques of shallow-frying

Searing

This involves starting the cooking of the item (plain flesh: not sausages, puddings, breaded items or batters) in a hot pan to develop colour and flavour. It is often stated that this procedure seals in the juices and therefore retains goodness and reduces weight loss. This is a myth which should be ignored.

Speed of cooking

The speed of cooking varies with the item being cooked, but as a general rule, the thicker the item the lower the frying temperature. A common mistake is to fry at too high a temperature resulting in over-cooking, fat breakdown and off-flavours.

Turning an item

Foods being shallow-fried should be turned with a palette knife: never stab or pierce foods with a fork *at any stage* of preparation or cooking. This applies to all foods including sausages.

Shallow-frying: key points

- Remember that certain shallow-fried foods, e.g. beef dishes, are served at a particular degree of cooking; i.e. rare, underdone, medium and well done. The decision on when to start cooking a particular item must be made in keeping with service requirements.
- Always pre-heat the frying utensil to reduce both fat absorption into the food and the risk of the food sticking to the pan.
- Place the foods with the longest cooking times into the pan first, e.g. chicken *legs* before *wings*.
- Ensure the presentation side of the food item is fried first so that the item does not become discoloured or marked with sediment.
- Keep the frying fat to a minimum and if possible dry-fry on a non-stick surface.
- Use lean cuts of meat and trim off excessive fat before cooking. Also drain the food to remove as much surface fat as possible prior to service. To reduce fat content, cook foods by grilling rather than shallow-frying. This applies to meat, poultry, game and made-up items such as sausages and hamburgers.

Deep-frying

Definition

Deep-frying is a dry heat method of cooking, where prepared food is cooked in pre-heated fat or oil. Deep-frying is a fast method of cooking because all the surfaces of the food being fried are cooked at the same time, with temperatures of up to 195 °C (383 °F) being used.

Reasons for deep-frying foods

Foods are deep-fried in order to:

1. make foods tender, by breaking down and softening protein, fat, starch, cellulose and fibre
2. make foods more palatable and digestible
3. make foods safer to eat, by destroying bacteria which can cause food poisoning
4. produce a particular quality in food, of colour, flavour and texture (e.g. apple fritters).

Methods of deep-frying foods

Partial cooking or blanching

This is the deep-frying of foods until tender, but without developing colour. The reason for blanching foods is that they can be stored on trays until required for service then fried quickly in hot fat until crisp and golden brown. Chips are usually blanched in this manner; fruit fritters and battered vegetables may also be blanched prior to service.

Complete cooking

This is the deep-frying of foods until fully cooked, where serving takes place immediately, to maintain a crisp, dry product.

Pressure frying

This is the frying of food under pressure in special fryers. Pressure fryers are usually automated and work on a timed cooking cycle. These fryers are fast at producing high quality fried foods and are safe to use.

Food items suitable for deep-frying

1. *White fish:* e.g. some small whole fish (haddock) and fillets of fish.
2. *Chicken or turkey:* portions of poultry.
3. *Made-up items and convenience foods:* e.g. Scotch eggs, savoury cutlets, croquettes and cromesquis.
4. *Vegetables:* raw (e.g. aubergines, courgettes); or cooked (e.g. prepared celery, fennel and cauliflower).
5. *Potatoes.*
6. *Fruits:* e.g. bananas, peaches, apple and pineapple slices.
7. *Batters and doughs:* e.g. choux paste (fritters), bun dough and dough-nuts.

Important: Some foods are less suitable for deep-frying because they contain fats or oils which will contaminate the frying medium. Examples of these foods are oily fish, fatty meats and meat products such as bacon, gammon, ham, sausages and meat puddings.

Equipment used when deep-frying foods

1. Fritures (old-fashioned frying vessels: see *Deep-frying: key points* below).
2. Free-standing fryers (electric or gas) with manual thermostat control.

3 Automatic fryers.
4 Continuous fryers.
5 Pressure fryers.

Basic techniques of deep-frying

Draining foods to be fried

Wet foods should be thoroughly drained and dried as much as possible before being cooked. Placing wet foods into hot fat is very dangerous as the fat reacts violently and rapidly increases in volume.

Coating foods to be fried

Many fried foods are coated with batter or breadcrumbs prior to frying. This not only produces a crisp, coloured surface but also reduces the juices and fat from the item entering and contaminating the frying medium.

Battered foods

Foods which are battered are passed through the batter (usually after coating with flour), and then placed directly into the hot fat. The food should be placed carefully into the fat to avoid splashes of fat which can cause burns. A basket should never be used when frying battered foods.

Breaded foods

Foods coated with breadcrumbs are usually fried on trays or in baskets.

Speed of cooking

Most foods are fried at a temperature which will cook, colour and crisp the food all at the same time. Avoid low temperature frying as this increases fat absorption in the food.

Draining fried foods after cooking

Fried foods should be drained thoroughly after cooking to remove surface fat. In addition it is standard practice in many establishments to serve fried food on dishpapers which absorb surface fat.

Hot storage of fried foods

To produce high quality fried food for the customer, the food should be served immediately after frying. Never use a lid to cover fried foods as this produces condensation and softens the crisp coating.

Deep-frying: key points

- Always wear sleeves long to avoid burns to the arms from splashes of fat.
- Always use a well-designed fryer with a thermostat and never an old-fashioned friture.
- Never exceed a maximum frying temperature of 195 °C (383 °F).
- Never fry too much food at once as this is not only dangerous but will reduce the frying temperature and increase fat absorption.
- Check the accuracy of the thermostat at regular intervals.
- Strain the fat regularly to remove food particles. Keep the number of fried foods offered on your menus to a minimum.

Baking

Definition

Baking is a dry heat method of cooking where prepared food and food products are cooked by convected heat in a pre-heated oven.

Reasons for baking foods

Foods are baked in order to:

1. make foods tender, by breaking down and softening protein, fat, starch, cellulose and fibre
2. make foods more palatable and digestible
3. make foods safer to eat, by destroying bacteria which can cause food poisoning
4. produce a particular quality in food, of colour, flavour and texture (e.g. Victoria sandwich).

Methods of baking foods

Baking fruits, vegetables and potatoes

This is a form of simple oven cooking where the food items are cooked in an oven until tender.

Baking within a *bain-marie* (water bath)

This involves placing the item to be baked in a water bath, so that low temperatures may be maintained during cooking. The baking of egg custard mixtures is an example of this type of cooking; where a gentle oven heat is maintained by a bain-marie, reducing the likelihood of the mixture curdling.

Baking flour products

This is often a more complex form of cooking than the methods given above. When baking flour products such as cakes, the dry heat of the oven is usually modified with steam which has developed within the oven from the cake mixture during baking. The oven conditions in this instance should not only provide the correct temperature but also the correct humidity.

Cooking eggs

Cooking shirred eggs (*Oeufs sur le plat*) is also a form of oven cooking.

Food items suitable for baking and oven cooking

1. *Fruits*: e.g. apples and pears.
2. *Potatoes*.
3. *Milk puddings and egg custard products*.
4. *Flour products:* e.g. cakes, sponges, pastries and yeast goods.
5. *Vegetables* prepared in vegetarian bakes.
6. *Meat and vegetable hotpots* which are oven cooked.
7. *Eggs* which are oven cooked.

Equipment used when baking foods

Types of equipment used include: general purpose ovens, pastry ovens, forced-air convection ovens, baking ovens with steam injection (for bread) and specialist ovens (e.g. pizza ovens).

Basic techniques of baking

Traying up items to be baked

Certain categories of food require the baking tray to be prepared in a particular way. Lightly greasing trays or tins with white fat (rather than butter or margarine which may cause the food to stick to the tray) is essential for cakes and sponges. However, with some foods it may be advisable to use silicone paper (e.g. brandy snaps) to avoid this problem. Allowance must be made when traying up foods which will expand during baking (e.g. choux paste and yeast goods).

Marking foods to be baked

Some products, such as short pastry items (e.g. flans, pies and tarts) have their top edges neatly marked to produce an attractive finish. This is sometimes referred to as *notching* and may be done with the thumb and forefinger or special tweezers.

Gilding or coating with eggwash

Many items which are to be baked, especially pastry and yeast goods, are lightly brushed with eggwash just prior to baking so that a good colour will develop on the surface of the item.

Speed of cooling

Most items are baked so that the product rises (if appropriate), develops colour and cooks through at the same time. It is therefore important that the oven is set to the correct temperature and pre-heated before inserting the food.

Proving

This is the final fermentation of yeast goods after they have been shaped and placed on the baking tray. It is usually carried out at 28–30 °C (82–86 °F) in a moist atmosphere to prevent the surface of the goods developing a skin.

Cooling

Many baked items are very delicate when hot (e.g. cakes and pastries) and should be cooled or allowed to cool slightly prior to use. This is usually done on a cooling wire designed for the purpose, which allows the air to circulate under the food and prevents condensation and softening of the product.

Baking: key points

- When preparing your time plan, remember that many baked goods have long preparation times; such as yeast goods, large cakes and gateaux. In addition they may have to be served cold and you will therefore need to allow for cooling time.
- Ensure that you allow sufficient time for the oven to reach the correct baking temperature. Do not create a situation where a sponge is ready for the oven but cannot be cooked because the oven is not hot enough.
- Take care to measure and weigh foods accurately. When baking cakes, even small errors can have disastrous consequences; when weighing baking powder ensure the exact amount is used or the cake may be useless.

Cold preparations

Definition

Cold preparations are cold items which have been prepared and assembled and are either raw, or cooked then cooled.

Reasons for making cold preparations

1. To make foods more palatable and digestible.
2. To produce a particular quality in food, of colour, flavour and texture (e.g. Florida cocktail).
3. To make foods visually attractive (eg whole decorated salmon).

Methods of producing cold preparations

Cold savoury items

These may be accompanied or finished with appropriate sauces or dressings. Categories include:

- *different types of hors d'oeuvre:* including single item, selection (varies) and cocktail types
- *different types of salad:* including simple, mixed, fish, meat and poultry salads
- *cold decorative items:* e.g. fish, shellfish, meat, poultry and vegetables
- *different types of sandwiches:* plain, rolled and pinwheel types.

Cold sweet items

These may be garnished and decorated. Categories include: cold mousses, bavarois and charlottes; table jellies; trifles and condés; fresh fruit salads and coupes; syllabub.

Food items suitable for cold preparations

1. *Fresh vegetables* and salad vegetables.
2. *Cold cooked fish, shellfish, meat, poultry and game.*
3. *Convenience and processed foods:* e.g. canned, jarred, frozen, smoked and foods in brine.
4. *Fresh fruits.*
5. *Dairy foods:* including cream, fromage frais and yoghurt.
6. *Baked goods:* such as sponges and finger biscuits.

Equipment used when producing cold preparations

1. Small equipment, e.g. bowls, whisks, spoons, cutters.
2. Motorised equipment, e.g. mixers, blenders, food processors.
3. Large equipment, e.g. refrigerators, chillers and freezers.

Basic techniques of cold preparations

1. *Dividing skills:* such as slicing, dicing, shredding, chopping.
2. *Combining skills:* such as mixing, binding, dressing, garnishing.
3. *Artistic/creative arrangement of food:* attractive presentation and decoration of different food materials.

Cold preparations: key points

Always use good hygienic practices when handling and storing cold foods and remember the golden rule; *keep it clean, keep it cool, keep it covered.*

UNIT 2D1

Preparing and cooking meat and poultry

ELEMENTS 1, 2 AND 3: Preparing and cooking meat and poultry

What do you have to do?

- Prepare meat and poultry correctly for individual dishes.
- Combine prepared meat with other ingredients ready for cooking where appropriate.
- Prepare cooking and preparation areas and equipment ready for use, then clean after use.
- Cook meat and poultry dishes according to customer and dish requirements.
- Plan your work, allocating your time to fit daily schedules and then carry out the work within the required time.
- Finish and present meat and poultry dishes according to customer and dish requirements.
- Carry out your work in an organised and efficient manner, taking account of priorities and any laid down procedures.

What do you need to know?

- The type, quality and quantity of meat and poultry required for each dish.
- The equipment you will need to use in preparing meat and poultry.
- How to prepare meat and poultry for cooking.
- How to satisfy health, safety and hygiene regulations concerning preparation and cooking areas and equipment, and how to deal with emergencies.
- How to cook meat and poultry dishes by roasting, boiling, grilling, frying or stewing.
- Why it is important to keep preparation, cooking and storage areas and equipment hygienic.
- The main contamination threats when preparing, cooking and storing both uncooked meat and poultry and finished dishes.
- Why time and temperature are important when cooking meat and poultry dishes.

Introduction

Meat and poultry are an important part of most people's diets, supplying much of the protein we need for the growth and repair of our bodies and providing a source of energy. Offal, especially liver and kidneys, is also a good source of protein and some vitamins and minerals (particularly Vitamin A and iron).

When you cook meat, think about the structure of the food you are handling; this will help you to understand why certain processes suit particular types of meat better than others. The lean flesh of meat is actually composed of fibrous muscles, bound together by connective tissues. The size and thickness of the fibres in the muscles determine the grain and texture of the meat: so younger animals (with less developed muscle fibres) provide a more tender meat.

The connective tissue binding these fibres also affects the tenderness of the meat. There are two kinds of connective tissues: one (*elastin*) must be broken up by pounding or mincing, while the second (*collagen*) can be broken down by the cooking process.

The amount, condition and distribution of fat on a meat carcase will also affect tenderness and flavour. Where fat is found between the muscle fibres, the meat is said to be *marbled*. This type of meat will be more tender, moist and flavoursome. These three qualities are also enhanced in all meats by a process of hanging, which matures the meat before the carcase is dissected.

Poultry is the name given to all domestic fowl bred for food (their meat and eggs). It is more easily digested than meat and contains less fat. The tenderness and flavour depends on the type and age of the bird: older laying hens are tougher than younger ones, and need to be cooked by a wet process (e.g. boiling) to enhance tenderness. You may often use frozen poultry: make sure you are familiar with all the points listed on page 42.

Health, safety and hygiene

This is particularly important when working with meat and poultry, as you will be handling some high risk foods and working with potentially dangerous equipment. Check back to Unit G1: *Maintaining a safe and secure working environment* and Unit G2: *Maintaining a professional and hygienic appearance* in the Core Units book to make sure you are familiar with the general points noted there.

Also note the following points:

- when moving from preparing cooked to uncooked food, or from one type of meat or poultry to another, always clean or change your equipment, working areas and utensils
- make sure you are familiar with the colour-coding system in your kitchen. Always use the correct colour-coded boards for preparing meat and poultry, and use different boards for preparing cooked and uncooked meat and poultry
- store uncooked meat and poultry in separate areas within refrigerators, or (preferably) within separate refrigerators. Use trays for storage to prevent dripping and check that the temperature remains below 5 °C (41 °F), preferably at 3 °C (37 °F)
- when handling high risk meat and poultry (e.g. pork, chicken, turkey), wash your working surfaces with a bactericidal detergent to kill bacteria
- you will need to use a variety of knives when preparing meat and poultry. Work safely, following the guidelines in Unit 2D10: *Handling and maintaining knives* in the Core Units book. When using a boning knife, wear a safety apron for your own protection.

Planning your time

When preparing meat and poultry, some preparations and cooking methods will take longer than others; some may need to be carried out at a particular time (e.g. immediately before service). Plan your work carefully: which dishes require a fixed starting time? Which ones are more flexible? Dishes with short cooking times are more flexible in terms of starting times than those that need to be cooked slowly over a long period.

Equipment

Before starting to prepare or cook meat or poultry, decide what equipment and utensils you will need to complete the process. Put them all out beforehand, making sure they are clean and ready to use. You may need to use motorised cutting equipment, such as mincers, slicers or food processors; check that these are also assembled correctly.

Knives
You will need to use a variety of cook's knives for trimming and cutting meat. The following knives have particular uses that you should be familiar with.

- Butcher's steak knife: this has a firm blade with a curved end; it is often used for slicing raw meat.
- Boning knife: this is a very strong, sharp knife which can be used with pressure to cut close to the bone in meat joints. It must always be used with great care: it is made to withstand any force you may have to use, but any slippage could cause serious injury. Always wear a protective apron when using this knife.
- Carving knife: used for carving cooked joints of meat.
- Cleaver or chopper: you will need to use this to chop through large bones (e.g. the chine bone when preparing sirloin of beef). You can also use the back of the blade when necessary to crack bones.

Other equipment

- Cutlet bat: you will need to use this for batting (see page 39).
- Poultry secateurs: these may be used instead of a cook's knife for cutting through poultry bones.
- Cook's fork: this is used to lift roast meats from the roasting tray. It has very sharp points, so care must be taken to avoid either piercing the meat or poultry, or injuring yourself.
- Carving fork: this is used in conjunction with the carving knife. It usually has a guard to prevent your fingers accidentally slipping down the blade and coming into contact with the carving knife.

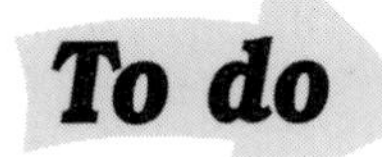

- Find out how long the hanging process is for the different types of meat. Why do these differ?
- Find out how the colour-coding system works in your kitchen. What do the different colours represent?
- Find all the pieces of equipment listed above. Make sure you know how to use them all correctly and safely.
- Look at several examples of joints from younger and older animals. How does the structure of the meat differ?
- Find out where and how meat and poultry are stored in your kitchen. What temperature are they stored at?

ELEMENT 1: Preparing meat for cooking

Beef *(Le Boeuf)*

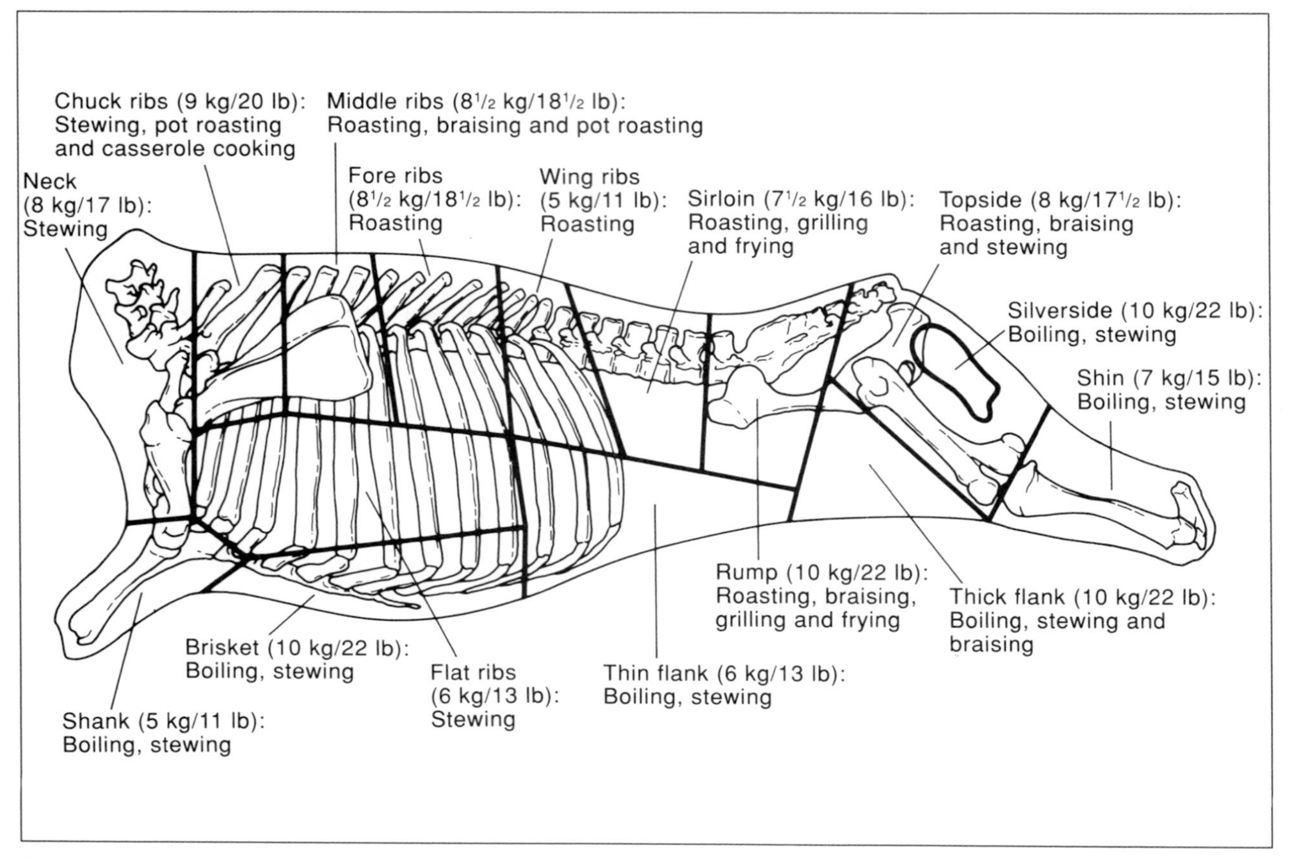

Beef carcase

Beef is hung for up to 14 days to increase tenderness and enhance flavour. It should smell fresh (almost odourless), and have a good red colour (or *bloom*). Beef cattle are slaughtered around the age of 18–21 months, and the meat from younger animals is generally more tender and brighter in colour; if the flesh has a deep red colour it has probably come from an older animal and may be tougher.

The colour of the meat is also affected by the length of time the meat has been hanging: the longer the meat has been hung, the darker the meat will be. You should be able to see flecks of fat (*marbling*), especially in the prime joints such as sirloin and wing rib. The outer layer of fat should be firm, fairly brittle and creamy-white. The most tender cuts come from the sirloin and rump.

Sirloin *(L'aloyau)*

The sirloin is used whole for roasting. When preparing whole, you will need to remove the *chine bone* (the T-shaped bone running from the bottom of the meat towards the fatty top).

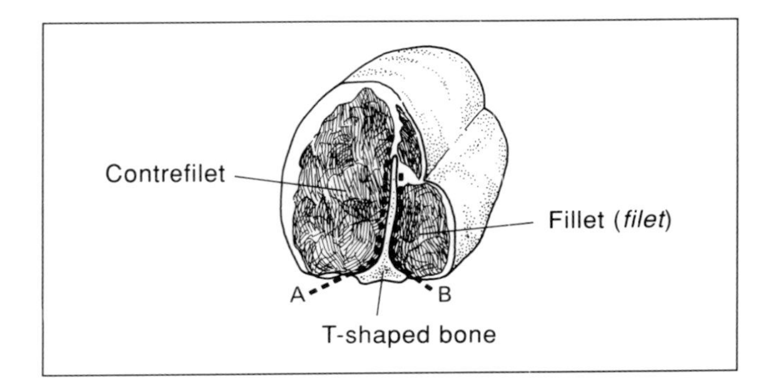

Preparation of sirloin

Preparing sirloin

1. Cut down between the meat of the fillet and the right-hand side of the chine bone, starting just above the top of the bone (**B**).
2. Cut down between the meat of the contrefilet and the left-hand side of the chine bone, again starting just above the top of the bone (**A**).
3. Pull back the cut away part of the contrefilet and saw or cleaver through the bone.
4. Remove the sinew (under the top layer of fat) and trim off any excess fat.

When roasting, the bone can be replaced and the sirloin tied with string. The fillet may be cut away entirely, as it can easily overcook.

Cuts from the whole sirloin (suitable for grilling or frying)

- *Contrefilet* — (see below)
- *Fillet* — (see below)
- *T-bone steak* — cut through the whole sirloin from the rib end; this contains a large piece of fillet
- *Porterhouse steak* — as for T-bone but cut from rump end of the sirloin, containing less fillet

Contrefilet

This is prepared from the sirloin, following the steps given above, but to remove the contrefilet from the chine bone completely, follow the steps given below.

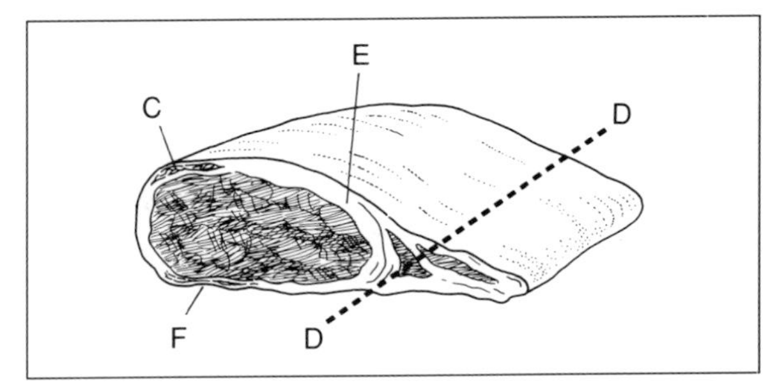

Preparation of the contrefilet

Preparing contrefilet

1. Cut away the bottom flap (containing little meat) along the line marked **D**.
2. Remove the sinew and nerve under the fat at **C**.
3. Trim any excess fat and sinew from **E** and **F**.

Cuts from the contrefilet (suitable for grilling or frying)

- *Sirloin (entrecôte) steak* — a piece 1–1.5 cm (1/2 in) thick from contrefilet, weight:150 g (6 oz)

- *Double sirloin (entrecôte double) steak* — a large sirloin steak: 2 cm (1 in) thick, approx. weight 300 g (12 oz)
- *Minute steak* — a sirloin steak flattened (batted) to approx. 3 mm (1/8 in) thick

Fillet *(Le filet)*

Cut the fillet from the sirloin following Step 1 of *Sirloin Preparation* (page 29), but cut the meat away from the sirloin entirely. Trim the meat by cutting away any sinew (running along the length of the fillet) and any excess fat.

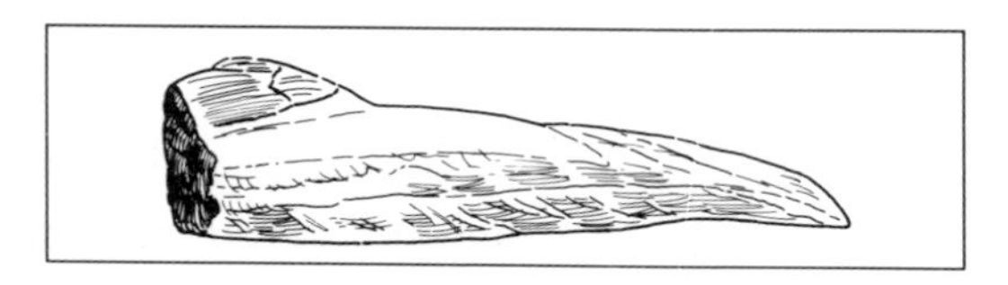

Fillet

Cuts from the fillet (suitable for grilling or frying)

- *Fillet steak* — cut from the middle of the fillet: 1.5–2 cm (1/2 in) thick
- *Chateaubriand* — cut from the head of the fillet as a double fillet steak: 3–10 cm (1–4 in) thick, normally served for two covers
- *Tournedos steak* — cut from the middle of the fillet: 2–4 cm (1–1 1/2 in) thick. Usually tied with string and barded with speck
- *Médaillons* — small slices or sticks cut from the tail end of the fillet
- *Bâtons* — small slices or sticks cut from the tail end of the fillet
- *Filets mignons* — cut from tail end of fillet, so small that if used as fillet steaks, two should be served per portion; can be diced finely, e.g. for Bolognese sauce
- *Paillard steak* — a flattened fillet steak

Wing and fore ribs

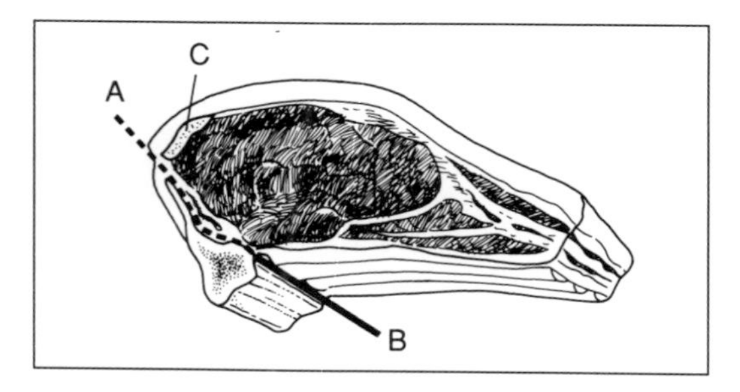

Wing and fore ribs

Remove the chine bone and sinew (under the top layer of fat) following the same method as for preparing sirloin (page 29). The fat and flesh from the bottom of the ribs may be trimmed away to expose the rib bones.

Wing ribs are roasted whole or cut into steaks for grilling or frying.

Rump *(La culotte)*
This is cut away from the hip bone, and any sinew or excess fat removed.

Cuts from the rump

- *Rump steak* — 2 cm (1 in) thick
- *Point steak* — cut from the triangular section of rump; 2 cm (1 in) thick

Topside and silverside *(La tranche tendre* and *La gite à la noix)*

These are from the hindquarters of the carcase. To prepare, carefully cut and remove the bone and trim off any sinew or excess fat. Topside is jointed and tied with string for roasting and braising, or cut into dice or steaks for stewing.

Thick flank *(La tranche grasse)*

This is also cut from the hindquarters of the carcase. To prepare, remove any excess fat and cut as required. The thick flank is used for braising and stewing.

Shin *(La jambe)*

Another cut from the hindquarter, shin is prepared by boning out then removing any excess sinew. It can be cut or chopped as required. It is generally used for stews.

Beef steaks (clockwise from top left): entrecôte steak, fillet steak, tournedos, T-bone steak, chateaubriand, minute steak

Lamb and mutton *(L'agneau* and *le mouton)*

Lamb meat comes from a sheep that is under 12 months old when slaughtered; if the animal is over 12 months old the meat is classed as mutton. Lamb flesh should be a dull red colour, while mutton flesh should be slightly darker: a dull brownish-red. Both should have a fine texture and grain. The fat should be clear white, brittle and flaky, and evenly distributed across the carcase. You would expect lamb bones to be pink and slightly porous.

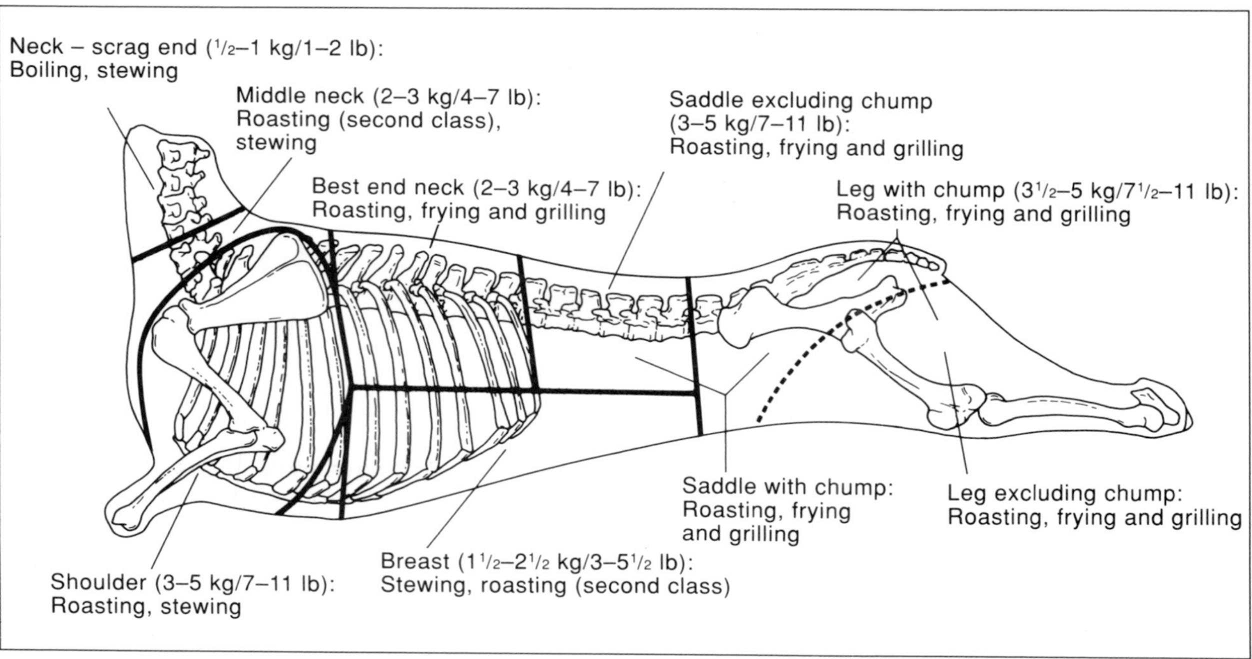

Lamb carcase

Best end of lamb *(Le carré)*

This can be roasted whole or cut into cutlets for grilling or frying.

1 Cut away the chine bone: cut down following the lines marked **A** on the illustration, then use a saw or cleaver to cut away the bone (**B** joining up to **A**).
2 You now have 2 cuts of best end. Remove the skin from both and the blade bone (**C**).
3 Remove the tough sinew lying just below the fat at the top of each best end.
4 Cut along the length of each best end, allowing you to trim away any flesh or fat between the bones. Scrape clean the bones.
5 If necessary, shorten the rib bones using a cleaver.

Cuts from the best end: cutlets (cut from between rib bones) and double cutlets (each having 2 rib bones).

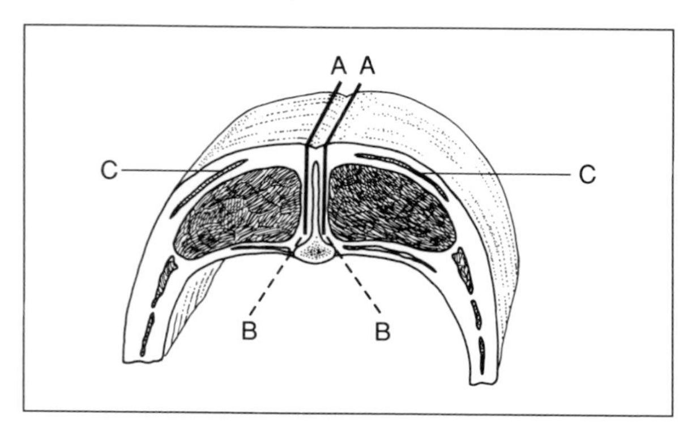

Best end of lamb

Loin *(La longe)*

This is half a saddle. Split the saddle through the middle (i.e. through the backbone) using a cleaver and then skin. The bone can be removed to allow the loin to be stuffed or used for cuts.

Saddle *(La selle)*

This can be roasted whole or cut into fillet, loin and chump chops for grilling and frying.

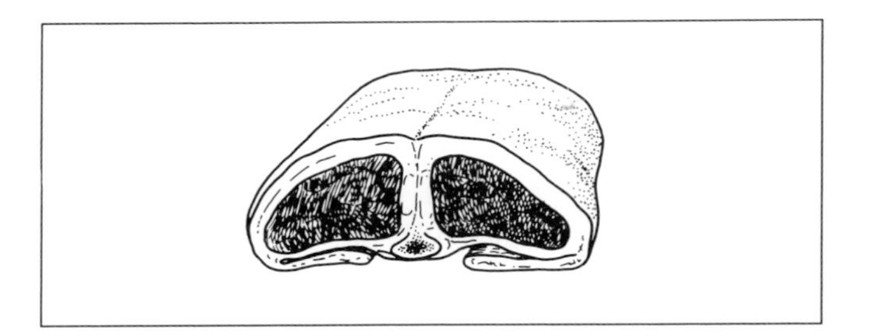

Saddle of lamb

1. Pull away the skin.
2. Trim off any sinew or excess fat, and remove the kidneys.
3. Trim the fatty ends from the flaps and tuck them under the saddle.
4. Remove the aitch bone.
5. Score the fat. Tie with string before cooking to keep the shape intact.

Cuts from the loin and saddle

- *Single loin chops* — cut across the unboned loin; each chop 100–150 g (4–6 oz) in weight
- *Noisettes (French style)* — cut from a boned loin at an angle of 45°; cuts are 2 cm (1 in) thick; flattened out and trimmed of excess fat
- *Rosettes* — cut from a boned saddle (i.e. across two loins); 2 cm (1 in) thick; ends rolled in and secured with string to achieve a flat heart shape
- *Barnsley chops* — cut from the unboned saddle (i.e. across two loins); 2 cm (1 in) thick

Leg *(Le gigot)*

This is generally partly boned for roasting. Leg of mutton is usually boiled.

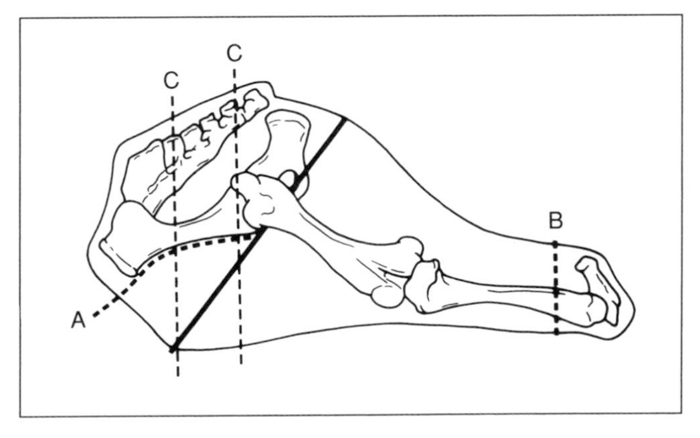

Preparation of a leg of lamb

1. Cut along line **A**, following the line of the aitchbone and through the ball and socket. Remove the aitchbone.
2. Saw off the bottom knuckle and bone.
3. Remove any excess fat, and tie with string before cooking.

Cuts from a leg of lamb

Gigot chops — cut from the centre of the leg
Chump chops — cut from the chump end of the leg

Shoulder *(L'épaule)*

Skin, then trim and clean the knucklebone, leaving 3 cm ($1^1/_2$ in) exposed. It can then be boned, stuffed and rolled if necessary.

Veal *(Le veau)*

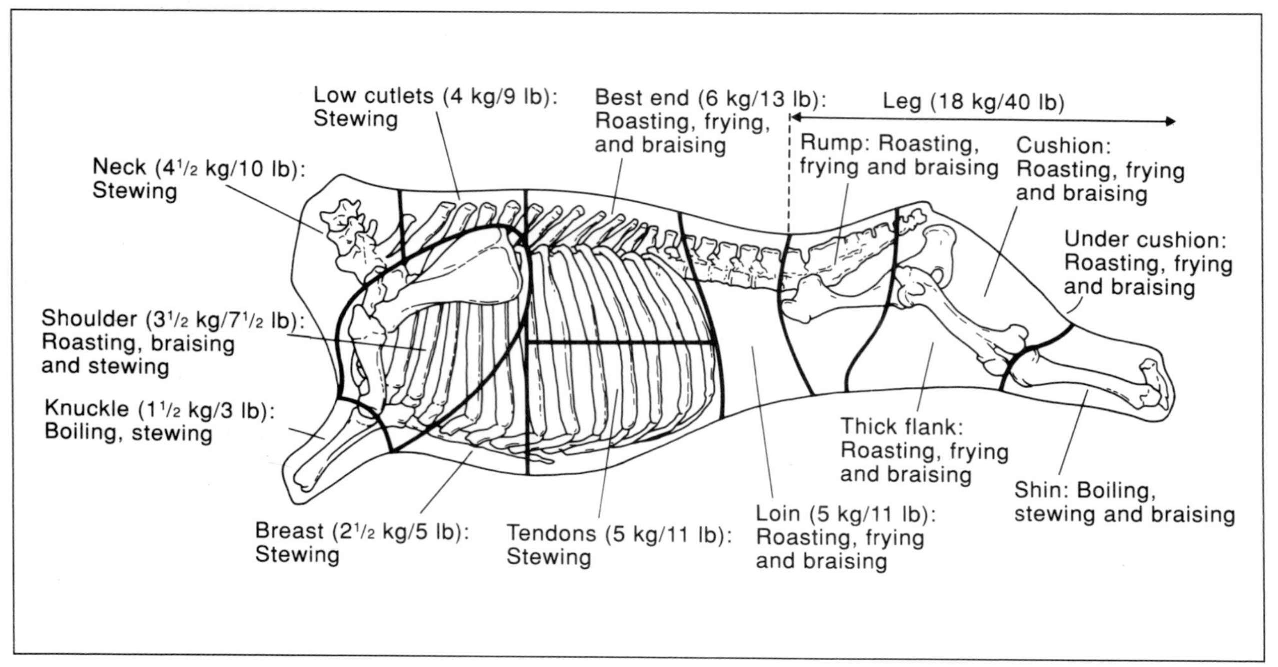

Veal carcase

Veal comes from the flesh of a calf (generally about three months old). The meat should be pale pink, with a fine, evenly-grained texture, and moist with no sign of stickiness. It should be covered by a fine layer of creamy-white fat. Any bones should be pinkish-white and quite soft.

Best end *(Le carré)*

This is the prime cut from a veal calf, and is prepared in the same way as for best end of lamb (page 32).

To prepare for roasting: bone out, trim, roll out stuffing and then tie with string to secure shape.

Cuts from the best end: veal cutlets (cut from between rib bones).

Saddle and loin *(La selle* and *La longe)*

A saddle of veal is prepared following the same method as for saddle of lamb (page 33). The loin can be cut from the saddle and roasted whole or cut into veal chops for grilling or frying.

Leg *(Le gigot)*

The leg is the leanest cut of veal. It is boned to produce three large joints: the cushion, under-cushion and thick flank. These can all be cooked whole or used to obtain the cuts listed on page 35.

Cuts from the cushion, under-cushion and thick flank

- *Escalopes* — large slices cut against the grain; 50–75 g (2–3 oz) in weight; batted if required
- *Grenadins* — small, thick slices cut across the grain; 10 cm (4 in) long and 2 cm (1 in) thick. Larded with speck to prevent dryness during cooking
- *Fricandeaus* — thick slices cut from along the muscle 17 cm (7 in) long and 2 cm (1 in) thick. Larded with speck as for grenadins
- *Médaillons* — small round slices cut against the grain; 5 cm (2 in) diameter and 4 mm (1/4 in) thick

To do

- Watch a sirloin being boned, and if possible the removal of the contrefilet and fillet.
- Find and identify at least three different types of beef steaks.
- Watch a best end of lamb being prepared.
- Find out if your kitchen currently holds any veal joints or cuts. If so, identify the joints or cuts and notice how the colour of bones and flesh differs from beef.

Pork *(Le porc)*

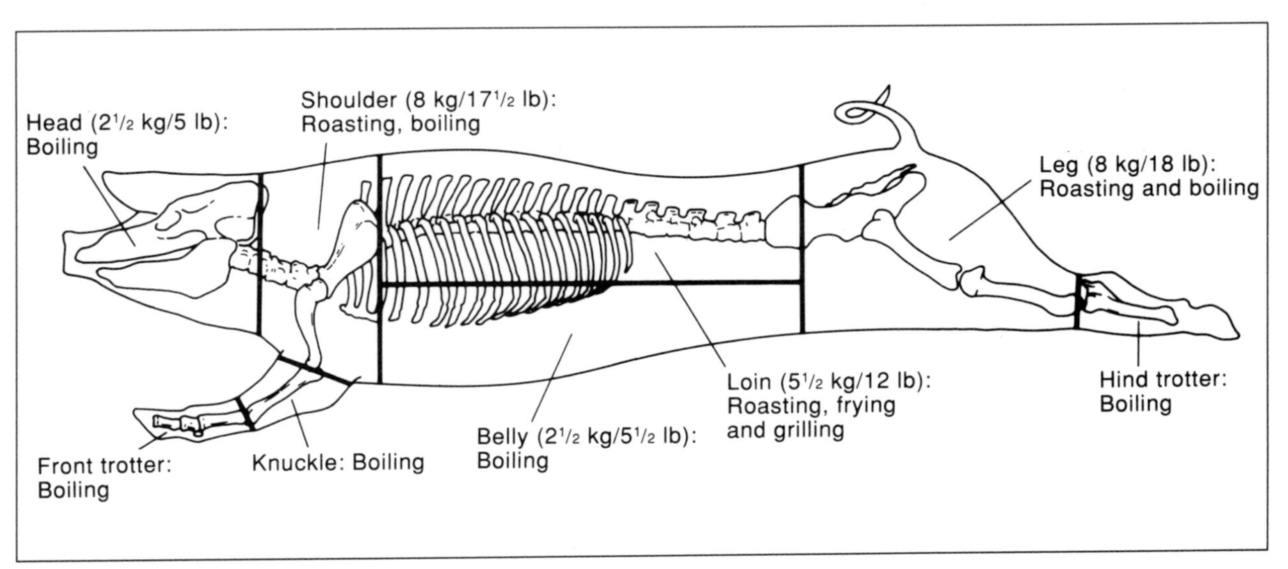

Pork carcase

Pork keeps less well than other meats, and needs very careful handling, preparing and cooking. It may contain parasitic worms, which are destroyed by thorough cooking. Always serve pork well done, *never* under done.

Pork joints should be well-fleshed without excessive fat. The flesh should be pale pink, firm, finely textured and not too moist. Look for smooth skin and fairly pliable bones. There should not be any unpleasant odours.

Loin *(La longe)*

This can be boned or left unboned. *If leaving unboned:* saw off the bottom of the chine bone (as for beef: *wing rib* on page 30), trim away any sinew and excess fat and score with diagonal strokes across the top. *To bone:* remove the filet mignon (tail of the fillet), cut away the bone completely, remove any sinew and excess fat. Neaten the flap and score as before. The filet mignon can be replaced before the meat is rolled and tied (for roasting). The loin can be roasted whole or cut into chops for grilling or frying.

Cuts from the loin

- *Chops* — cut from the unboned back loin; 2 cm (1 in) thick
- *Cutlets* — cut from the unboned fore loin
- *Escalopes* — prepared from the fillet; flattened steaks

Leg *(Le cuissot)*

Prepare as for leg of lamb (page 33), removing the trotter. The leg of pork can be roasted or boiled.

Shoulder *(L'épaule)*

Bone the shoulder by removing the rib bones, shoulder blade, shoulder bone and backbone. Remove any sinew and excess fat then score across the skin. Roll the joint (stuffed if required) and tie with string. The shoulder can be roasted, cut into spare ribs or used for making sausages or pies.

Cuts from the shoulder: spare ribs.

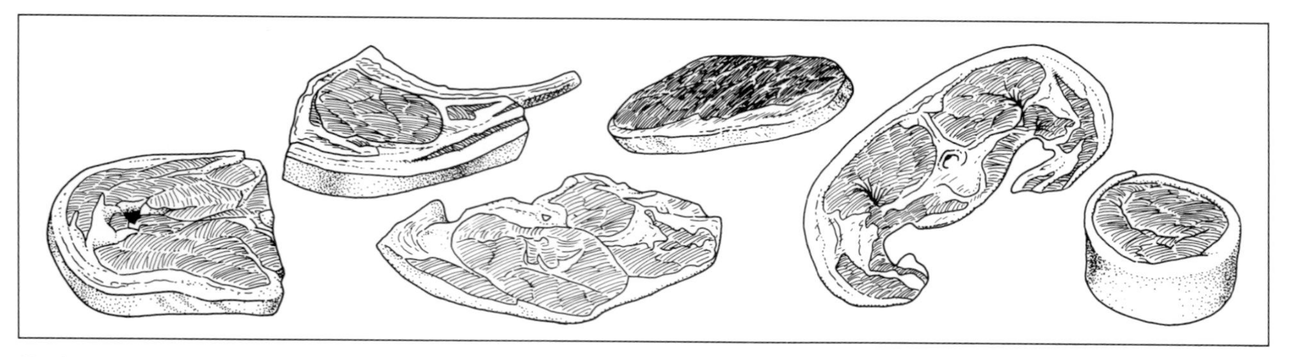

Pork and lamb chops
From left: loin chop, cutlet, chump chop, noisette, Barnsley chop, rosette

Bacon *(Le lard)*

Bacon differs from pork in that it comes from a different, usually larger breed of pig (*a baconer pig*) and the meat is cured (salted in brine) and sometimes smoked. *Green bacon* is a name often used to describe unsmoked bacon. Ham comes from the hind leg of a baconer pig, which has been cut away from the carcase and cured or pickled in brine, and sometimes smoked.

There should be no sign of stickiness and no unpleasant smell. The rind should be thick, smooth and free from wrinkles. Check that the fat is white, smooth and not excessive in proportion to the lean meat (flesh) which should be deep pink in colour and firm.

Joints and cuts of bacon

Back bacon is cut from the loin of bacon pigs while streaky bacon is cut from the belly, and gammon steaks from the hind leg. The hock and collar are cut from the shoulder and neck (these are tough but richly flavoured cuts).

Offal

Offal is the term used to describe edible parts of an animal taken from the inside of the carcase. The most commonly used offal are kidneys and liver, though this category also includes sweetbreads, tongue, brains, heart, tripe and trotters.

Liver *(Le foie)*

Liver should look fresh, moist and smooth and have a pleasant smell and colour. The liver from lamb, mutton, veal, pork and chicken can be used for cooking. Calf's liver is considered to be the most tender and flavour-some, while lamb's liver is also very tender but with a stronger flavour. The difference in quality is demonstrated by the vast price differentials.

1. Cut away any tubes and gristle.
2. Remove the membrane (the thin outer skin).
3. Cut as for recipe; usually at a slant into thin slices.

Kidneys *(Le rognon)*

The kidneys from lamb, mutton, beef (calf and ox) and veal can be used for cooking, although calf's kidneys are far superior in quality. Ox kidney requires prolonged cooking and is often blanched prior to cooking to remove excessive uric acid. All kidneys should be moist, with any fat being brittle, and should smell pleasant.

1. Remove the membrane (skin) and cut away all fat and tubes.
2. Cut through the centre of the kidney without cutting right through: you should be able to open it out but keep it in one piece.
3. Trim any fat or gristle found.
4. Cut or skewer whole as for recipe; if cutting, cut at a slant into thin slices or dice.

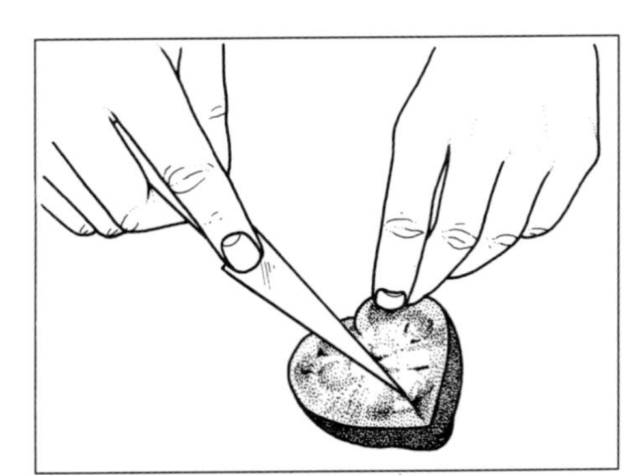
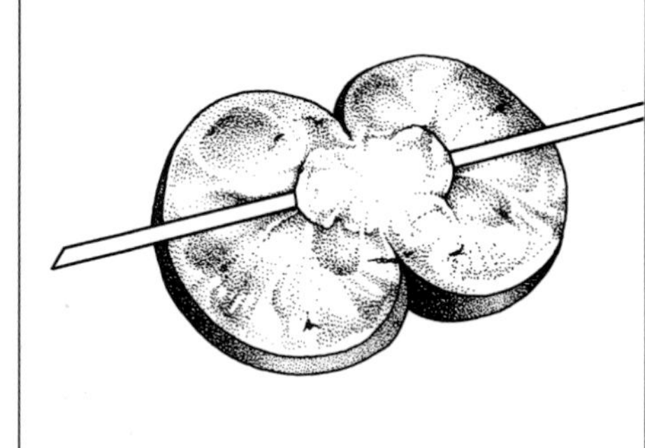

Left: slitting the kidney
Right: skewering the kidney

Tongue *(La langue)*

Ox's tongues and lamb's tongues may be used in cooking. They should smell pleasant, and not need too much trimming from the root end. Both types may be used fresh, but ox tongues are also often used salted. They are suitable for boiling or braising.

1 Remove any bone and gristle from the throat end and underside of the tongue.
2 If pickled, soak in cold water for 4–6 hours. If fresh, soak for 24 hours.

Sweetbreads *(Le ris)*

Neck and heart breads are commonly used. The heart bread is generally considered to be superior to neck breads. Both kinds should be fleshy, large and creamy-white with a pleasant smell. They can be braised or fried. As for kidneys, calf's sweetbreads are recognised as being of the highest quality, while ox-breads are not normally used except in the food processing industry.

1 Soak in cold, salted water (2–3 hours) to remove any visible blood (*disgorge*).
2 Wash thoroughly under cold water.
3 Blanch and trim.

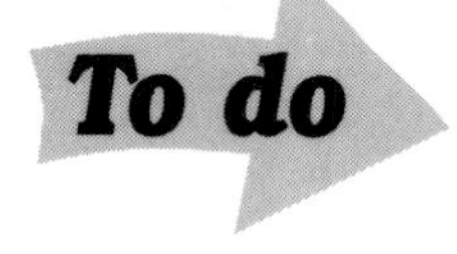

- Find and identify at least three types of pork steaks or chops.
- Examine some pork and bacon cuts. What differences do you notice in colour, texture, fat quantity and smell?
- Check today's menu to see if any offal dishes are to be served. If so, watch them being prepared.

Preparation methods for meat

Trimming

Meat is often trimmed (using a cook's knife) to cut away any unwanted fat or sinew, or simply to obtain a required shape. For example, when cutting minute steaks (*entrecôte minute*) from the contrefilet, you would need to cut the contrefilet into slices, flatten the slices, then finally trim the steaks to a uniform shape.

Dicing

Meat is often diced when used as an ingredient in stews. The meat is trimmed and cut into even-sized cubes to ensure even cooking.

Stuffing

Large joints of meat are often stuffed and cooked. Note the points listed below:

- the stuffing usually consists of a combination of some of the following ingredients: breadcrumbs, sausage meat, chopped suet, onion, herbs, an egg (to bind the stuffing) and seasoning. There are some notable exceptions, such as Beef Wellington, which is stuffed with goose liver (*foie gras*)

- stuffings for meat are inserted raw into the boned meat; the ingredients are combined and inserted, to be cooked as the meat cooks. Some ingredients, particularly vegetables, may be cooked lightly before being combined with other ingredients (e.g. onions, which are always sautéed first)
- stuffing cuts of meat: this can be applied either to large fillets of meat (e.g. Beef Wellington) or smaller cuts of lean meat (e.g. Beef olives). Trim and flatten out the meat (if necessary), season, then lay the stuffing lengthways down the centre of the meat. Roll (see below) and tie with string. Cook as for the recipe
- stuffing joints of meat: some dishes require a joint of meat to be boned in such a way as to leave the joint intact, but provide an inner cavity for the stuffing, e.g. Roast leg of lamb in pastry (*Gigot d'agneau en croûte*). Here the stuffing is simply inserted into the pocket left by the removal of the main leg bone and cooked with the meat.

Trussing or tying

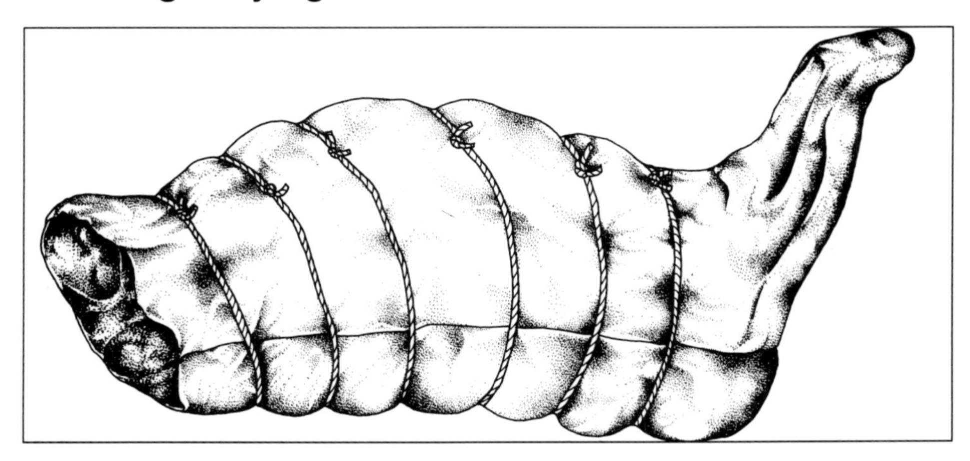

A prepared and tied shoulder of lamb

Trussing or tying is a way of keeping a piece of meat together with string to hold the meat in shape during cooking. The term *tying* is correctly used when referring to meat, while *trussing* is used when referring to poultry. You might tie or truss poultry and large joints of meat for roasting or braising, although some steaks are also tied with string to give a particular shape, e.g. tournedos. The string is always removed before serving. (See *Trussing*, page 43).

Rolling

Rolling takes place once a joint or cut has been boned and/or stuffed, to achieve the required shape. Simply roll the meat (using your fingers) into the correct shape, making sure that you are rolling the entire length at the same time, to achieve an even tightness. Secure the rolled meat with string to hold the shape while cooking.

Batting

Batting is a way of tenderising and flattening out meat by beating it with a heavy implement; usually a cutlet bat. You can do this most hygienically by placing the meat between polythene sheets before batting.

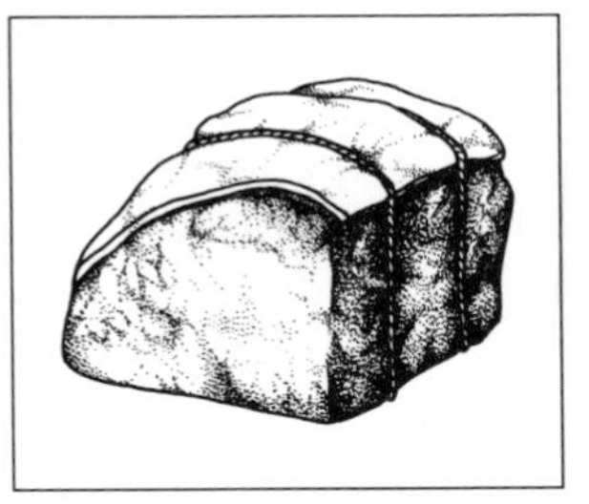

Barded topside of beef

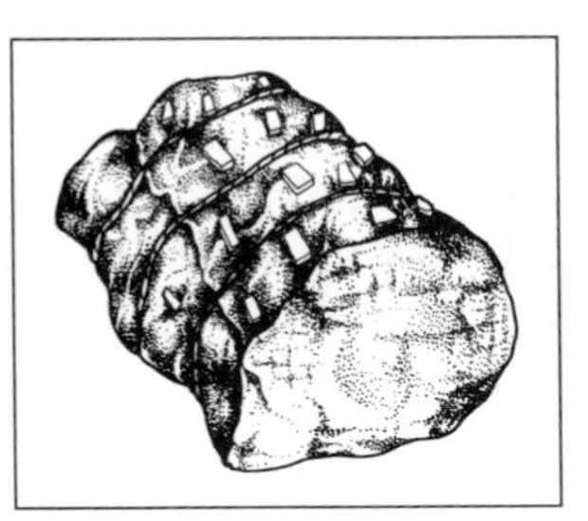

Larded filet of beef

Barding

Meats are liable to dry out slightly when cooked by some of the dry process methods, especially roasting. One way to combat this is to cover the surface of the meat with slices of pork or bacon fat (*speck*), securing them to the top of the joint; this is called *barding*. The fat will act to protect the surface of the meat, keeping it moist during cooking. It will also spread through the meat during cooking, helping to make it more tender and adding flavour.

Meat and poultry that has been barded will not usually need to be basted during cooking. This means that you will not need to open the oven door frequently while cooking, and so a constant temperature is maintained.

Larding

Larding is a similar process to barding and also helps to prevent meat from drying out. In this case, strips of fat are inserted into the meat to prevent the interior of the meat from drying out.

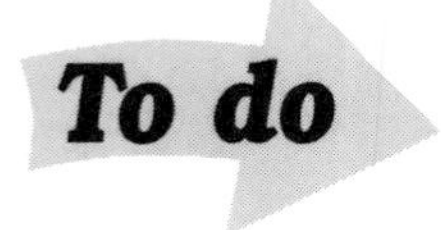

- Find as many different reasons as you can for trimming meat.
- Watch a stuffed dish being prepared. Note how the cavity is formed to take the stuffing (e.g. where a joint has been boned and rolled).
- Find two stuffing recipes to accompany each of the following: lamb, pork, veal.
- Watch a joint being barded. Find out why that type of meat needs barding and which cooking process is to be used.

Essential knowledge

The main contamination threats when preparing, cooking and storing meat are caused by the following:

- storing uncooked and cooked meat and/or poultry together: this allows food poisoning bacteria to pass between the foods
- using the same preparation areas, equipment and utensils for preparing cooked and uncooked meat and/or poultry: this also allows food poisoning bacteria to pass between the foods
- using unhygienic preparation areas, equipment and utensils for preparing meat and poultry: food poisoning bacteria can pass from these to the foods
- leaving food uncovered: this allows pests carrying bacteria to contaminate the food
- inadequate personal hygiene: food poisoning bacteria can pass from your hands, mouth, nose or any open wounds to uncooked meat
- incorrect storage temperature: any meat or poultry stored at a temperature above 10 °C (50 °F) will become contaminated
- incorrect waste disposal.

Essential knowledge

You must always keep preparation, storage and cooking areas and equipment in a hygienic condition because:
- this prevents the transfer of food poisoning bacteria from working areas or equipment to food
- this prevents pest infestation in storage areas
- you are obliged by law to comply with food hygiene regulations.

What have you learned?

1 What are the main contamination threats when preparing and storing uncooked meat?
2 Why is it important to keep preparation and storage areas and equipment hygienic when preparing meat?
3 What features should you look for when buying lamb or mutton to establish freshness and quality?
4 How does bacon differ from pork?
5 What is the difference between *barding* and *larding*?

ELEMENT 2: Preparing poultry for cooking

Introduction

The term *poultry* is used to describe any domestic birds bred specifically to be eaten or for their eggs. Chicken, *poussin* (baby chicken), turkey, duck, goose and guinea fowl are all forms of poultry. Wild duck, pheasant, grouse and snipe on the other hand are classified as *feathered game,* as they are wild birds which may be eaten.

Quality

Fresh chicken and turkey should have an unmarked skin (no cuts, rubbed portions or blood spots) and a white colour (unless corn fed, when they will look yellowish). Look for a straight, broad, well-fleshed breast and a pliable breastbone. The flesh should be firm but pliable, and there should not be too much fat, especially in the abdominal cavity.

When buying duck or goose, look for bright yellow feet and bills, where the bill breaks easily. The webbed feet should tear easily. Check that the breast is plump and the lower back is flexible.

Poultry weight and portion size

English	French	Weight	Number of portions
Single baby chicken	*Poussin*	360–500 g (12 oz–1 lb)	1
Double baby chicken	*Poussin double*	500–750 g (1–1½ lb)	2
Small roasting chicken	*Poulet de grain*	0.75–1 kg (1½–2¼ lb)	1½–2
Medium roasting chicken	*Poulet reine*	1–2 kg (1¼–4½ lb)	2–4
Large roasting chicken	*Poularde*	2–3 kg (4½–6½ lb)	4–6
Turkey	*Dinde*	3.5–20 kg (7–43 lb)	**
Goose	*L'oie*	4–7 kg (9–16 lb)	**
Duck	*Le canard*	2–3 kg (4½–6½ lb)	3–4

**Allow 200 g (½ lb) raw weight per portion.

Poultry is often bought frozen. In this case, check that the packaging is in a good condition before opening. Do not use any poultry showing signs of freezer burn, i.e. dry white patches on the skin. Defrost the bird gradually, by transferring it from the freezer to the cold room, to prevent any damage occurring during thawing.

Preparation methods for poultry

The following preparation methods are for chicken and turkey. However, the general preparation for roasted duck and goose would follow the same procedures.

Washing

Poultry may need washing after plucking, singeing and gutting. If these procedures are carried out carefully you may need only to wipe around the inside of the bird with a clean cloth. However, if the intestines break inside the carcase when you are trying to remove them, the bird should always be washed.

Method
Hold the bird under cold running water and wash well, inside and out. Dry the carcase thoroughly using either a clean cloth or absorbent kitchen paper.

Skinning

If a bird is to be skinned, this is usually carried out while jointing. To skin a jointed chicken (e.g. the breast once the legs and wings have been removed):

1. Hold the neck end of the breast with your left hand.
2. Pull back the skin with your right hand, gently easing it away from the flesh. Pull the skin down to the back end of the breast.
3. Cut away the skin from the bird using a cook's knife.

Trimming

While preparing poultry you will often need to trim parts of the carcase to neaten the shape (removing any straggly pieces), cut away skin or remove any sinew. Use a cook's knife and work carefully: remember that chicken is a high risk commodity.

Jointing

Chicken or turkey may be jointed before or after cooking using the same method:

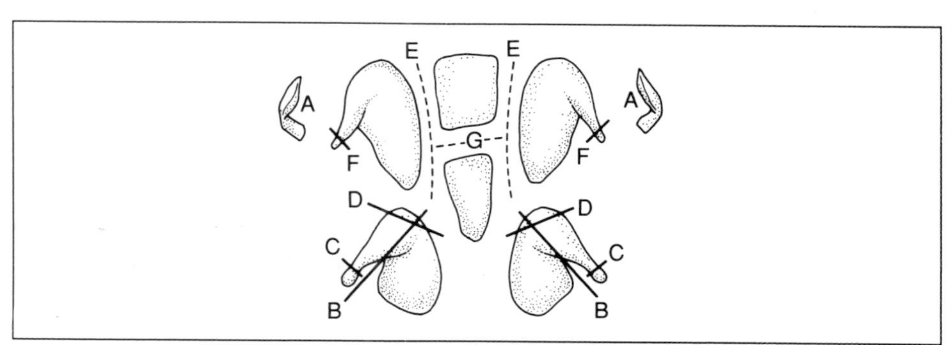

Jointing a bird

1 Remove the wishbone (cutting the sinews to release it).
2 Cut off the winglets (at the tip of the wings) and trim (**A** and **F**).
3 Remove the legs from the carcase and then cut the knuckles from each leg (**C** and **D**).
4 Cut each leg in half through the joint to divide into thighs and drumsticks (**B**), then remove the knuckle ends.
5 Remove the wings carefully, cutting parallel to the breastbone and through the joint (**E**).
6 Stand the carcase upright (with the parson's nose on top) and cut down between the carcase and the breast to remove the breast flesh.
7 Cut the breast in two.
8 If presenting the joints on the carcase, cut it into three pieces and trim.

Various cuts are achieved from the jointing process, including: legs, thighs, breast, wings, winglets, supremes and ballotines.

Trussing

Chicken and turkey are trussed for boiling and roasting. This keeps the bird in shape (especially important if it is to be carved in the restaurant), and allows the bird to cook evenly (the legs and wings can quickly become overcooked if they are left stretching away from the bird).

Method

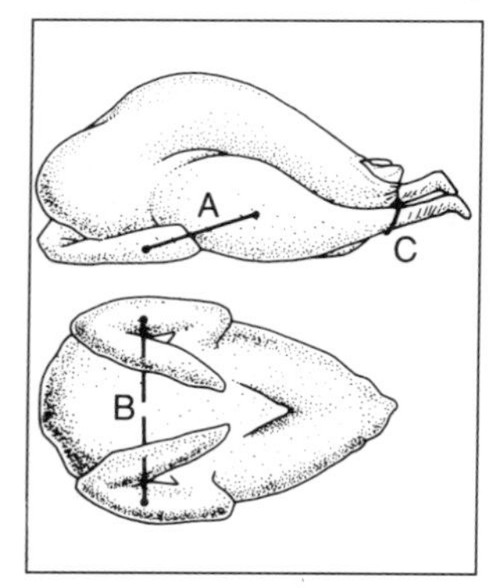

Trussing a bird

1 Prepare a 16–20 cm (6½–8 in) trussing needle, threading it with string.
2 Remove the wishbone: cut the sinew holding it to the carcase and remove whole.
3 Cut off wing tips and all the claws except the centre one on each leg: shorten these by half.
4 Turn the bird onto its back, raise the legs on each side and push them down onto the front of the bird.
5 Hold the legs in position using your left-hand thumb and middle fingers, and push the needle and string through the bottom of the leg (at the joint of the drumstick and thigh).
6 Push the needle right through the body and out the other side at the same position (i.e. at the joint of the drumstick and thigh).
7 Place any stuffing in the neck cavity and pull down the neck flap to seal the stuffing.
8 Turn the bird on its side and push the needle though the middle of the wing joint (on the same side as the leg you have just pulled the string through).
9 Now pass the needle in and out of the neck flap going under the backbone, in a stitch attaching the neck flap to the top of the bird.
10 Push the needle out through the middle of the other wing. The string will now be back on the side you started from (Step 5).
11 Unthread the needle, pull the string gently to tighten the shape of the bird and knot the two ends of the string. Trim off any string beyond the knot to neaten.
12 Thread the needle with more string. Turn the bird leg-side up (breast bone down) and hold the legs down and together.
13 Push the needle through the bottom of the carcase, just below the ends of the drumsticks, and out the other side.
14 Loop the string over the tops of the legs, then push the needle back into the carcase under the leg to come out at your starting position for this piece of string.
15 Tighten the string, knot and tidy the ends.

Batting

Chicken may be grilled whole, but needs to be opened out and flattened. This preparation is known as *chicken spatchcock.* To do this:

1. Remove the wish bone from the cleaned, prepared chicken.
2. Cut off any claws (and winglets if required).
3. Insert a large cook's knife into the abdominal cavity and slice through the backbone, cutting from the neck to the parson's nose.
4. Open out the carcase and lightly bat out (flatten) the bird using a cutlet bat.
5. Remove the ribs and trim the backbone. Place on a clean tray for grilling.

Barding

Chicken and turkey benefit from barding when roasting, to combat the drying-out effect of the process. To do this: cover the breast of the bird with strips of bacon or pork fat before placing in the oven.

Dicing

Poultry (usually chicken) may be cooked and diced as an ingredient for a *salpicon* (savoury filling). Salpicons are used in the preparation of vol-au-vents, crêpes, bouchées, tartlets, etc. Cook the chicken, usually by boiling, then skin and remove the flesh from the bones. Dice the chicken flesh by cutting it into small cubes of around 1 cm (1/2 in) wide.

Diced chicken may also be used when preparing chicken salad.

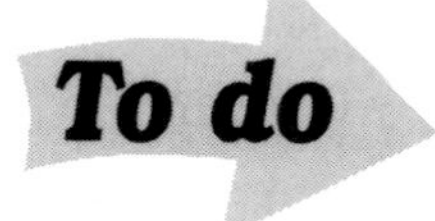

- Find out whether your kitchen uses frozen poultry. If it does, ask your supervisor where the frozen birds should be placed to thaw, and how long different sizes of birds might be expected to take.
- Watch a bird being jointed. Ask what the resulting cuts are to be used for.
- Watch a bird being trussed. Notice the equipment and techniques used, and how long the process takes.
- Ask your supervisor how to truss a bird using the 'one string method'. Find out which method is normally used in your kitchen.

Marinading

Meat and chicken are often marinaded to tenderise the food and add flavour. The marinade may also cause colour changes, depending on the ingredients used. The meat or poultry may be marinaded just before cooking for a short time (e.g. 20 minutes) or steeped in the marinade for anything from 1–12 hours. The marinade itself usually consists of some kind of liquid (e.g. wine, oil, lemon juice), vegetables (e.g. onions, garlic) and herbs. You can increase the absorption of the marinade by pricking the food, but remember that this will also increase fluid loss from the food during cooking.

Method

1. Mix up the ingredients for the marinade in a porcelain or stainless steel bowl, and add the chicken (usually jointed). Do not use an aluminium dish: this can cause discoloration.

2 Turn the chicken pieces until all the surfaces are covered with the marinade and keep chilled for the time specified in the recipe. While chilling, turn the chicken pieces occasionally, basting them in the marinade to ensure maximum absorption.
3 When ready to cook, remove the chicken pieces from the marinade, drain and place under the grill or salamander.
4 Occasionally baste with the marinade during cooking.

What have you learned?

1 Why should you be especially careful when handling and cooking pork and chicken?
2 Why might you truss (or tie) a chicken before cooking?
3 How would you detect *freezer burn* in a frozen chicken?
4 What are the main contamination threats when preparing and storing uncooked poultry?
5 Why is it important to keep preparation and storage areas and equipment hygienic?

ELEMENT 3: Cooking meat and poultry dishes

Customer requirements

You will need to be aware of customer preferences in the cooking of certain meat dishes, notably beef steaks. These may be ordered:

- rare (blue or *bleu*)
- underdone (*saignant*)
- medium (*à point*)
- well done (*bien cuit*).

Test for cooking using grilling tongs (do not pierce the meat). Generally, the rarer the steak, the more springy it will feel; rare and underdone steaks will also show traces of blood (i.e. red juices will appear on the plate).

Never serve pork or chicken underdone: this carries health risks.

Roasting

Roasting is a dry process, where food is cooked in an oven or on a spit with a small amount of fat or oil (see page 13). The process is particularly suitable for meat and poultry, resulting in tender joints, and allowing the meat juices to be used for gravy.

Time and temperature

- The type of meat and the size and shape of the joint will affect cooking time and temperature. When considering the size and shape, note particularly the thickness of the joint: the thicker it is, the longer it will take to cook. Long, flat shapes cook faster than short, thick ones.
- The quality of the meat or poultry also affects cooking time. Older, tougher joints will need to cook for longer than younger, tender ones. A boned joint will also take longer than an unboned one, as bones conduct heat well, carrying it to the centre of the joint.
- The higher the temperature, the quicker the food will cook. However, cooking at a high temperature can burn the outside while leaving the

inside uncooked, and cause shrinkage and fluid loss. Most joints are therefore started at a high temperature (225 °C/440 °F) for a short time (to develop colour) and then the cooking completed at a lower temperature (150-175 °C /300–350 °F).
This is particularly suitable for larger, tougher joints and those that must be cooked thoroughly (e.g. pork).

- If cooking large joints of meat or those that need to be well done, you may need to roast them at a high temperature for the entire cooking period. This can cause the meat to dry out; make sure the joint is barded or larded sufficiently.
- Cooking joints at a low temperature can also cause drying out because the cooking time is increased as the temperature is lowered. Turkey is cooked at a low temperature (165 °C /325 °F), as it benefits from the very slight colouring achieved, and needs to be thoroughly cooked.
- Stuffed joints or poultry should be cooked at approximately 165 °C (325 °F). This allows the heat to penetrate sufficiently to the stuffing, killing any bacteria there. A higher temperature would cook the outside too quickly, and a lower one would allow any bacteria to multiply.
- Remember that forced-air convection ovens cook more quickly than the older style ovens, so you will need to check what kind of oven you will be using when estimating cooking time.
- The top section of an oven is the hottest, and the lowest section the coolest. Roast joints in the centre part if possible.
- Do not baste too frequently: every time you open the door the temperature will drop.

Testing for cooking

Use a temperature probe to test for cooking: insert it into the thickest part of the joint before placing the food in the oven.The internal temperatures reached should be as follows:

- chicken and turkey: 75 °C (167 °F)
- rare meat (beef, mutton): 55–60 °C (130–140 °F)
- medium meat (beef, mutton, lamb): 66–71 °C (150–160 °F)
- well done meat (beef, veal, lamb, pork): 78–80 °C (172–176 °F).

If you are unable to use a temperature probe, test for cooking by pressing against the surface of the meat and checking the colour of any juices that run out: red juices indicate that the meat is underdone, pink juices indicate a medium degree of cooking and clear juices mean that the meat is cooked thoroughly.

Essential knowledge

Always cook meat and poultry for the correct time and at the correct temperature. Failure to do so may result in:

- food poisoning. High risk foods such as chicken and pork must always be thoroughly cooked, to kill off bacteria or worms that may be present
- an incorrectly cooked meat or poultry dish. The finished dish depends on your achieving the right degree of cooking at each stage
- food shrinkage. Some foods will shrink if cooked for too long a time or at too high a temperature
- loss of moisture. Food will become dry if cooked for too long or too fast using a dry process
- an uncooked stuffing. The temperature must be high enough to cook the stuffing and prevent bacteria from multiplying, while not so high that the outside (meat) is burned.

Method

1. Prepare the joint and any mirepoix (bed of vegetables or bones, see *Points to consider* below).
2. Place the mirepoix into the roasting tray and lay the joint on top.
3. Brush any unbarded joints with oil, then place into the oven. If you are using a combination of temperatures for the joint, start cooking at 225 °C (440 °F) to brown, then reduce the temperature to 150–175 °C (300–350 °F).
4. Cook until the correct internal temperature is reached (see page 46).
5. Remove the joint from the oven and allow to rest in a warm place for 5–15 minutes (depending on the size of joint).
6. Cut away any string and carve.

Points to consider when roasting

- Never place a joint of meat directly onto a roasting tray: the bottom of the meat will start to fry. Raise the meat either by placing it onto a metal trivet resting on the bottom of the tray, or by placing it on top of roughly chopped vegetables. This vegetable base is known as a *mire-poix*, or bed of roots.
- Always cook poultry with the bird resting on its side, e.g. 20 minutes on the left leg, 20 minutes on the right leg, 20 minutes on its back. Meat joints should be roasted with the barded side uppermost.
- Bear in mind any customer requirements concerning the degree of cooking.
- Check that the oven shelves are in the correct position, allowing enough room for the size of joint you are roasting.
- Note all points listed under *Time and temperature* (page 45). Do not forget to pre-heat the oven.
- Check that the tray is large enough to hold the joint or bird, and deep enough to catch the juices during cooking.
- You will need to baste the meat or poultry every 20–30 minutes.
- Roasted items have some *carry-over* cooking time. When the roasted item is taken from the oven, it will continue to cook for a short amount of time (5–15 minutes depending on size). Keep the joint or bird in a warm place while this takes place. This reduces the risk of the person carving being burned and makes the joint or bird easier to carve. All poultry should be kept vent upwards during carry-over cooking to allow cooling to happen quickly, and therefore preventing any bacteria from multiplying.
- Take care when lifting large roasts from the oven. Remember that the fat is extremely hot, so always protect your hands and wear long-sleeved clothes to shield yourself from spitting fat.

Meat and poultry suitable for roasting

Beef: sirloin (boned or unboned), wing rib, fillet, fore rib, rump, middle rib, topside (a second class roasting joint).
Veal: saddle, best end, loin, breast, shoulder, leg, cushion, under cushion, thick flank.
Lamb: best end, saddle, leg, breast, shoulder.
Pork: spare rib, loin, leg, shoulder.
Poultry: chicken and turkey.

Example recipes: Roast turkey with chestnut stuffing, Roast saddle of lamb.

Boiling

Boiling is a wet cooking process, where food is cooked in boiling liquid (see page 5). It is suitable for many types of meat and poultry for several reasons:

- the food does not dry out or shrivel
- the food can be flavoured by the cooking liquor and its additives
- the cooking liquor is enhanced by the juices from the food and can be used as a base for stocks, soups, etc.
- the high temperature of the water will kill most forms of bacteria
- tough meat can become more tender when cooked in this way. It is therefore particularly suitable for tough, older cuts of meat and for pickled or salted meats.

Bacon cuts suitable for boiling

Method

1. Place the meat or bird into a saucepan, cover with cold water and bring to the boil. (This stage may have to be repeated two or three times for some salted meats, see *Points to consider* below.)
2. Allow the food to simmer, occasionally skimming off any impurities that have risen to the surface (to prevent these from discolouring or affecting the flavour of food).
3. Towards the end of the cooking time, add any vegetables, dumplings, etc. so that the amount of time they need to cook coincides with the further amount of time the meat or poultry needs to cook; i.e. if you are adding a vegetable that needs to simmer for 30 minutes, add it to the pan when the main item has 30 minutes left to cook.
4. When cooked, remove the food from the pan, keeping back the cooking liquor if necessary.
5. *Meat:* remove any excess fat then slice across the grain. *Poultry:* skin and cut into joints or as needed for recipe.

Points to consider when boiling

- The joint should be an even shape so that heat penetrates evenly through it during cooking.
- Choose a large enough saucepan to hold the joint of meat or poultry and any food to be cooked in the same liquid (e.g. dumplings).
- Consider whether the meat or poultry should be started in cold or boiling water. Starting meat in cold water will remove any undesirable fat or blood, and any excess salt from cured or pickled meats. For very salty meats, like gammon, you will need to repeat Step 1 several times, starting with fresh cold water each time.

- Make sure that you have allowed sufficient cooking times for starting with cold water in your time plan.
- Check in advance whether you need to preserve the cooking liquor for sauces, stocks, etc.

Meat and poultry suitable for boiling

The following cuts of meat are particularly suited to boiling because they are either fairly tough (and will become more tender through the process) or salted (and the salt will be drawn out by the process): leg of mutton, silverside and brisket (salted or fresh), thick flank of beef, tripe, tongue, ox tail, pork belly, gammon joints, hock and collar bacon joints, ham, boiling fowls (hens).

Example recipes: Boiled beef and dumplings, Boiled leg of mutton (or lamb) with caper sauce.

To do

- Watch steaks being prepared rare, underdone, medium and well done. Notice the differences in the appearance of the cooked steaks.
- Find out how long and at what temperatures you would expect to roast: a leg of lamb, topside of beef, a pork loin.
- Find three example recipes for boiled meat dishes. Notice the differences. How long does each need to be boiled for? Is the meat started in cold or hot water? Are the final dishes served hot or cold?

Grilling, barbecuing, tandoori cooking

Grilling

Grilled meat cuts

This is a dry cooking process, cooking food by radiated heat generated by infra-red waves (see page 15). The process is suitable for cooking many types of meat and poultry, especially where health considerations are important; no fat need be added and some of the fat from the meat itself will melt and drip away. The process can have a drying-out effect however, and the process is not suitable for very lean cuts of meat (e.g. veal cutlets); most meats and poultry will benefit from a light brushing of oil.

Grilled food also benefits from the effects of *browning*: i.e. the food surface develops a good colour (becoming brown) and a full flavour.

Method

1. Brush the grill bars with oil then pre-heat the grill.
2. Prepare the meat or poultry cuts and brush lightly with oil or any marinade.
3. Place them onto the hot grilling trays or bars.
4. Turn the food when half-cooked.
5. Complete cooking and serve immediately.

Grilling on skewers (kebabs)

Kebabs are usually marinaded before grilling.

1. Prepare and dice the meat and any vegetables.

2 Place them into the bowl containing the marinade and allow to stand as dictated by the recipe.
3 Thread the meat and vegetables onto the skewers.
4 Place the skewers under a pre-heated grill, turning and basting with the marinade as necessary until cooked.

Points to consider when grilling

- Browning should occur at the same rate as internal cooking. Adjust the distance between the food and the grill as necessary to avoid the surface burning before the inside is cooked.
- Meat that is marbled with fat will remain more tender than meat with surface fat. All grilled meats should be tender with a reasonable amount of fat.
- Try to avoid adding salt to food that is to be grilled: it is thought to slow colour development.
- Cooking time: this depends on the shape and size of the food, the distance of the food from the grill bars, the heat of the grill and customer requirements. Small, thin items will cook more quickly than large, thick ones which may need to be browned at a high temperature and then finished under reduced heat.
- As with roasting and shallow-frying, meat (especially beef steaks) may be ordered to a particular degree of cooking (see *Roasting*, page 45). Keep the customer requirement in mind when deciding on length of cooking time. *Never* serve pork or chicken underdone.
- Make sure that your time plan has allowed for any marinading time.
- Do not allow foods to dry out while cooking: brush with oil (or marinade) as necessary.
- Never pierce food during cooking; turn when necessary using tongs or a palette knife.
- Serve grilled food immediately; it will dry out quickly.

Barbecuing

While grilling can take place over or under the heat source (or even between two heat sources), barbecuing always involves cooking meat *over* the heat source. The process is very similar to grilling, and the same cuts of meat and poultry are suitable for both processes.

Brush the bars of the barbecue with oil before cooking, to prevent the food from sticking to them. Notice which parts of the barbecue are hotter or cooler than others; you may need to start food cooking on the hotter parts and then transfer it to complete cooking on the cooler parts. Note all the points listed under *Grilling* above.

Tandoori cooking

This is a dry process form of cookery, where food is cooked by a combination of radiant, conductive and convective heat. The clay oven is fired by gas or wood charcoal, providing a direct heat source at the foot of the oven. However, since the ovens are made from clay, which radiates heat evenly, the oven is no hotter near the base than at the top. Meat, poultry or fish is threaded onto very long skewers which are then inserted into the oven from the top. Food items at the top of the skewer (nearest the top of the oven) will cook at the same rate as those at the bottom of the skewer. No fat or oil is used in the process, which is advantageous for health reasons.

The food is cooked in a similar way to barbecued food (over the charcoal or gas), but also by convective heat from being inside an enclosed space (the oven). This means that food tastes similar to barbecued food, but cooks more quickly, as the overall temperature is much higher. A tandoori oven on medium to high heat has a temperature of approximately 375 °C (700 °F). This means that you can cook much larger pieces of meat or poultry than would be possible on a barbecue: a whole chicken could be successfully cooked in a short period of time (approximately 30 minutes). The temperature of the charcoal fired ovens is controlled by the air supply: the more air fed into the oven, the higher the temperature. Gas fired ovens can be manually controlled.

Preparing tandoori chicken

Meat and poultry cooked in a tandoori oven is generally marinaded first, as the marinade helps to secure the food to the skewers. The distinctive flavour of tandoori-cooked food comes from both the marinade and the cooking process. The food is marinaded for several hours before being cooked in the oven. The spices for a tandoori marinade need to be well-cooked at a high temperature; if the food was to be grilled rather than baked in a tandoori oven, the marinade would need to be cooked before being applied to the food items.

Most forms of meat and poultry can be successfully cooked in this way.

Example recipes: Chicken tikka, Chicken tandoori.

Meat and poultry suitable for grilling, barbecuing and tandoori cooking

Beef: steaks, e.g. entrecôte, double entrecôte, minute, T-bone, Porterhouse, rump, chateaubriand, fillet, tournedos; mince from tougher cuts (as hamburgers, sausages, etc.); diced beef cuts (for kebabs).

Lamb: cutlets, Barnsley chops, chump chops, loin chops, noisettes, rosettes.

Pork: pork chops, pork cutlets, bacon slices, gammon steaks.

Poultry: chicken and turkey breasts, whole young chickens (*poussins*).

Offal: from young animals, e.g. lamb, veal. (Note that liver is passed through seasoned flour then brushed with oil before grilling).

Example recipes: Grilled fillet steak garni *(Filet de boeuf garni)*, Mixed grill.

Shallow-frying, griddling, stir-frying

These are dry cooking processes, where the food is heated through contact with a hot cooking surface and hot fat or oil (see page 17). Shallow-frying generally takes place in a frying pan, sauté pan or sauteuse, while griddling takes place on a solid metal plate. One advantage of griddling rather than shallow-frying meat is that frying can then take place on a ridged-surface griddle, which allows the fat to drain away. This is useful for people wishing to keep down the fat content of their diet.

Stir-frying uses the same principles as shallow-frying, but takes place in a wok: a large, bowl-shaped cooking pan which allows food to be tossed quickly with minimum spillage over the sides of the pan (see page 18). The food is cooked very quickly at a high heat, and therefore only suits very small cuts of meat or poultry (e.g. thin strips) which can be thoroughly cooked in a very short time.

Meat and poultry are suitable for shallow-frying for the following reasons:

- browning occurs, enhancing the colour and flavour of the meat and poultry cuts
- the food cooks quickly, which suits the more tender cuts of meat
- the tougher cuts of meat can be minced or chopped then shaped and fried
- meat with a low fat content (e.g. veal) which would dry out if grilled, will be kept moist using this process.

Method

1. Trim any excess fat from the prepared meat cuts. Add any seasoning.
2. Lightly brush the cooking pan with oil and place on the stove to heat.
3. Add the meat or poultry items carefully to the hot pan and fry, presentation side down. Do not add too many items to the pan at once: the temperature will drop and the food may start to simmer in its own juices rather than fry in the oil.
4. Turn the items when they are approximately half-cooked. (Small drops of blood will appear on the top surface at this point.)
5. Cook to the required degree (noting customer and dish requirements).
6. Remove from the pan and drain off any excess fat. Keep warm if preparing a sauce; otherwise serve immediately.

Cooking *pané* (with breadcrumbs)
Some items may be shallow-fried with a coating of breadcrumbs. To prepare food in this way: dip the item into seasoned flour, then into egg-wash and finally into breadcrumbs. Fry following the method given above.

Points to consider

- Is the food tender and small enough to be cooked successfully by shallow-frying?
- Have you made ready all the equipment and ingredients you need before starting? (Items cook very quickly when shallow-fried, and you should avoid having to reduce the heat while you fetch or prepare other items or equipment.)
- Most meat cuts need to be fried at a temperature of 125–175 °C (250–350 °F), with the thicker cuts towards the lower temperature and thinner cuts towards the higher temperature. *Never* undercook pork or chicken. You may need to start some (e.g. thick) items at a high temperature and then lower the heat to complete cooking.
- Is the pan at the right temperature? Do not cook the items at too high a temperature: the food will burn and dry out. If the temperature is too low, the items will absorb too much oil.
- What are the customer requirements for degree of cooking? Should the item be rare, underdone, medium or well-done (see *Customer requirements,* page 45)?
- Are some of the items to be cooked larger than others? If so, these will take longer to cook, and will need to be added to the pan before the smaller cuts.

Suitable meat and poultry cuts

Beef: e.g. steaks (entrecôte, double entrecôte, minute, tournedos, fillet, escalope, médaillons, bâtons, rump, point, filet mignons).
Veal: cutlets, chops, escalopes.
Lamb: cutlets, gigot chops, chump chops, Barnsley chops, noisettes, rosettes, fillet.

Pork: chops, cutlets, escalopes, back bacon, gammon steaks.
Offal: liver (calf, lamb, chicken), kidneys (calf, lamb), sweetbreads (calf; with flour and breadcrumbs).
Poultry: chicken or turkey breast, chicken legs (thighs and drumsticks) and wings.

Example recipes: Chicken Maryland, Pork escalopes with Calvados sauce.

Braising or stewing

Braising and stewing are wet processes, where food is partially covered by liquid and cooked either in the oven or on the stove (see pages 9–13). When braising meat or poultry, the resulting cooking liquor is often used as a base for a sauce, while the cooking liquor of stews becomes part of the dish itself.

Both braising and stewing are lengthy processes, and therefore suit the tougher cuts of meat very well. Whole joints of meat or poultry can be braised (they may need *barding*, see page 40), while the particularly tough cuts can be diced and stewed. Both methods usually require the meat to be browned first (either by shallow-frying or flash-roasting), to enhance colour and flavour. However, some white stews (i.e. of tougher cuts of white meats and poultry) often require you to *blanch* rather than *brown* the meat or poultry pieces before stewing.

Brown stews

Here the food (meat or poultry and vegetables) is *browned* before stewing and an appropriate brown stock is used. The meat and vegetables are coated with flour during the shallow-frying stage to thicken the cooking liquor while cooking, so that the food cooks in a sauce which becomes part of the finished dish.

White stews

Blanquette
Here the meat or poultry is *blanched* before stewing. This removes any impurities from the meat or poultry which would discolour the finished dish. The vegetables are added raw, and the cooking liquor is used as a base for the sauce (velouté) made once the food is cooked.

Fricassée
Here both the meat or poultry and vegetables are first shallow-fried without browning (*sweated*). Flour is added at this stage so that the meat and vegetables cook in a thickened sauce which forms part of the finished dish.

Both forms of white stew use white stock and may have a liaison (egg yolk and cream beaten together) added before service to enhance colour and flavour and to correct consistency (the liaison will thicken the stew).

Ingredients: stewing and braising

Cooking liquid
This can include stock, wine, jus lié, tomato purée, seasoning, herbs.

Vegetables
When *braising*, these are chopped and placed under the main item. This base of vegetables is known as a *mirepoix* (or *bed of roots*). When *stewing*, the chopped vegetables are mixed in with the meat or poultry and cooking liquid (see *Brown stews* and *White stews*, page 53). Vegetables commonly used are carrots, onion, celery, leeks and garlic. Any vegetables used must be able to retain their structure (i.e. not disintegrate) during the long cooking process.

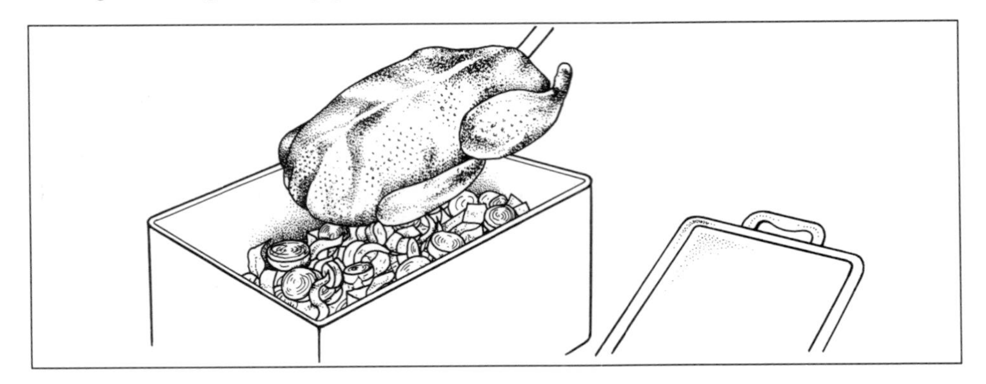

A braising pan with mirepoix

Method: braising

1. Prepare the meat or poultry and roughly chop the vegetables.
2. Brown the meat and vegetables by shallow-frying. Large joints may be flash-roasted (i.e. placed for a short time in a very hot oven).
3. Arrange the vegetables on the bottom of the dish, place the main item on top and add the cooking liquid to cover two-thirds of the main item.
4. Place the dish on top of the stove and bring it to the boil.
5. Skim off any surface fat or impurities, cover the dish with a lid and transfer to the oven. The oven temperature should be 180 °C (360 °F) approximately.
6. When cooked, remove the main item and keep it warm while you strain the cooking liquor.
7. Thicken or reduce the cooking liquor to form a sauce.
8. Slice the main item if required, place it into a serving dish and coat with the sauce.

Method: stewing

1. Dice the prepared meat into 2 cm (3/4 in) cubes.
2. Roughly chop any vegetables.
3. *If cooking a brown stew:* shallow-fry the meat in a very hot pan until browned, then add the vegetables and continue cooking for 6–8 minutes. Add a small amount of flour and stir in to coat the meat and vegetables, cooking for a further 4 minutes.
   If cooking a white stew: blanch the main item and place it with raw vegetables into a pan.
4. Add any tomato purée and the cooking liquid, pouring it in a little at a time and stirring continuously.
5. Bring the pan to the boil and then lower to a simmer. You can continue to simmer the food either on top of the stove or in the oven. This stage may take anything from 1–3 hours. The oven temperature should be 180 °C (360 °F) approximately.
6. While cooking, skim off any excess fat or impurities and top up the cooking liquid if necessary.
7. Complete cooking. *If cooking a blanquette:* whisk in the liaison and do not allow the sauce to re-boil or curdling will take place. *If cooking a*

fricassée: remove the meat when cooked, reduce the cooking liquor to pouring consistency, strain the liquor, add the liaison, then coat the meat with the sauce. *All stews:* check consistency and seasoning, transfer to a serving dish and serve immediately.

Points to consider when braising or stewing

- Both stewing and braising are long processes. Make sure you have allowed enough time for them in your time plan.
- Make sure that you have allowed enough time for any meat or poultry preparation (including soaking).
- Always find out whether the stocks you will need are already made.
- When considering whether to oven cook or cook on the stove, check which area will have the most free space at that time.
- Make sure that you know what quantity of stew or the braised dish makes up one portion.

Meat and poultry suitable for braising and stewing

Stewing

Beef: shin, topside, silverside, thin and thick flank, chuck steak
Lamb: middle neck, breast, shoulder, chump
Pork: shoulder, belly, spare rib
Veal: neck, shoulder, breast, knuckle
Poultry: chicken
Offal: ox kidney

Braising

Beef: topside, thick flank, middle rib, chuck
Lamb: shoulder, chump
Pork: shoulder
Veal: neck, shoulder, best end, saddles, leg cuts
Poultry: turkey winglets, boned and stuffed chicken legs
Offal: ox liver, heart and tongue; lamp tongue, heart and sweetbreads; veal sweetbreads and tongues.

Example recipes: Braised stuffed shoulder of veal (*Épaule de veau farcie*), Braised sweetbreads (*Ris de veau braisé*), Coq au vin, Beef Burgundy style (*Boeuf bourguignon*).

- Watch some beef cuts being grilled. Notice the temperature, the distance between the meat and the grill bars, the length of cooking time and the appearance of the cooked meat.
- Find: a plat à sauter, a frying pan, a sauteuse and a wok. Notice the differences, and find out when you would expect to use each.
- Watch a brown stew and a white stew being prepared. Notice how the ingredients, methods and finished dishes differ.

What have you learned?

1. Why are time and temperature important when cooking meat and poultry dishes?
2. Why is it important to keep preparation and storage areas and equipment hygienic while cooking meat and poultry?
3. What is *carry over* cooking time? Why is this important?
4. How does braising differ from stewing?
5. Which beef cuts are suitable for frying?
6. When would you use a mirepoix?

Extend your knowledge

1. Find out about the different types of food poisoning, their causes and effects.
2. Find out how meat and poultry can be successfully frozen, and why this is done.
3. Find out the buying costs for the different types of meat. Notice the menu price of dishes using these different meats.
4. Find out about the different types of game that can be cooked, and what processes are commonly used.

UNIT 2D2

Preparing and cooking fish dishes

ELEMENTS 1 AND 2: Preparing and cooking fish dishes

What do you have to do?

- Prepare fish correctly and as appropriate for individual dishes.
- Combine prepared fish with other ingredients ready for cooking where appropriate.
- Prepare preparation and cooking areas and equipment ready for use, then clean correctly after use.
- Finish and present fish dishes according to customer and dish requirements.
- Cook fish dishes according to customer and dish requirements.
- Plan your work, allocating your time to fit daily schedules, and complete the work within the required time.
- Carry out your work in an organised and efficient manner, taking account of priorities and any laid down procedures or establishment policy.

What do you need to know?

- The type, quality and quantity of fish required for each dish.
- What equipment you will need to use in preparing fish.
- Why time and temperature are important when preparing and cooking fish.
- Why it is important to keep preparation, cooking, storage areas and equipment hygienic.
- What the main contamination threats are when preparing, cooking and storing raw fish and cooked fish dishes.
- How to satisfy health, safety and hygiene regulations, concerning preparation and cooking areas and equipment, and how to deal with emergencies.

Introduction

The catering industry in the UK is well supplied with a wide range of both salt and freshwater fish. A highly organised fishing industry distributes its catch in prime condition, either fresh, frozen or chilled. Extensive farming of freshwater fish has also increased the availability of certain species, so that certain types of fish are no longer regarded as luxury commodities. However, exotic species of fish from other parts of the world are also readily available and appear regularly on menus.

Fish, as a food, is an important contributor to a well-balanced diet, being high in protein, rich in certain vitamins yet low in fat and easy to digest. It does, however, require careful handling and cooking to maximise its benefits to the consumer.

The correct place of fish on the formal menu is as a separate course directly before and as a contrast to the meat course. Today fish is often used in other sections of the menu, for instance as an hors d'oeuvre or as part of a salad, savoury or farinaceous dish. In modern menus fish is frequently offered as a main course choice suitable for luncheon or dinner and this reflects its popularity with the modern palate.

ELEMENT 1: Preparing fish for cooking

Classification

Fish are normally classified for culinary purposes as *round* or *flat* fish. This very broad classification may be sub-divided into *white* and *oily* fish, encompassing the freshwater and sea-water varieties.

The following tables give examples of some fish in general use and a comparison of white and oily fish.

Fish classifications

Flat white	Round white	Round oily
Brill	Cod	Anchovy
Dover sole	Haddock	Eel
Halibut	Hake	Herring
Lemon sole	Huss	Mackerel
Plaice	Whiting	Red mullet
Turbot	Monkfish	Salmon
		Salmon trout
		Sardine
		Trout
		Tuna
		Whitebait

A comparison of white and oily fish

White fish	Oily fish
White flesh	Dark flesh
Oil is stored in the liver	Oil is distributed throughout the flesh
May be round or flat	Always round
Vitamins A and D found only in the liver	Vitamins A and D dispersed throughout the flesh

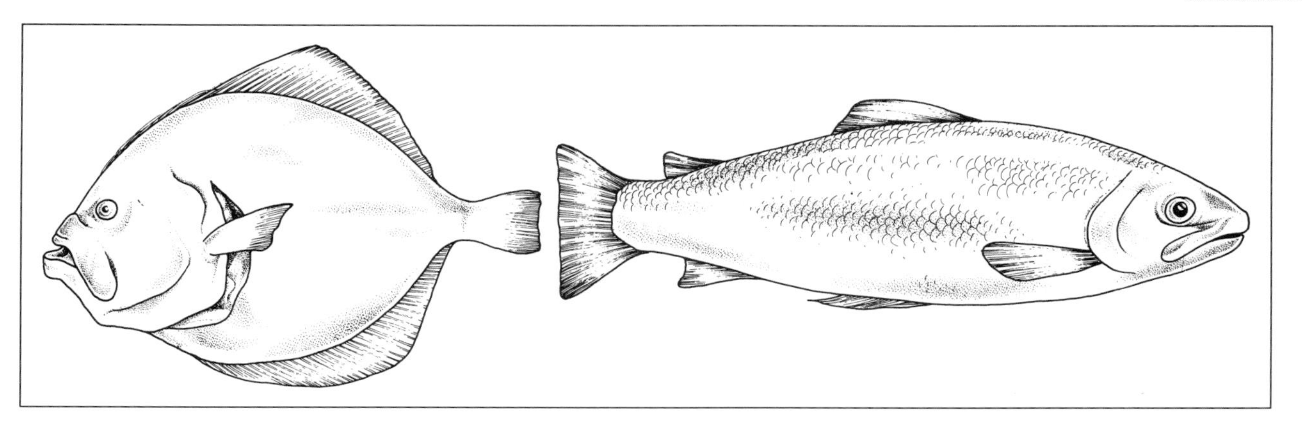

Left: 'flat' fish
Right: 'round' fish

Quality points

Absolute freshness is essential when purchasing fresh fish to ensure a high quality dish. The flavour and appearance of fresh fish quickly deteriorates if it is stored incorrectly or not used within a short space of time. This means that fish must be purchased daily from a reputable supplier if standards and quality are to be maintained.

Frozen fish, if properly processed, defrosted and prepared, is a useful substitute when the fresh item is not available. Note, however, that it is not suitable for all recipes. This is because the cellular structure of the fish changes as it freezes. When preparing mousselines for example, it is difficult to achieve good results with fish that has been frozen because the protein needed to bind the cream with the fish is broken down during the freezing process.

Factors indicating freshness in whole wet fish

- Scales: plentiful, firmly attached, moist and shiny.
- Eyes: bright, clear and full (not sunken).
- Skin: bright, with a sheen and evidence of slime.
- Gills: bright when lifted, deep pink in colour, not sunken or dry.
- Flesh: firm to the touch (the flesh should spring back to its original shape when pressed).
- Smell: wholesome and pleasant, no hint of ammonia or any offensive odour.

Factors indicating freshness in cuts of wet fish

- The flesh should be firm to the touch.
- The fish should have a clean and pleasant smell.
- There should not be any areas of discoloration (through bruising or blood clots).

Factors indicating quality in frozen fish

- There should not be any evidence of dehydration or freezer burn.
- The packaging, if applicable, must be undamaged.
- There should be a minimum fluid loss during thawing.
- Once thawed the flesh should still feel firm.

Storing fresh fish

Fresh fish should be purchased and used daily where possible. However, due to fluctuations in business you may need to store fish for short periods.

When storing fish:

- wash the fresh fish thoroughly under cold running water and store in a refrigerator designed specifically for storing fish
- place the fish in a container with holes at the base. Cover the fish with crushed ice to prevent any surfaces from drying, allowing the drips from the ice to drain away through the holes in the container. Change the ice daily
- store the fish immediately after delivery at a temperature of 1–2 °C (34–36 °F); just above freezing point
- store whole fish separately from fillets to minimise the risks of cross-contamination
- store frozen fish at –18°C (0 °F) or below and use it in strict rotation
- store ready-prepared, cooked fish products separately from raw products to minimise any risk of transference of harmful bacteria from raw to cooked foods
- follow a strict rotation policy on all purchases of fish to ensure maximum freshness.

Storing and thawing frozen fish

Frozen fish must be purchased commercially frozen and then stored in an appropriate freezer at –18°C (0 °F). The length of storage will vary according to the product and the type of packaging (certain products have recommended storage times of one, two or three months, which is stated on the packaging).

Frozen fish must be be defrosted in a refrigerator (perferably overnight) to ensure that it is evenly thawed prior to cooking. Forced defrosting (i.e. in water or at room temperature) is a potentially dangerous practice as bacteria multiply in a warm environment.

Once defrosted, frozen fish must be used quickly, but should be stored as for fresh fish before use.

Under no circumstances should defrosted frozen fish be re-frozen. This is potentially hazardous to health.

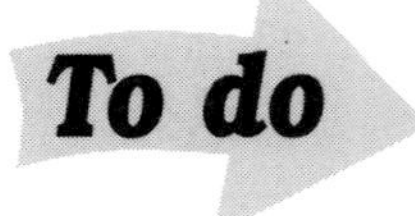

- Identify and categorise four flat white fish, four round white fish and four oily fish currently stored in your establishment.
- Prepare a written checklist for inspecting a delivery of fresh fish.
- Check a delivery of fish against your checklist and discuss the results with your supervisor.
- Check the current operating temperature of your fish refrigerator using a thermometer.
- List the items currently in stock in the fish refrigerator and check their dates of purchase to ensure a strict rotation policy is being observed.

Health, safety and hygiene

Observation of the basic rules of food safety and hygiene (as outlined in Units G1 and G2 of the Core Units book) are critical when handling fish as you will be handling some high risk foods.

Note the following points before you start to prepare or cook any fish dishes:

- always use the correct colour-coded board for preparing raw fish, using a different board for cooked items
- do not work directly on the table. Keep your chopping board clean by using fresh disposable wipes
- use equipment reserved for the preparation of raw fish. Where this is not possible, wash and sanitise equipment before use and immediately after use
- work with separate bowls for fish offal, bones and usable fish; never mix them together as the risk of contamination from the offal is high
- store fish correctly at the correct temperatures (see *Storing fresh fish,* page 59)
- work away from areas where cooked foodstuffs are being handled
- keep your preparation area clean
- wash your equipment, knives and hands regularly and use a bactericidal detergent or sanitising agent to kill bacteria
- dispose of all swabs immediately after use to prevent contamination from soiled wipes.

Essential knowledge

It is important to keep preparation and storage areas and equipment hygienic in order to:

- prevent the transfer of food poisoning bacteria to food
- prevent pest infestation in storage areas
- ensure that standards of cleanliness are maintained
- comply with food hygiene regulations.

Planning your time

When preparing fish, certain preparations and cooking methods are more time-consuming than others. Some dishes may require only minutes and can be undertaken at a specific time (i.e. cooked to order *à la carte* style), while others may either require lengthy preparation before cooking or use fish stocks or court-bouillons which must be prepared in advance.

Identify the basic culinary preparations required to complete the dish to the required establishment standard *before starting any procedure.* Plan your approach to the dish carefully, addressing the longest and most time-consuming jobs first. Make sure that you have everything in place before starting to prepare a dish. Familiarise yourself with all ingredients and cooking and storage methods necessary.

Equipment

Before starting to prepare fish, decide what equipment and utensils you will need to complete the process. Place them ready to use, making sure they are clean. You may need to use mechanical equipment such as mincers or food processors; do not use these unless you have received instruction from your supervisor on how to use them and you are familiar with the safety procedures outlined in Units G1 and G2 in the Core Units book.

Knives

You will need a variety of cook's knives to prepare fish efficiently. Read Unit 2D10: *Handling and maintaining knives* in the Core Units book and make sure that you are familiar with the types of knives and the safety points listed there. Remember that you are less likely to have an accident using a sharp knife than when using a blunt one.

You will need to be familiar with the following types of knives and cutting implements when preparing fish:

- *filleting knife:* a thin, flexible blade ideal for following bones closely
- *fish scissors:* used for trimming fins and tails in fish preparation
- *fish steak knife:* a long, firm blade with a curved end ideal for slicing through the skin, flesh and bones of fish
- *carving knife:* a long, thin, slightly flexible blade used for portioning both raw and cooked fish.

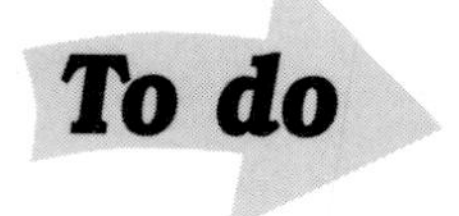

- Prepare a simple written *do's and don'ts* list of procedures when working with fresh fish.
- Prepare two working time plans (see *Planning your time,* page 61): one for fish cooked *à la carte* style (to order), and a second for a fish cooked table d'hôte style or for a banquet or function.
- Identify:
  - all tools from a knife set that may be used in fish preparation
  - three pieces of mechanical equipment that may be used in fish preparation.

Preparation methods

Washing

It is essential to wash all fresh fish under cold running water in order to:

- remove any coating of slime which may be present
- facilitate ease of handling

Washing should be carried out:

- before any preparations commence
- during the preparation process (particularly when removing scales)
- after removing the intestines, to ensure that no traces of blood are left on the bone.

Fish should be washed at all stages of preparation and cutting and immediately before cooking. (The most notable exception is when preparing fresh river trout for the classic dish *Truite au bleu* [Blue trout], where the natural skin slime is essential to maintain the distinctive colour achieved in cooking. In this case only the belly cavity should be washed).

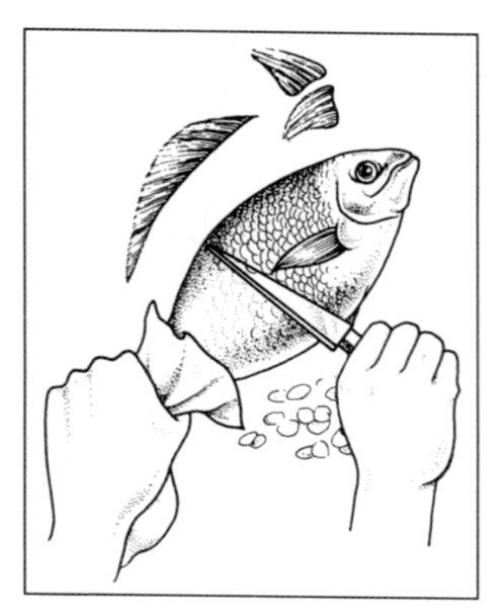

Removing the fins and scales: cut off the fins using fish scissors. Hold the fish by its tail, preferably over the sink or clean paper, then scrape away the scales. Work towards the head.

Trimming

This is the removal of surplus items such as the gills, fins, eyes (where necessary) and head (if appropriate). Trimming is also an essential part of the fish filleting process, necessary to achieve a neat, well presented piece of fish ready for cooking.

Fish scissors are used to trim fish. The technique involves cutting along the natural line of the fish fin where it joins the body. If scissors are not available a cook's knife may be used.

Essential knowledge

The main contamination threats when preparing and storing uncooked fish are as follows:

- contamination through incorrect storage temperatures. Fish must be stored at temperatures of 1–2 °C (34–36 °F) and packed in ice which is changed daily
- cross-contamination between cooked and uncooked food during storage. Fish must be stored in separate refrigerators to any other foods and the cooked and uncooked fish kept away from each other
- cross-contamination from unhygienic equipment, utensils and preparation areas
- contamination from inadequate personal hygiene, i.e. from the nose, mouth, open cuts and sores or unclean hands to the fish
- contamination through incorrect thawing procedures when fish is prepared from frozen
- contamination through the fish being left uncovered
- contamination through incorrect disposal of waste.

Gutting

Many of the fish commonly found in culinary use such as cod, haddock, sole, plaice and whiting are delivered already opened and gutted; inspect these carefully to ensure they are thoroughly clean. Further minor cleaning may be necessary, such as scraping any congealed blood from under the spinal vertebrae or removing any hard membrane from the inner surface of the abdominal cavity. Many round fish such as herrings, mackerel, salmon and trout may be delivered whole and these require gutting. A general procedure may be followed as given below.

Method

1. Remove the gills and fins using fish scissors (see page 62).
2. Make an incision from the anal vent along the belly of the fish to two-thirds the length of the fish.
3. Remove the gut or intestines, pulling them out with your fingers or the handle of a spoon (for large fish use the hooked handle of a small ladle).
4. Remove any congealed blood lying under the vertebrae or backbone and wash the fish thoroughly.

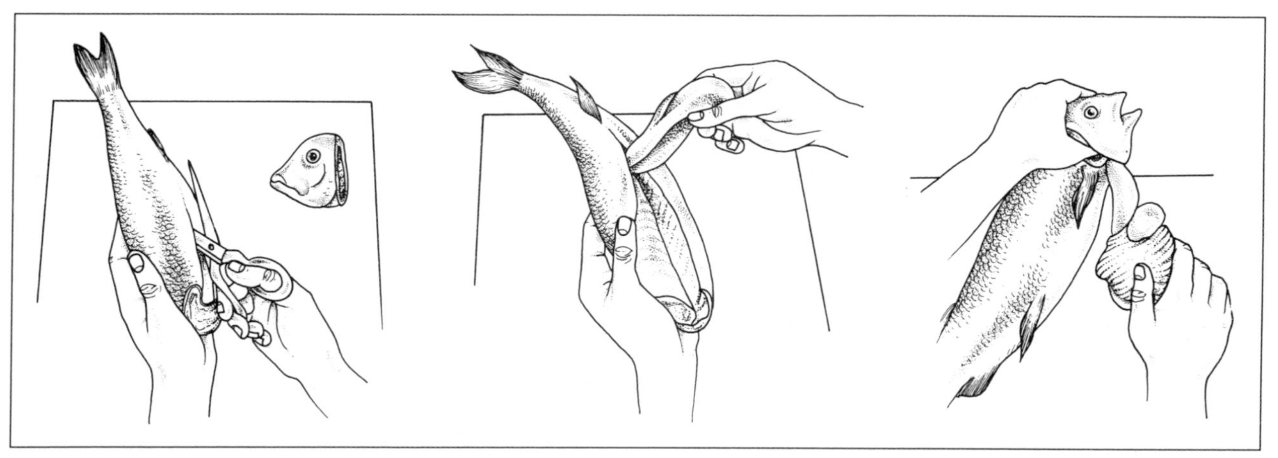

Gutting a round fish

Certain preparations may require the belly to be left intact. If this is the case the gut may be removed through the gills slits. This requires skill and care in handling the fish, and should not be attempted unsupervised.

Filleting

Filleting is the neat removal of the flesh of the fish from its skeleton to yield sections of fish flesh free from skin and bone, making the fish easier to eat. Correctly removed fillets are evenly shaped and smoothly cut without ridges. There should not be any flesh remaining on the skeleton of the fish.

Flat fish yield four fillets (two from each side) which are known as *quarter-cuts*. Small flat fish are sometimes prepared into two fillets (one from each side) which are known as *cross-cuts*. Round fish only yield two fillets (one from each side of the vertebrae). These also require further trimming to remove all bones; many of the smaller bones are removed using stainless steel pliers (kept specifically for that purpose) or tweezers.

Filleting a flat fish

1. Lay the fish flat on the appropriate chopping board.
2. Cut down the natural centre line of the fish to the bone. Make your initial incision as close to the head as possible to minimise loss of flesh; finish at the tail.
3. Work from the centre, cutting with smooth sweeping strokes to the left, keeping the knife pressed against the bone. Detach the fillet from the bone, lifting the fillet away from the bone as you cut.
4. Repeat the process for the second fillet.
5. When the fillets have been completely removed, turn the fish over and repeat on the underside to yield the four fillets.

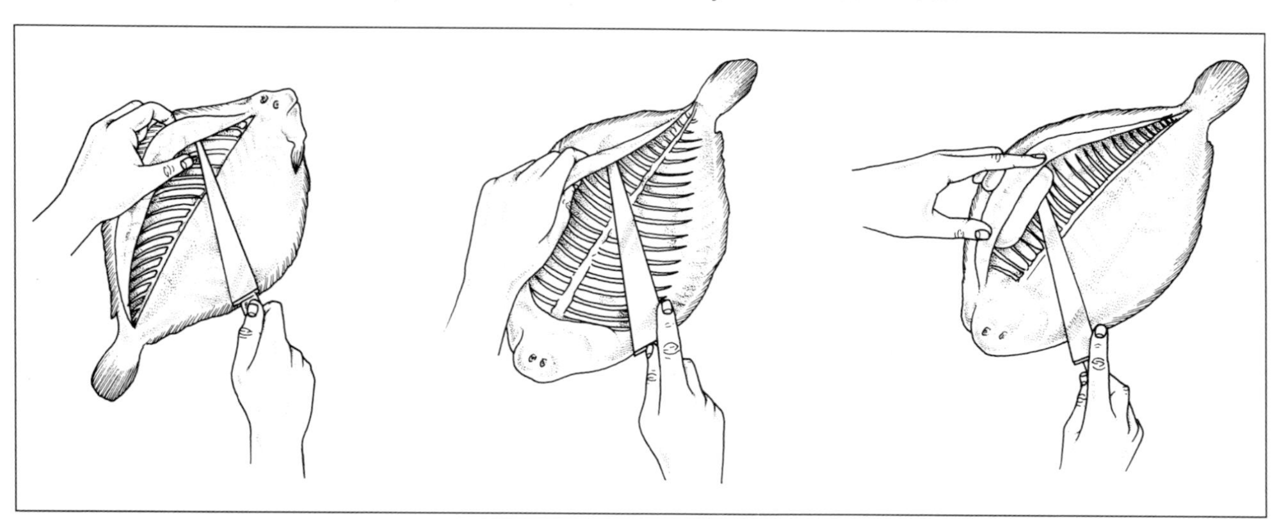

Filleting a flat fish

Filleting a round fish

1. Lay the round fish on the appropriate chopping board (the fish must be prepared for filleting: i.e. cleaned, gutted and trimmed).
2. Working from the head to tail, cut down the backbone following the vertebrae and over the rib cage. Lift the fillet as you cut so that you have more control over the action of the knife.
3. Detach the fillet at the tail and the head.
4. Turn the fish over and repeat to yield two fillets. (The head may be removed before filleting to make it easier to detach the fillets.)

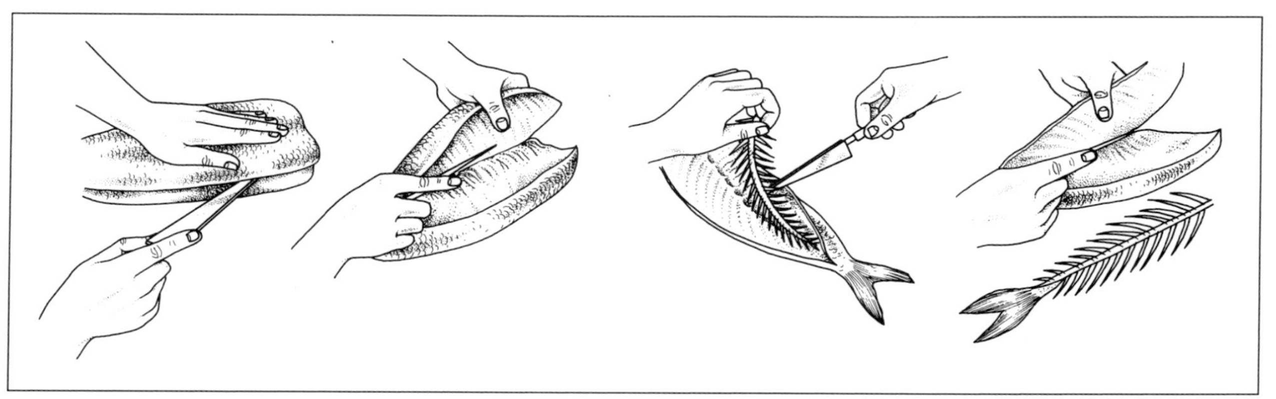

Filleting a round fish
through the backbone

Boning a round fish without separating the fillets
This process is necessary where the recipe requires the fillets to still be attached for presentation purposes. It is commonly used with small round fish such as trout or herring or where the fish is to be filled with a stuffing (farce), when the head and tail are also left attached.

Method

1 Clean the fish in the normal manner but remove the intestine through the gill slits at the back of the head.
2 Lay the fish on the appropriate chopping board with the tail close to you.
3 Using the point of a filleting knife, open the back halfway through on the left side of the fish, following the bone closely. When this is complete, turn the fish over and proceed in the same way on the right side, separating the flesh from the bone halfway through the fish.
4 Hold the backbone between two fingers, then draw the tips of your fingers along the backbone from head to tail, separating the flesh from the bone.
5 Using fish scissors, cut the backbone at the head and the tail.
6 Pull the bone out by gently lifting it up with your right hand while supporting the flesh with your left hand.

This is a difficult procedure requiring some practice and supervision to perfect the technique. The benefits of this method are in the final presentation of the fish dish, the most common example being *Truite meunière Cleopartre*.

To do

- Wash and descale a round fish and a dover sole. How do the two preparations differ?
- Trim a flat fish and a round fish before filleting. Compare the trimmings for quantity, style and ease of removal.
- Gut and clean a fresh salmon or trout. Weigh the fish prior to gutting and again after; compare the weight loss.
- Fillet any flat fish and any round fish. Note the difference in the techniques used and compare the length of time taken to fillet both.
- Prepare a herring or trout leaving the fillets intact and identify a suitable culinary use.

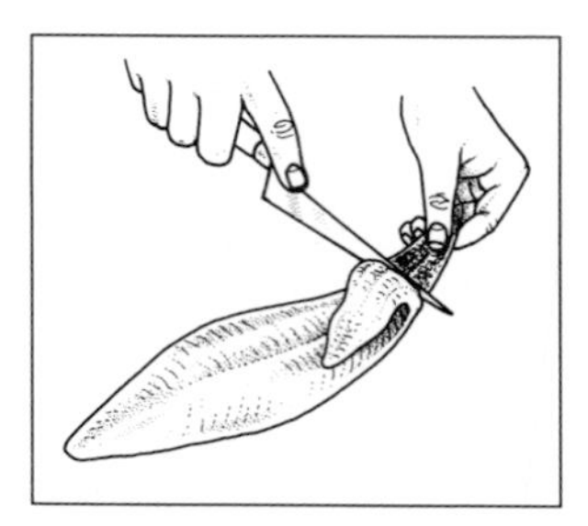

Skinning a fillet of fish

Skinning

It is normal culinary practice to remove the skin from most fillets of fish where this can be done without harming the flesh. The skin should also be removed from whole flat fish, and in all cases the dark skin must be removed before cooking.

Skinning a fillet of fish

1. Place the fillet skin-side down on the appropriate chopping board.
2. Cut through the flesh to the skin at the tail or tip end of the fillet.
3. Hold the fillet firmly by the tail with one hand.
4. Turn the fish filleting knife to an angle of 45°and push and cut forwards with the knife using a sawing motion, at the same time pulling the skin of the fish with the other hand until the skin and flesh are detached from each other.
5. Keep the blade as close to the skin as possible, but do not cut through it: the skin of the fish must be taut at all times to minimise loss of flesh in the skinning process.
6. Trim carefully as required to achieve a neat and sharp appearance.

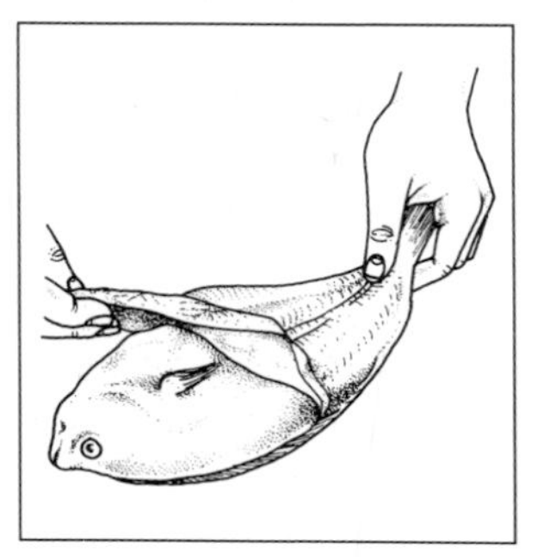

Skinning a whole Dover sole

Flat fish (e.g. lemon sole or plaice) may be skinned whole from head to tail using your thumb and fingers to detach the skin without damaging the flesh.

Dover sole is skinned tail to head with the edges loosened first by the thumb to minimise the risk of tearing the flesh. Dover sole is always skinned before filleting.

Portioning fish

The amount of fish allowed per person will depend on the following:

- the type of fish being served
- the degree of preparation
- the method of cookery and, if appropriate, the sauce or garniture which will accompany it
- the cost of the raw commodity
- the pricing structure or policy of the establishment
- the customer requirements
- the menu type the fish appears on; i.e. table d'hôte, à la carte or banquet
- whether it represents a separate course as part of a menu or a main course choice.

You will need to discuss portion size with your supervisor in order to avoid making costly mistakes, taking into account the points listed above. Fish is portioned not only for effective cost control but importantly for *standardisation of cooking times*. By dissecting the fish into similar muscle tissue forms we control and select the most appropriate method of cookery to maximise the quality and flavour of a delicate product.

The *three P's* are essential for achieving excellent results in fish preparation:

- Purchasing
- Preparation and
- Portioning.

The quality of every finished fish dish depends upon the quality of preparation prior to cooking.

Stuffing fish

Whole round fish

When a round fish has been boned with the head and tail intact (see page 65), the cavity in the belly forms an ideal pocket for filling with a stuffing to enhance the flavour of the fish and gives an added dimension to the concept of fish cookery.

The filling or *farce* may be a simple mixture of breadcrumbs, herbs and spices; a duxelle (such as a simple mushroom duxelle); or a more sophisticated fish mousseline or quenelle mixture (where a purée of fish is bound with egg white, cream and seasoning to form a light forcemeat).

The basic rules are simple:

- ensure the belly cavity is thoroughly clean and dry
- do not over-fill the fish with any farce or stuffing as this may cause the fish to burst or distort in shape
- allow a longer cooking time for stuffed fish to ensure heat penetration and correct cooking
- select the cooking method appropriate to the type of farce used.

Flat fish

Flat fish may also be boned and stuffed, normally with a mousseline type of forcemeat. This is a complex procedure and requires skill and practice to achieve high quality results both in preparation and in cooking.

Fish fillets

Fillets of fish may also be stuffed. This is simply done by placing a little filling on the skin side of the seasoned fillet and rolling the fillet from head to tail. This is called a *paupiette* (see *Fish cuts and preparations* below).

Fish cuts and preparations

Whole fish may be divided into smaller pieces to accommodate various cookery processes, to maximise cost efficiency and to standardise cooking times (see *Portioning fish,* page 66). Importantly, these cuts often also add to the overall presentation of the dish and usually make the fish easier for the customer to eat.

Cuts of fish. From top left (clockwise): paupiette, en goujons, tronçons, délice, darne

Common cuts identified in fish cookery and menu terminology include:

- fillet or *filet*: the flesh of the fish free from skin and bone, presented as long flat fillets
- délice: a fillet of fish which has been trimmed and neatly folded for presentation prior to cooking
- paupiette: a small fillet of fish lightly flattened, usually spread with a stuffing or fish farce, rolled from head to tail and wrapped in a buttered paper (to retain the shape and protect the flesh from drying out during the cooking process)
- goujons: strips of fish cut at an angle diagonally from a fillet approximately 6–8 cm (2½–3 in) long and 15 mm thick
- supreme: a section of fish cut across and on the slant from a large fillet of fish
- médaillon: cut as for a supreme (above) but further trimmed to a neat oval or round shape
- *steaks*
  darne: a section of fish cut across and through the bone of a whole round fish
- tronçon: a section of fish cut from a large flat fish through the bone.

Note that darnes and tronçons are cooked *with* the skin. The skin is removed before service.

To do

- Skin the fillets of any small flat fish (e.g. plaice or lemon sole) and any fillets of large fish (e.g. salmon). Compare the technique to that of skinning a dover sole.
- Identify four fish that are filleted then skinned and one fish that is skinned then filleted.
- With your supervisor, prepare and cut *darnes* and *tronçons* of fish. Weigh them and record the weight. Prepare and cut *fillets* or *délice* of plaice or sole, then weigh them and record the weight.
- Look at four or five fish dishes and compare the amounts of fish required for a portion. Identify the raw weight costs of the fish used.
- Prepare a fish forcemeat from whiting using a fish mousseline recipe, then prepare paupiettes of lemon sole or plaice. Record the length of time it takes you after the initial filleting stage.

What have you learned?

1. What are the main contamination threats when preparing and storing uncooked fish?
2. Why is it important to keep preparation and storage areas and equipment hygienic?
3. Why is fish a contributor to a well-balanced diet?
4. How may fish be simply classified for culinary purposes?
5. How could you identify absolute freshness in fish purchases?
6. At what temperature should you store:
   - fresh fish
   - frozen fish?
7. What factors are involved when determining portion size?
8. Why must fish be used in a strict rotation policy?

The cookery processes

All of the standard methods of cookery may be applied to most fish. To determine which cookery process to adopt you will need to consider the following:

- any dish, recipe specification, or menu demands
- any customer requirements and relevant establishment policies
- the type of menu on which the dish is to feature
- the quality, type and cut of fish being used
- the quantity of fish being cooked
- the equipment and resources you have available.

You will also need to plan your work to determine the shortest possible time between cooking and serving.

Points to consider

Having selected the most suitable cookery method for achieving the desired result, you will need to consider the points given below in order to ensure the quality of the dish.

- Excessive cooking will render any fish flesh dry and tasteless. Fish is composed of delicate muscle tissue and requires relatively little cooking in comparison to meat and poultry.
- Insufficient cooking may not destroy harmful bacteria that may be present, therefore care in timing and temperature control are essential to achieve a safe product.
- Handling should be kept to a minimum both before and after cooking as contamination can occur at either stage. Ensure that cooked fish is stored above 63°C (145 °C) and that cooked cold fish is stored below 5 °C (41 °C), and preferably as near to zero as possible without causing damage.
- Excessive time in hot storage will cause the fish to dry out.

Care in preparation and storage prior to cooking, and attention to detail to ensure correct cooking must constantly be observed if consistency in quality is to be achieved.

Essential knowledge

The main contamination threats when preparing, cooking and storing fish dishes are as follows:

- contamination through incorrect storage temperatures. Fish stored in a warm environment for a prolonged period may provide a breeding ground for bacteria
- cross-contamination between cooked and uncooked food during storage
- cross-contamination from unhygienic equipment, utensils and preparation areas. Cooking, storage areas and equipment must be correctly cleaned before, during and immediately after use
- contamination from inadequate personal hygiene, i.e. from the nose, mouth, open cuts and sores or unclean hands to the fish
- contamination through incorrect thawing procedures when fish is prepared from frozen
- contamination through the finished fish dishes being left uncovered
- contamination through incorrect disposal of waste.

Baking (oven-cooking) fish

Here fish is cooked by the controlled dry heat inside an oven. The delicate nature of fish flesh means that it requires some protection from the direct heat of the oven. This is achieved by retaining and protecting the natural moisture.

To achieve this you may use one of the following methods:
- cooking the fish in an ovenproof dish with a lid
- cooking it in a foil or greaseproof parcel or bag (this method is called *en papillote*)
- wrapping then cooking it in a protective pastry casing (known as cooking *en croûte*). Puff pastry, filo pastry or brioche dough may be used
- baking it in a protective clothing, e.g.in a savoury custard for quiche or barquettes or within a savoury soufflé
- stuffing the fish with a moist mixture prior to baking and brushing or basting with oil or butter during cooking
- baking in a dish with an accompanying cooking liquor or complete sauce.

Fish suitable for baking
Whole fish: white or oily, round or flat. Cod, hake, bass, haddock, tuna, salmon and pike are amongst the fish most suited to the application of baking.

Cuts of fish: e.g. suprêmes, fillets or darnes.

Menu examples: *Rouget en papillote*, Baked codling fillet with parsley sauce, *Coulibiac de saumon*.

Grilling fish

Grilling is a dry method of cookery using the concentrated heat radiated by gas flames, electrical elements or glowing charcoal.

The heat may be directed:
- from *above* the food; as a salamander or overhead grill
- from *below* the food; as in a charcoal or simulated charcoal grill
- from *above and below* the food; as in some infra-red grills.

Key points

- Grilling radiates a fierce heat so the surface of the fish requires protection to prevent drying. In general, exposed areas of fish flesh will need to be passed through seasoned flour and brushed with oil or butter before cooking. Where the natural skin covers the flesh the importance of using flour is reduced, although some chefs will flour the fish as a matter of course.
- When fish is grilled by the overhead method (i.e. by salamander), it is quite natural to mark the flesh of the fish with a very hot poker. This is to simulate the appearance of grill bar marks, enhancing the appearance of the fish when performed neatly. The culinary term for this is *quadrillage* (see illustration, left).
- When grilling whole fish, score the thickest part of the flesh with 2–3 parallel cuts through the skin on either side. This allows the heat to penetrate for uniform cooking. This scoring process is known as *ciseler* (to score).

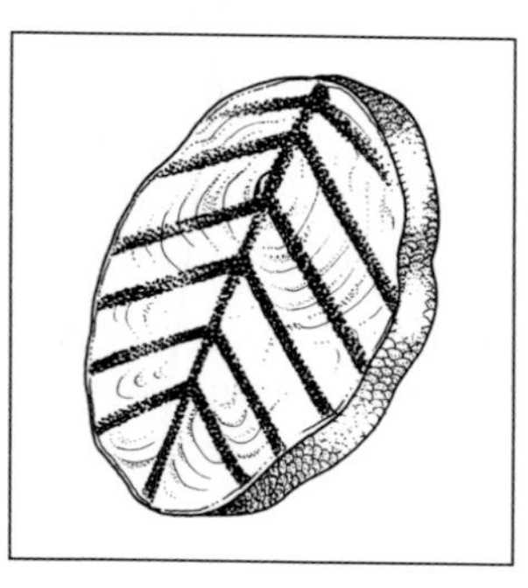

A darne of salmon marked with a hot poker (*quadrillage*) ready for grilling

- Whole fish or thick cuts should be turned halfway through the cooking process to ensure even cooking. You may need to use special fish grilling wires to turn the fish. This reduces the risk of damaging the flesh and may be used in conjunction with a salamander or charcoal grill.
- Grilled fish is normally accompanied with a chilled savoury compound butter, e.g. parsley, red wine or anchovy butters or a warm butter sauce such as béarnaise or choron.
- Lemon and fresh parsley are the standard garnish.
- The bones of cuts of fish are normally removed after cooking and prior to service, although this will depend on establishment policy.
- Grilling is recognised as a quick method of cookery and is employed where the food is cooked to order. Grilled fish should be served to the customer as soon as possible after cooking is completed. Prolonged hot storage will allow the flesh to dry and spoil the dish.

Grilling breaded fish

A more gentle grilling procedure is applied to cuts of fish which are coated with breadcrumbs for protection prior to cooking. Unlike many breaded items, grilled fish uses flour and *melted* butter to adhere the crumb to the surface and a salamander is used to achieve a gentle, even cooking and colouring. Fish grilled by this method should not be turned during cooking as this would spoil the appearance. On menus this method of grilling often appears as *St Germain* or *Caprice.*

Fish suitable for grilling

Small to medium whole fish: round, flat, oily or white.

Cuts of fish: this includes darnes, tronçons and supremes.

Menu examples: Darne of salmon with béarnaise sauce (*Darne de saumon grillée, sauce béarnaise*), Fillets of plaice in breadcrumbs with béarnaise sauce and noisette potatoes (*Filets de plie St Germain*).

Barbecuing fish

This method of grilling is often performed outdoors using the underheat method. The fish is barbecued over charcoal and wood, and the more informal way in which these dishes are eaten requires them to be less complex and relatively easy to cook.

Prior to cooking, barbecued items are placed in marinade to tenderise and improve flavour. The remaining marinade is brushed onto the fish during the cooking to keep it moist and improve appearance and taste.

Whole fish may have the belly cavity filled with branches of fresh herbs to give a distinctive flavour during cooking. They may also be wrapped in foil and baked over the barbecue.

Fish suitable for barbecuing

Small and medium whole fish, cuts of fish.

Example dishes: kebabs or brochettes (pieces of fish mixed with vegeta-bles on skewers and grilled).

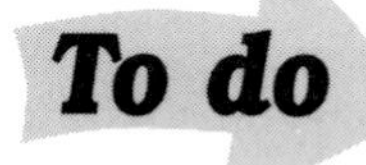

- Watch your supervisor or chef baking a small cut of fish in puff pastry and in an ovenproof dish. Compare the preparation times and cooking times and examine the results for texture and flavour.
- Under supervision, grill a whole cleaned trout, recording the time it takes to cook. Examine the skin of the cooked fish for excessive dryness. If dry, how could you have protected it to preserve quality?
- Under supervision, grill a fillet of fish prepared by the melted butter and breadcrumb method. Note the position in the salamander required to achieve even cooking. Compare the cooking time to the conventional grilling method.
- Prepare a simple marinade for a fish kebab, then marinade and barbecue the kebab. Record the effects of the marinade in terms of texture and flavour and colour.

Deep-frying fish

Deep-frying is a process which cooks the food by entirely immersing it in hot oil or fat. The frying temperature should be 175 °C–185 °C (347–365 °F) so as to bring sufficient heat into contact with the entire surface of the food at one time, enabling a golden brown and crisp surface to develop. The flesh of the fish will need a protective coating during cooking because of the high cooking temperature; this acts to protect the delicate flesh and to enhance the appearance and eating quality.

Protective coatings for fish

- Flour, eggwash and breadcrumbs (*pané à l'anglaise*). See page 52.
- Fresh breadcrumbs (*mie de pain*).
- Dry breadcrumbs or brown breadcrumbs (*chapelure*). Commercial golden-yellow rusk may be used in place of dried breadcrumbs.
- Batter (*pâte à frire*).
- Milk and flour (*pané à la française*).

Batter (*pâte à frire*)
This uses a combination of flour, liquids (milk, water), eggs (à l'orly), and an aerating agent (to give a lighter texture). The aerating agent may be yeast, baking powder or beaten egg whites; adding beer to some batters will produce a fermentation similar to that of yeast. To prepare fish in batter, flour the fish (in seasoned flour) then pass it through the batter, removing any surplus before deep-frying.

Milk and flour (*pané à la française*)
Here the fish is passed through milk and into seasoned flour to produce a thicker coating than by simply passing through flour. This gives better protection in cooking and assists in the quick colouring of the fish.

To improve the flavour of deep-fried fish it is advisable to marinade the cuts of fish for 1.5–2 hours prior to cooking; the fish must remain in the fridge during this period. A simple marinade of cooking oil, lemon juice, crushed parsley stalk, salt and white mill pepper, will enhance the quality and flavour of any deep fried white fish dish.

Protective coatings for deep-fried fish, if administered correctly and cooked at the appropriate temperature, will:

- enhance the flavour
- help retain the shape of the fish during cooking
- enhance the colour and presentation
- reduce moisture loss in the flesh
- prevent excess absorption of the cooking medium.

Keep handling after cooking to a minimum; at this stage contamination can take place by bacteria being passed from the food handler to the fish. Use tongs and wire baskets or frying spiders.

Key points

In order to present deep-fried fish at its optimum quality, observe the following:

- *draining:* allow the fish to drain well. The surface should be grease-free for presentation
- *seasoning:* season with a little salt immediately after cooking and draining away from the fryer. Season whilst hot so the salt sticks to the surface of the food
- *storage:* fried fish should be cooked and served as soon as possible. Do not hold quantities of fried fish for service as prolonged storage will cause the surface to lose its crisp texture. Never cover deep-fried fish as the moisture created will also cause the food to lose its crispness
- *garnishes:* keep these simple for deep-fried fish, using lemon crowns, lemon wedges (free from pips and pith) and sprigs of fried parsley
- *sauces:* these may be hot (as in tomato sauce for Fish Orly) or cold (as in sauce tartare for Fish à l'anglaise).

Safety precautions

Owing to the potential hazards when working with any deep-fryer it is essential to familiarise yourself with basic *do's and don't's* and to have a thorough knowledge of procedures for accidents and fires.

Follow the correct establishment procedures for filtering and changing oil in all fryers.

Never:

- over-fill the fryer with oil or fat
- over-heat the oil. If it begins to smoke, switch off the fryer and inform your supervisor
- plunge uncoated wet items into hot fat
- attempt to extinguish a fryer fire with water or sand
- leave fryers unattended whilst in use.

Make sure you know:

- where to find and how to sound the fire alarm
- who to inform in the case of an accident or fire and how they may be contacted quickly
- what type of fire extinguisher to use for a fryer fire, where the nearest is kept and how to use it
- where the switches are to turn off the gas or electricity to the fryer.

Fish suitable for deep-frying

Whole small round and flat fish, suprêmes, fillets and goujons are suitable for deep frying. Fish with a high fat content such as mackerel, tuna or salmon, do not suit this process as they taste unacceptably greasy after absorbing a percentage of the frying medium.

Menu examples: Fillets of lemon sole in batter (*Filets de limande a l'orly, sauce tomate*), Strips of plaice breaded and fried (*Goujons de plie frite, sauce tartare*), Devilled whitebait (*Blanchaille diablé*).

To do

- Prepare four fillets of plaice, marinade in simple frying marinade and deep-fry:
  - in a frying batter made with yeast
  - in a frying batter made with egg white
  - pané a l'anglaise (flour, eggwash and breadcrumbs)
  - pané a la française (milk and flour).
- Note the differences in colour and variable cooking times.
- Examine for flavour and texture. Which do you prefer and why?

Shallow-frying fish

Shallow-frying involves cooking the fish in a small amount of clarified butter or oil at high temperatures using a frying pan or special fish meunière pan.

Cooking *à la meunière*

This method is applied to small whole fish (trout, red mullet, dover sole) or cuts of fish (fillets, suprêmes, médaillons, goujons) and produces quality results in white or oily fish.

After preparation the fish is passed through seasoned flour immediately before cooking and placed in the hot clarified butter presentation side first then cooked until it has developed a light golden brown colour. The fish is turned during cooking to even the process. Note that you may need to reduce the heat during cooking when cooking larger cuts and allow a prolonged period on the stove to ensure thorough cooking.

The amount of colour developed during shallow-frying fish is very important. Under-coloured fish may look insipid and unappetising, while over-coloured fish may look as if it is burnt (and will probably be dry and over-cooked).

Finishing *à la meunière* dishes

A nut brown butter or *beurre noisette* is added to finish these dishes. To prepare beurre noisette, melt approximately 30 g (1 oz) butter per person in a clean pan and allow to cook to a light brown colour, sharpen with a squeeze of lemon juice and add a pinch of freshly chopped parsley at the last moment. Pour over the cooked fish whilst still bubbling.

Alternatively, season the cooked fish and squeeze lemon juice directly onto the surface. Sprinkle with chopped parsley and pour over the heated brown butter.

Fish meuniere dishes are normally garnished with peeled lemon rings free from pips and pith. However, garnishes may vary from simple to complex depending on the requirements of the dish.

Simple garnishes

- *A la meunière Doria:* peeled lemon, nut brown butter and blanched, turned cucumbers.
- *A la meunière aux amandes:* peeled lemon, nut brown butter, chopped or flaked almonds.

Complex garnishes

- *A la meunière Murat:* peeled lemon, nut brown butter, small turned potatoes, strips of artichoke bottom.
- *A la belle meunière:* peeled lemon, nut brown butter, slice of peeled tomato, shallow-fried herring roe, turned mushroom head.

Key points

- Shallow-frying is a quick process which lends itself to *cooking to order.*
- After cooking the fish should be served quickly as prolonged storage will allow the fish to become dry, or if stored with beurre noisette, soggy and unappetising. Timing in the execution of shallow–frying is critical.
- make sure all equipment and commodities are ready to hand and all prior preparations are in place. Complete the task in a single action, cooking and presenting for service.

Fish suitable for shallow-frying

Small whole fish (trout, red mullet, dover sole) or cuts of fish (fillets, suprêmes, médaillons, goujons). White or oily fish.

Menu examples: Filet de plie grenobloise, Shallow-fried Dover sole with cèpes, *Suprême de turbot meunière à l'orange et poivrons vert.*

Poaching fish

Poaching is a process widely employed in fish cookery. It involves cooking fish in liquids at a temperature below boiling (approximately 90–95 °C/194–203 °F), creating very little or no movement in the cooking liquor.

Because of the gentle nature of this process, it is a most effective way of cooking most varieties of fish, particularly those with a delicate flavour or texture. Invariably the cooking liquid contributes to the quality of the finished dish by:

- imparting flavour to the food, and/or
- forming the basis of the resulting sauce.

Poaching may be divided into two categories: *deep poaching* and *shallow poaching.*

Deep poaching

During deep poaching the fish is totally immersed in a particular type of liquid: an *aciduated court bouillon.*

Vinegar court bouillon

This is used for salmon, salmon trout, trout and shellfish. It is made from: water, vinegar, carrots, onions, thyme, bay leaf, parsley stalks, salt and peppercorns.

Method: Bring the ingredients to the boil. Simmer for 25 minutes then strain, cool and use as required.

White wine bouillon
This is used for freshwater or oily fish. It is made from: dry white wine, water, lemon juice, sliced onions, thyme, bay leaf, parsley stalks, salt and peppercorns.

Method: Bring the ingredients to the boil. Simmer for 25 minutes then strain, cool and use as required.

Equipment for deep poaching

The fish may be deep poached in a special fish kettle such as:

- a salmon kettle or saumonière
- a trout kettle or truitière
- turbot kettle or turbotière

Key points

- The scales of the fish are removed but the skin is left on for protection during cooking.
- *Whole fish* is started in a cold court bouillon, to minimise the risk of distortion.
- *Cuts of fish* are started in a hot court bouillon to reduce the cooking time and prevent the natural juices escaping and coagulating into a white coating on the cut surface of the fish.
- Remember the importance of temperature control when deep poaching fish: excessive boiling will cause the flesh to shrink and distort and affect the texture of the flesh.
- Insufficient cooking may not destroy any harmful bacteria that may be present: check that poaching liquor temperatures reach 90–95 °C (194–203 °F).
- The cooking process is completed on top of the stove.
- The liquor contributes to the flavour of the food during cooking only.
- After deep poaching, fish may be served with lemon, plain boiled turned potatoes, picked parsley and a suitable sauce (often a warm butter sauce, e.g. sauce hollandaise) or a beurre fondue.

Fish suitable for deep poaching
Larger whole fish and cuts of oily fish (especially where the fish is to be dressed for cold buffet use).

Menu examples: Saumon froid en belle-vue; Darne de saumon poché.

Whole fish cooked in a deep court bouillon

Shallow poaching

Fish is shallow poached when it is to be coated with a sauce, and the cooking liquor is invariably used for finishing the sauce.

Shallow poaching may be identified by definite procedures:

- poaching may take place in a sauteuse, shallow saucepan, deep sided tray or in a special fish poaching pan: a *plaque à poisson*
- the dish is lightly greased and any garnish (e.g. shallots, mushrooms, tomatoes) that requires cooking is placed on the bottom of the dish
- the seasoned fish is placed in the dish but kept slightly apart so that the poaching liquor heats evenly
- the poaching liquor is normally fish stock and dry wine (see Unit 2D4: *Preparing and cooking stocks, sauces and soups*). The stock is added at a cool temperature and to a level of two-thirds the height of the food
- the exposed areas of fish are covered with a buttered paper lid or cartouche, to prevent the fish from drying out during cooking
- pre-warming takes place on top of the stove (do not boil as this will cause shrinkage and distortion) and the poaching is completed in the oven on a medium heat to ensure even heat distribution
- once cooked, the fish is removed from the cooking liquor and kept covered to prevent the surface from drying. The cooking liquor is returned to the stove and reduced. The reduced liquor is then either:
  a) added to the pre-prepared velouté sauce to improve flavour and quantity
  b) reduced further (to a syrupy consistency), before cream is added and the liquid cooked to the desired sauce thickness and finished with unsalted butter to become the finished or complete sauce
- when dishing for service, place a little of the complete sauce on the serving dish to facilitate ease of service and to absorb any excess liquor, then arrange the fish neatly on the dish and coat it with the finished sauce.

Glazing shallow poached fish dishes

Recipe requirements may demand that certain fish dishes are *glazed.* This refers to browning the surface of the completed sauce within its service dish under a salamander to enhance the appearance of the dish.

To achieve a good glaze, egg yolk (cooked to sabayon) and unsalted butter may be added to the velouté sauce. In the case of the fully reduced cooking liquor (Method **b** above), the consistency of the syrup once the cream and butter is added should be sufficient to achieve a good glaze without any further additions.

It is advisable to check the glazing quality of the sauce prior to completing the dish; this may be done by pouring a little sauce onto a plate and testing glazing under the salamander. If an even glaze does not develop, rectify by adding sabayon.

Surface browning may also be achieved by the use of cheese, as dictated by the recipe. Breadcrumbs may also be used, but note that this method is called *gratinating* and should not be confused with *glazing.*

To do

- Under supervision, prepare two fillets of fish for cooking meunière with a simple garnish and two fillets of fish meunière with a complex garnish.
- Make a note of the time differentials in preparing the two dishes. Notice how the dishes differ in terms of colour, balance and presentation.
- Watch your chef or supervisor preparing a court bouillon.
- Under supervision, cook one glazed and one non-glazed dish of shallow poached fish. Compare the finishing of the sauces and the method employed to achieve a glaze.

Health and hygiene points

Owing to the relatively low temperatures at which fish is poached, it is important that the fish spends as little time as possible in a warm environment. Keep handling after cooking to a minimum, using the correct utensils. Arrange your timing so that cooked fish does not have to be held for long periods of time.

Fish suitable for shallow poaching
Small whole fish (sole, plaice, baby turbot) or cuts of fish (fillets, délice, suprêmes, paupiettes). This method lends itself to delicate or fragile types of fish flesh that would break up if deep poached.

Menu examples: Glazed shallow poached dishes: Bonne femme, Brevale, Bercy, Marguery, Veronique. Non-glazed shallow poached dishes: Duglère, d'Antin, Dieppoise, Suchet, Vin-blanc.

Essential knowledge

Time and temperature are important when cooking fish dishes in order to:

- prevent food poisoning. Insufficient cooking can cause food poisoning
- prevent shrinkage of fish
- ensure a correctly cooked fish or fish dish. Fish is a delicate food item which can easily be spoiled if cooked for too long or at the incorrect temperature
- prevent fat absorption during the frying processes.

Steaming fish

Small whole fish, fillets or other small cuts of fish can be cooked with excellent results by steaming. It can be carried out in a low-pressure steaming unit, high pressure steaming oven or a pressureless convection steamer. All units give good results but cooking times vary, depending on the type of steamer you employ.

Steaming has the advantage of being easy to carry out, while offering rapid cooking with little loss of flavour, colour or nutrients. It is particularly useful when preparing large quantities for banquets as it reduces the number of times the fish is handled and allows the sauces to be made in advance. This reduces the amount of time the fish needs to be kept an a warm environment prior to serving; i.e. the fish is cooked immediately before it is required for service.

Many shallow poached dishes can be executed by the steaming process, where excellent results are obtained if the accompanying sauce is of the required quality.

Method
Prepare the fish for steaming by moistening it with a little fish stock, lemon juice or wine (or as recipe instruction). Season with a little salt and white pepper. Steam until just cooked (remembering that steaming is a rapid method of cookery). Prepare the accompanying sauce separately.

Fish suitable for steaming fish
Small whole fish, fillets or other small cuts of fish.

Menu examples: As for any shallow or short poached dishes.

Corrective action checklist for cooking fish

Fault	Corrective action
The fish is cooking too quickly	Reduce the heat, move to cooler part of the stove or oven.
The fish is not colouring	Increase the heat, cook less food at one time. Check whether you floured the fish (if required).
Service is unavoidably delayed	Cover with moist paper, keep hot but moist: brush with fish stock. Note that you should not add any sauce until required for service.
The sauce is too thick	Thin with the appropriate liquor.
The sauce is too thin	Reduce or correct the consistency prior to serving.
The fish is dry	Check whether you have cooked the fish for too long or at too high a temperature. Has it been badly handled or incorrectly stored?
Poached fish has a tough texture	Check whether you have cooked the fish too quickly. Has it boiled? At what temperature did you begin the cooking?
The fish sticks to the cooking dish	Check whether you greased the cooking dish.
The cream in the sauce separates	Check that the cream used was fresh. Did you shake and stir the sauce during reduction to ensure even heat distribution?
Steamed fish has a poor flavour	Check that the steamer was cleaned correctly before being used for cooking.

Storing fish dishes

In order to prevent bacterial growth, hot fish dishes should be stored above 65 °C (149 °F); cold fish dishes should be stored below 5 °C (41 °F).

Dealing with unexpected situations

If you have problems when preparing fish dishes, refer to the diagnostic corrective action list (above). This should help you to deal with any unexpected situations.

Make sure that you are familiar with all health and safety procedures; this will help you to respond appropriately to any emergencies. Keep a note of any emergency phone numbers (both within and outside of your organisation) that you will need to ring to obtain immediate assistance.

What have you learned?

1. What factors determine your choice of cookery process for cooking fish?
2. What effect will under-cooking or over-cooking have on fish dishes?
3. What are the main contamination threats when preparing, cooking and storing fish dishes?
4. Why are time and temperature important when cooking fish dishes?
5. Why does fish require protection during baking or grilling?
6. Name three protective coatings for deep frying fish?
7. Why are oily fish not suitable for deep frying?
8. What is the difference between deep poaching and shallow or short poaching?
9. How is the cooking liquor used from shallow poached dishes?

Extend your knowledge

1. Research and prepare a list of fish used in other countries of the world. Many are readily available in this country. How would you cook them?
2. Find out how fish may be preserved other than by freezing. Note their culinary uses.
3. Visit a fish market and observe the variety of species, the handling and storage techniques and the current market prices. How and why does the fish reach you in first class condition?

UNIT 2D3

Preparing and cooking egg custard based desserts

ELEMENT 1: Preparing and cooking egg custard based desserts

What do you have to do?

- Prepare appropriate equipment ready for use.
- Prepare and cook the egg dessert ingredients correctly for each individual dish.
- Handle and mix the prepared ingredients according to the product requirements.
- Carry out your work in an organised and efficient manner taking account of priorities and laid down procedures.
- Cook the desserts for the correct time and at the correct temperature according to dish requirements.
- Finish and present desserts according to customer and dish requirements.
- Store finished desserts hygienically and according to laid down procedures.
- Dispose of waste hygienically and correctly.

What do you need to know?

- The type, quality and quantity of ingredients required for each dessert.
- What the main contamination threats are when preparing, cooking and storing egg custard based desserts.
- How to satisfy health, safety and hygiene regulations concerning preparation areas and equipment, both before and after use.
- Why it is important to keep preparation and storage areas and equipment hygienic.
- Why time and temperature are important in the preparation of egg custard based desserts.
- How to plan your time to meet daily schedules.
- How to deal with unexpected situations.

Introduction

Egg custard based desserts provide a tasty, easily digestible and light finish to a meal. They cover a range of dessert items, such as trifle, crème caramel, bread and butter pudding, diplomat and cabinet puddings, and queen of puddings.

The basic ingredients are eggs, milk, sugar and vanilla, although some recipes use additional ingredients. Egg custard may be set (by cooking) or left in a liquid form, and variations on the basic egg custard mixture include confectioner's custard (pastry cream or *crème pâtissière*) and egg custard sauce (*sauce anglaise*).

Raw egg custard is used with fruit, sponge and pastry ingredients to form a variety of interesting classical dessert and patisserie dishes and products. Savoury raw egg custard is used to make quiches, tartlettes and flans for dinner savoury courses, buffet and snack items. Confectioner's custard (*crème pâtissière*) is a versatile filling used for many pastries and desserts, and as a base in the production of more complicated dishes, such as soufflés, crêpe soufflés, gateaux, flans and tartlettes. Sauce anglaise is used for ice cream recipes and as an accompaniment for tarts, pies, puddings or fruit-based desserts.

In dishes where the egg custard is set, the sauce or custard is thickened by the coagulation (gelling) of egg protein, which occurs at a particular temperature. The correct temperature is dictated by the parts of the egg used: egg white coagulates at approximately 60 °C (140 °F), yolks at 70 °C (158 °F) and a mixture of yolks and whites at 66 °C (151 °F). If the egg proteins are overheated, they shrink and release their liquid content, causing undesirable bubbles on the surface of the set custard. The time and temperature for cooking are therefore crucial in ensuring the success of an egg custard based dessert.

This chapter shows you how to make a basic egg custard and its two main derivatives: confectioner's custard (*crème pâtissière*) and egg custard sauce (*sauce anglaise*). General points to note at each stage of general preparation and cooking are also given, followed by recipes for egg custard based desserts. The final section looks at egg based dessert items such as meringues, pancakes and omelettes.

Ingredients

Eggs

Eggs are one of the most basic of ingredients available to the chef, and they have been a part of our diet for a very long time. They are used in many ways and for many reasons: as a binding agent or to produce a glaze (egg yolks); as a foaming agent (egg whites); or as an ingredient in desserts, cakes, creams and custards (yolks and whites).

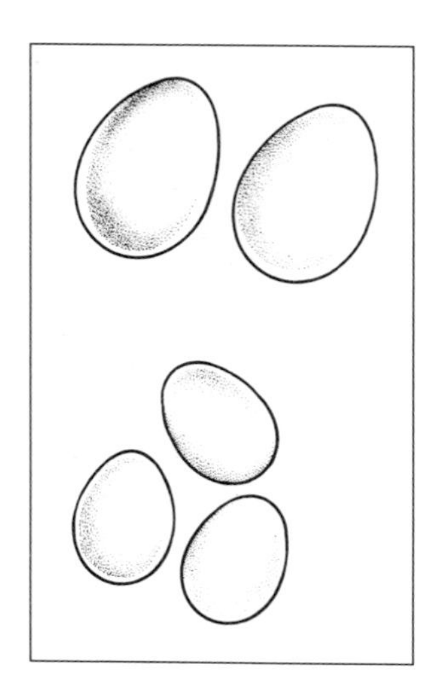

Hen eggs
Top: chicken eggs
Bottom: bantam eggs

Hen eggs are available fresh, frozen, freeze dried and liquid pasteurised, although most kitchens use only fresh eggs and dried egg white (for meringues and icing). Bakeries, however, use the whole range of egg products for producing meringues, royal icing, marshmallows, nougat and the japonaise range of products.

Eggs are graded by size and quality. Most eggs today are factory farmed for supply to large manufacturers, food wholesale companies and supermarket chains, although local free range suppliers can be found in most areas. Free range eggs have become more popular as people have become more aware of the benefits of fresh food bought directly from the farm or local supplier. Many hotels, restaurants and catering outlets buy fresh eggs twice weekly from grocery multiples, and cash and carry or consortium networks who supply large and small food chains.

Eggs have a high nutritional value. They are made up of egg yolk and egg white, both of which contain protein. Egg yolk protein is called *lecithin* and egg white protein is called *albumen*. Taken together, the yolk and white form a rich protein food with a high essential amino acid content. One egg provides 11 per cent of the average daily protein requirement and 6 per cent of the maximum recommended fat intake per day. Eggs are a rich source of vitamins essential to our health: particularly vitamins A, B and D. They also provide two important minerals: calcium and iron; and include the trace elements iodine, phosphorus, selenium and zinc. It is worth noting that lecithin (in the yolks) acts as a natural emulsifier, which means it can be used to combine fat and water.

Milk

Milk is one of the most fundamental commodities used by the pastry chef. Cows' milk contains protein, carbohydrate, fat, minerals, vitamins and water. It is available in a variety of types: pasteurised, homogenised, sterilised, UHT, semi-skimmed and skimmed, dried, evaporated and condensed. It is generally purchased daily in pint bottles or cartons, except where large quantities are needed, when it is bought in plastic bag/box dispensers known as *Pergals*. These are available in three and five gallon units, which fit into a chilled dispenser unit.

Composition of fresh pasteurised cows' milk

Water	87.0%
Protein (casein)	3.5%
Fat	3.5%
Carbohydrate (lactose)	5.0%
Minerals (especially calcium, potassium, sodium and magnesium)	0.7%
Trace components (iron and vitamins A, B-complex, C, D and E)	0.3%

Cream

Cream can be used to enrich egg custard based desserts, either as an ingredient (replacing some of the milk) or as a decorating medium, where it is whipped and served on trifles, Charlottes and individual dessert items. Single, double, whipping and clotted cream are used for a wide range of pastry recipes. Double or whipping cream can be used either in liquid form, semi-whipped (as for Bavarian creams) or fully whipped to a piping consistency (for decorating trifles, gateaux and dessert products).

Minimum legal fat content of cream

Single cream	18%
Whipped cream	35%
Double cream	48%
Clotted cream	55%

Cream may also be aerated to make it lighter and increase its volume. Bakeries often use cream aerator machines to achieve this, where the liquid cream is forced through a nozzle dispenser and mixed with air under pressure. Aerated cream ages more quickly as a result of this process, and any products containing it must always be refrigerated or chilled during display.

All cream products must be kept in a refrigerator until required for use for health and hygiene reasons, but also because cold cream will whip more effectively. Always use stainless steel, high quality catering plastic, glass or china bowls for whisking cream; never put cream into copper or aluminium containers.

Hygienic handling of dairy products

- Milk, cream and other dairy products like cheese and eggs present perfect opportunities for the cultivation of micro-organisms if stored in warm conditions. While produced or manufactured dairy products are safe, problems can arise in the subsequent handling and storage as they pass through the food chain. Contaminated milk and cream can cause food poisoning.
- Thoroughly sterilise all equipment used for preparing milk and cream products.
- Be aware of the risks and problems caused by unhygienic methods, particularly when you are under pressure. Do not take shortcuts in hygienic practices.
- Handle all food cleanly and safely according to the health, safety and hygiene regulations laid down in your establishment and by law.
- Remember that cream and milk absorb odours. Never place them near more pungent foods, such as fish or onion.

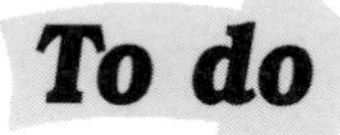

- Find out where the fresh cream is stored in your establishment and check the temperature of the unit.
- Find out the difference between fresh and synthetic (non-dairy) cream, and between single, double and whipping cream.
- Find out what types of bowls your establishment uses for whipping cream.
- Find out what size and quality of eggs are used in your kitchen. Where are they kept?

Sugar

All forms of sugar can be used in pastry cookery: cubed, granulated, caster, icing and the range of brown sugars. Caster sugar is most widely used for making egg based desserts, because it has fine grains which dissolve easily and have good creaming properties.

If you are making a dessert that involves boiling sugar, such as when making caramel for crème caramel, use cubed or granulated sugar.

Vanilla *(vanillin)*

The vanilla pod is the matured seed pod of a tropical orchid found in Asia, North America and The East and West Indies. It is black, approximately 4–6 inches long, and has a sweet, attractive perfume. You can use it in pod form, either by soaking it in milk and allowing the flavour to diffuse into the milk, or by placing it in a jar of sugar where the sugar will become flavoured over time. Vanilla flavour is widely used because it is cheaper, but it does not provide the really distinctive flavour of fresh vanilla pods.

Flour

There are many types of flour available. Two of these, soft and strong flour, may be used in the production of crème pâtissière. Soft flour, sometimes called plain or weak flour, is milled from wheat and has a lower protein content than strong bread flour. It also gives a softer texture to crème pâtissière when used alone, and so is often used together with strong flour when making flans, for example, where a set custard is required. Here the strong (or hard) flour is used in equal quantity to the soft flour and the two flours are sieved together. The resulting mixture is called a *medium strength flour.*

Flour is also used as a thickening agent for custard sauces. Cornflour, milled from maize, is used in custard powder. It is approximately 85 per cent pure starch (which provides its thickening quality) and when used to thicken clear liquids does not produce a totally clear sauce. If a clear sauce is required, arrowroot may be used in its place. Taken from the underground stems of tropical plants, arrowroot is 100 per cent pure starch. When added to a clear liquid, all of the starch absorbs all of the liquid. This allows the light to pass through and results in a clear shiny sauce.

Preparing equipment

Keep the following points in mind when preparing and cooking egg custard based desserts:

- check that ovens are pre-set and turned on
- if using an electric solid top hob, check that it is switched on
- make sure that moulds and dishes are thoroughly clean, but do not scour during cleaning
- dry all moulds and dishes before storing away. Never stack wet moulds and dishes as this causes bacteria to grow
- use silicone or greaseproof paper for *cartouches* (round paper lids)
- keep mixers, ovens and ranges clean. Make sure you clean them each day after production.

Equipment used for preparing egg custard based desserts (from left): saucepan, chinois, whisk, pastry brush, mixing bowl

Health, safety and hygiene

Note all points given in Units G1 and G2 of the Core Units book concerning general attention to health, safety and hygiene. When preparing egg custard based dessert products the following points are particularly important:

- make sure that all dairy products are handled and stored at the correct temperature
- strain all liquid egg custards to remove any shell contamination
- if using a bain-marie, take care when moving hot water trays around on top of the stove, and when placing them into or removing them from the oven
- wipe up all spillages as they occur
- transport hot desserts in a safe and hygienic manner
- handle ice cream with care and always dispose of any defrosted ice cream.

Essential knowledge

It is important to keep preparation and storage areas and equipment hygienic, in order to:

- comply with food hygiene regulations
- prevent the transfer of food poisoning bacteria to food
- prevent pest infestation in preparation and storage areas
- prevent contamination of food commodity items by foreign bodies.

Safety points when cooking sugar

- Keep a bowl of iced water readily available.
- Never work in a confined space or a crowded area of the kitchen.
- Carry or transport boiling sugar solutions carefully. Use an oven cloth and make sure that a clear space is available on the work bench for the pan.
- Make sure saucepan handles do not protrude over the edge of the stove where they can be knocked.
- Wash down the sides of the pan containing the boiling sugar using a clean pastry brush when necessary.
- If splashes of boiling sugar do land on your skin, place the burned area into iced water immediately.
- When pouring hot sugar solutions, pour it at arm's length and away from work colleagues.
- When adding the water to caramel for Crème caramel, hold the pan away from your face. The steam can cause serious burns.

Essential knowledge

The main contamination threats when preparing and producing egg custard and egg custard desserts are as follows:

- cross-contamination can occur between cooked and uncooked food during storage
- food poisoning bacteria can be transferred through preparation areas, equipment and utensils if the same ones are used for preparing or finishing cooked, baked goods and raw, uncooked dairy and dessert foods
- food poisoning bacteria may be transferred from yourself to the food. Open cuts, sores, sneezing, colds, sore throats or dirty hands are all possible sources
- contamination will occur if flour or any foods are allowed to come into contact with rodents (such as mice or rats), or insects (house flies, cockroaches, silver fish, beetles). Fly screens should be fitted to all windows
- food poisoning bacteria may be transferred through dirty surfaces and equipment. Unhygienic equipment (utensils and tables, trays, strainers and cooking trays) and preparation areas (particularly egg and cream based mixers) can lead to contamination
- contamination can occur through products being left opened or uncovered. Foreign bodies can fall into mixes and uncovered creams and sauces, such as eggshell, etc.
- cross-contamination can occur if equipment is not cleaned correctly between operations
- contamination can occur if frozen and dried eggs are de-frosted or reconstituted incorrectly (always check the manufacturer's instructions)
- incorrect waste disposal can lead to contamination.

Egg custards and variations

Liquid egg custard (sweet)

This is made from eggs, milk, sugar and vanilla flavouring (vanilla pod or flavour).

Method

1 Place the milk and vanilla pod into a pan and bring to the boil. Remove the pan from the stove and allow the contents to cool for a few minutes. Remove the vanilla pod.
2 Whisk the eggs and sugar together in a mixing bowl.
3 Pour the hot milk into the eggs and sugar, whisking all the time at a steady speed to prevent a foam occurring.
4 When thoroughly mixed, pass (strain) the liquid through a fine conical strainer (*chinois*) to remove any shell particles.

Optional: This mixture can be enriched by replacing some of the milk with cream, but note that this does increase the unit cost of the dessert.

Liquid egg custard (savoury)

This is made from eggs, milk, salt and cayenne pepper.

Method

1 Place the milk and seasoning into a pan and bring to the boil. Remove the pan from the stove and allow the contents to cool for a few minutes.
2 Whisk the eggs.
3 Pour the hot milk over the whisked eggs gradually, whisking all the time at a steady speed to prevent a foam occurring.
4 When thoroughly mixed, pass (strain) the liquid through a fine conical strainer to remove any shell particles.

Optional: This mixture can be enriched by replacing some of the milk with cream, but note that this does increase the unit cost of the dessert.

Crème pâtissière (confectioner's custard or pastry cream)

This cream or thick set sauce is the pastry chef's equivalent of the chef's bechamel sauce: it forms the basis of many other creams and is widely used as a filling for pastries, flans, tartlettes and barquettes, and as a base for soufflés. It is the continental equivalent of fresh cream.

Crème pâtissière is made from milk, egg yolks, flour, sugar and vanilla flavouring.

Method

1 Place the milk and vanilla pod or flavour into a pan and bring to the boil. Remove the pan from the stove and allow the contents to cool for a few minutes. Remove the vanilla pod from the pan if necessary.
2 Combine the egg yolk and sugar and mix well. Stir in the flour.
3 Pour the milk into the mixture and mix well.
4 Place the mixture into a clean saucepan, return to the stove and cook the cream sauce until it boils (approximately 1 minute) or until the sauce is thick.
5 When cooked, sprinkle with icing sugar or cover with a buttered cartouche to prevent a skin forming. When cool, store in the refrigerator.

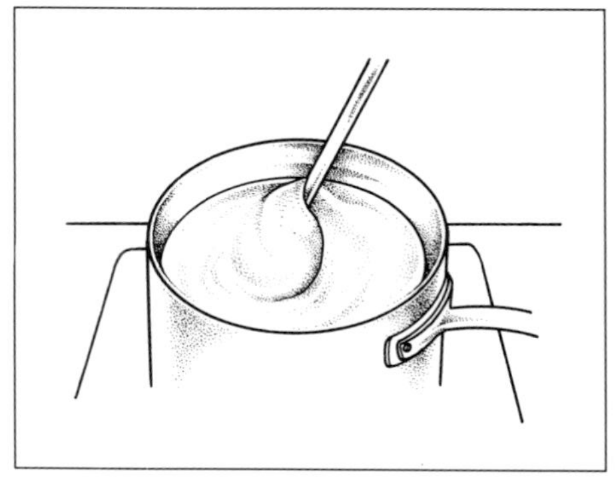

Cooking crème pâtissière
Left: pour the mixture into a saucepan and bring to the boil
Right: the cooked mixture is thicker and smoother

Optional: If a thicker set cream for cutting is required, such as for *Flan aux bananes*, increase the flour content of the recipe or use a medium strength flour (50 per cent soft and 50 per cent strong). Crème pâtissière can be flavoured with chocolate, almond, coffee or liqueurs (such as Kirsch, Cointreau, brandy, etc.).

Crème mousseline

Crème pâtissière can be made into Crème mousseline by adding butter. Add one third of the total butter used into the hot crème pâtissière and allow this to cool. Meanwhile, cream the remaining two-thirds of butter. When the crème pâtissière mixture is cool add it to the well-creamed butter.

Crème moussline is used as a filling for desserts, biscuits and sponges. It can be stored for up to four days in the refrigerator, but always needs to be mixed well before use.

Fresh egg custard sauce (*Crème ou sauce anglaise*)

Traditional English custard is a medium-thick custard made by using custard powder. Many catering operations use this product as a sauce garnish (for tarts, pies and puddings) or for trifles, desserts and ice cream recipes. In France, custard (sauce anglaise) is a thinner sauce made from fresh eggs, milk, sugar and vanilla.

Method

1. Place the milk and vanilla pod into a pan and bring to the boil. Remove the pan from the stove and allow the contents to cool for a few minutes. Remove the vanilla pod.
2. Whisk the eggs and sugar together in a mixing bowl.
3. Pour the cooled milk into the eggs and sugar, mixing well. The milk needs to be slightly cooled to prevent it from over-cooking the yolks, which can scramble if the milk is too hot.
4. Put the mixture into a clean pan and cook it on top of the stove until the custard is thick enough to coat the back of a spoon. Do not allow it to boil.
5. If necessary, keep the cooked sauce anglaise hot in a bain-marie until required for service.

Notes

- If you over-cook the sauce anglaise, do not throw it away: if you put it through a liquidiser it will reform and can then be used.
- This sauce is very susceptible to bacterial growth due to the low coagulation temperature of the egg yolks. Always make it fresh daily.

Convenience custard

Custard powder is also used to make custard. This is a flavoured cornflour based powder that is mixed with a little cold milk and sugar and then stirred into boiling milk. Always read the product instructions for the ratio of powder to milk. If the milk is not hot enough when the dissolved powder is added, the sauce will taste of the uncooked custard powder and will probably burn before thickening.

Points to remember during preparation and cooking

Preparing the dessert ingredients

- Check that your scales are accurate and meet existing regulations.
- When ingredient items are low in stock inform the chef or pâtissière, who will re-order the necessary items.
- Work cleanly. Replace lids on storage bins to prevent contamination. Cover any food items such as unused egg custard, custard or cooked creams and sauces with cling film.
- Check the date of milk, cream and eggs, and use the oldest stock first. Check whether other refrigerators have ingredients to be used up before opening new units of stock.
- Always wash your hands between handling fresh eggs or dairy products and other pastry ingredients.

Mixing the prepared dessert ingredients

- Check the type of dessert being produced to make sure you have weighed the ingredients correctly.
- If you are preparing crème pâtissière, remember to stir it well while cooking. Do not over-cook or the cream will burn and will be wasted.
- If using dried milk powder only use 60 g (2 oz) per pint of warm water. Milk powder contains fat which will not easily dissolve in cold water.

Cooking, finishing and storing the desserts

- When the products are baked or cooked remove them from the oven, handle them carefully and place them safely onto the bench. Hot liquids and sauces removed from the stove should be placed on a wooden triangle to cool, or placed into a bain-marie to keep hot.
- Remove cooked products carefully and cleanly, making sure you have cleared a space to place the hot goods for hygienic storage while cooling.
- Cool egg based dessert products should be stored in the refrigerator and covered with cling film to prevent contamination from other stored products.
- Never use cream to decorate a product that is still warm.
- Ensure the cooked products are cool before finishing, creaming, filling or decorating with media except desserts that are to be served immediately.
- Present and store the finished cooked products according to the *Food Hygiene (Amendments) Regulations 1993.*
- See also *Creaming and Decorating* on page 93.

Essential knowledge

Time and temperature are important in the preparation of egg dessert products in order to:
- ensure the custard, cream or sauce is prepared according to the recipe requirements
- ensure the custard, cream or sauce is cooked, cooled or set at the right temperature for the stated length of time in order to prevent faults in the finished desserts
- ensure the dessert is mixed, cooked and cooled for the required amount of time in order to prevent faults in the dessert products
- ensure that the desserts are correctly cooked or cooled prior to service, storage or the next stage of production
- prevent food poisoning from direct and indirect contamination sources
- eliminate the possibility of shrinkage and wastage.

To do

- Find out which method your chef favours for washing down the sides of the pan containing the boiling sugar solution.
- Check the temperatures of your refrigerators and freezers using a digital thermometer. Do they comply with existing regulations?
- Watch your chef making pastry cream (crème pâtissière).
- Find a recipe for convenience custard powder. How much is needed to produce 5 litres (9 pints) of custard sauce?
- Practice making a piping bag and cartouche with your eyes shut.

Egg custard based desserts

Egg custard dessert creams and sauces

Crème ou sauce anglaise, Crème pâtissière, Crème bavarois, Crème Charlotte, Crème St Honoure, Crème Chiboust, Crème Danoise, Crème mousseline, Sauce mousseline, Sauce sabayon, Sauce aux amandes

Egg custard desserts: sweet

Baked egg custard, Crème caramel, Crème renversée, Crème St Claire, Crème viennoise, Crème beau-rivage, Diplomat pudding, Cabinet pudding, Crème regence, Crème espagnole, Petits pot de crème vanille, Petits pot de crème chocolat, Bread and butter pudding, Queen of puddings, Tarte à l'orange, Tarte au citron, Flan normande, Baked egg custard flan, Custard tarts, Manchester flan, Crème brulée, Oeufs à la neige (Sauce anglaise based dessert), Ile flottante (Sauce anglaise based dessert)

Egg custard desserts: savoury

Quiche Lorraine, Quiche aux oignons, Tartelette Lorraine, Flan Evesham, Tartlette Evesham, Flan forestaire, Tartlette forestaire, Tarte au fromage

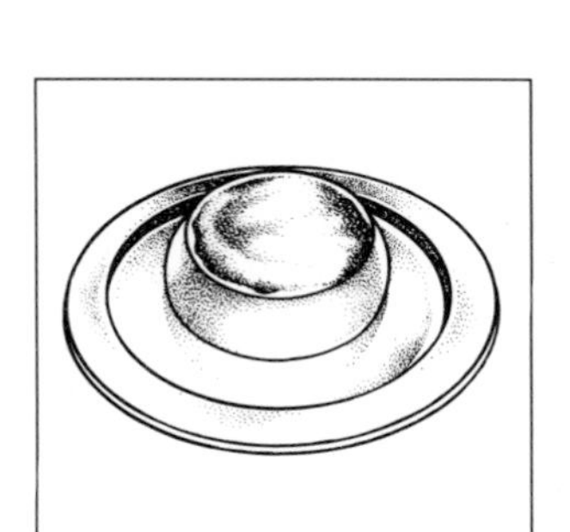

Baked egg custard

Crème caramel

For this dessert you will need to make a sweet liquid egg custard and caramel. Made from sugar and water, caramel is a tasty, nutty, sugar sauce often used as a flavouring agent for ice cream and desserts.

Method

1 Place the water into a clean saucepan (preferably copper) and add the sugar.
2 Mix together and heat gently to dissolve the sugar crystals.
3 When dissolved, boil over a full heat and cook until the sugar turns a rich chestnut colour. Wash down the sides of the pan during boiling using a clean, wet pastry brush to control recrystallisation.
4 When the sugar begins to smoke lightly, pour on some boiling water and shake to form the sauce.
5 Pour while hot into dariole moulds and allow to cool and set.
6 Pour and fill the dariole moulds with a sweet, liquid egg custard and cook in a bain-marie (in a tray half-filled with hot water) in the oven for 40–45 minutes or until set at approx 177 °F (350 °C).
7 When cool, press the edge of the cooked cream to allow air between the cream and the mould. Turn out onto a plate and pour over any sauce remaining in the base of the mould. To test the cream: insert a small knife into the top of the cream; if liquid rises, more cooking time is required. If the knife comes out clean, the custard is cooked.

Bavarois (Bavarian cream)

The Bavarian cream has become a standard dessert. Today it is used extensively for charlotte russe and royal, and bavarois desserts made from either a sauce anglaise or fruit base. Every pastry and chef trainee should know how to produce this dessert.

Sauce anglaise based bavarois

The ingredients are eggs, caster sugar, vanilla, milk, cream and gelatine. You may like to use leaf sheet gelatine (8–10 sheets per 28 g/1 oz) for this recipe. It is more expensive than powdered gelatine, but it is easier to use and works well in this type of product. Leaf gelatine is available in gold, silver and bronze grades. Remember that the use of gelatine makes this product unsuitable for vegetarians. Vegetable gum may be used as a substitute.

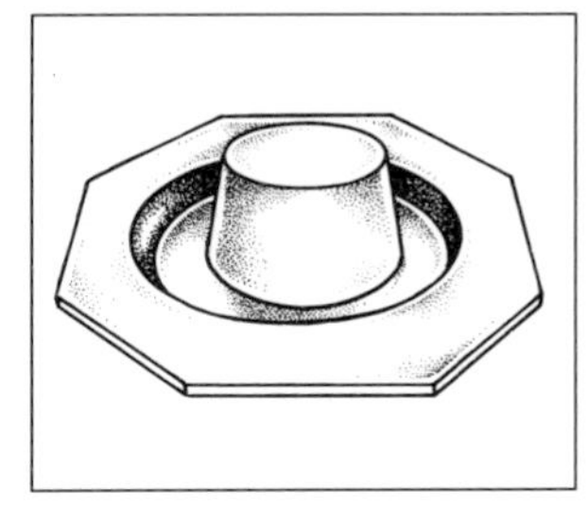

Bavarois

Method

1 Place the gelatine in cold water to soften. (*Optional:* when making a fruit bavarois, the gelatine may be dissolved in lemon juice rather than water.)
2 Place the vanilla (pod or essence) and milk into a clean pan and bring to the boil. Remove the pod if necessary.
3 Mix the egg yolks and sugar together well.
4 Pour on the boiled milk, stir well and return to the stove in a clean pan.
5 Cook the custard sauce until it thickens slightly and coats the back of the spoon.
6 Remove from the stove and cool slightly, then add the drained and softened gelatine.
7 Strain the custard into a clean stainless steel, glass or china bowl, then leave to cool or set the bowl on a bowl of iced water to cool.
8 Semi-whip the cream.
9 When the custard begins to thicken and set, fold in the cream in three stages. Then place the custard into moulds or glasses (depending on the dessert being made).

If you are making a fruit based bavarois using the sauce anglaise method, replace some of the the milk with fruit purée. For example, if the recipe requires 568 ml (1 pt) of milk and you are making a raspberry bavarois, use 50 per cent milk and 50 per cent raspberry purée. Add the purée only when the custard is setting (Step 9). If you add it too early, the acid of the fruit will *denature* (soften) the gelatine strands and prevent bonding, resulting in a soft bavarois that will not de-mould successfully.

Fruit based bavarois
Bavarois can be made using fruit purée, caster sugar, gelatine and cream. Mix the sugar with the fruit purée and bring to the boil. Stir in the soaked, drained gelatine leaves and stir well. Strain the mixture into a stainless steel bowl and leave to cool. Semi-whip the cream and fold it into the setting fruit mixture. Use as required by the recipe.

Trifle

Trifle

The trifle is a standard dessert that has been popular for many years. It can be made in a number of ways: the custard can be made either from fresh egg yolks, cream or milk, sugar and vanilla or by using custard powder with milk and sugar. Sherry trifle, the most common form of trifle, is made by soaking layers of jam and sponge in sherry, then pouring over hot custard (of medium thickness). Once it has cooled, it can be decorated with whipped fresh cream, glacé cherries, flaked toasted almonds, angelica and small macaroon biscuits. Piped chocolate motifs provide an artistic finish to this traditional dessert. Trifle is individually portioned in glass dishes or goblets, or multi-portioned in glass bowls.

Mousses

Mousses are traditionally produced by the bavarian cream method, using cream to replace some of the milk in the sauce anglaise. Whipped cream, fruit and/or light Italian meringue is also added, with an appropriate liqueur or other flavouring commodity (e.g. chocolate, coffee, etc.).

Prepare the basic sauce anglaise, then fold in the whipped cream and cool Italian meringue in two or three stages: add some of the cream, then some of the meringue, and so on, to produce a very light mousse dessert. Pour into moulds and chill in the refrigerator for 2–4 hours or pour directly into a fine glass or dish and chill. Mousses can be decorated in several ways: e.g. with nuts, piped chocolate or crystallised fruits and flowers (roses, violets and mimosa).

Mousses can also be produced using the fruit purée method as described for fruit based bavarois (above). Remember that high acid fruits will soften the gelatine and might cause problems with moulding. Add the purée only when the custard is setting.

Custard based desserts

Ice creams

Ice cream is made from a sauce anglaise base with cream which is then frozen. Freezing generally takes place in a turbine, which rotates and freezes the water content, thickening the ice cream as the ice crystals grow. Vanilla ice cream is made by infusing the seeds of the vanilla pod in the milk when making the sauce anglaise (the seeds are extracted by splitting open the pod with a small knife lengthways). The sauce is then cooked, cooled and mixed with fresh liquid cream.

You can add fruit pulp or use the array of frozen fruit purées from frozen food suppliers. These purée products are excellent for mousses, bavarian creams or as a fruit sauce (*coulis*).

Long term storage of ice cream should be at −18 to −20 °C (−0.4 to −4 °F) but the temperature for service should be −5 to −6 °C (21–23 °F), otherwise the ice cream would be too hard to serve. Ice cream that has defrosted should be disposed of as the risk of contamination is very high.

Strict regulations govern the production and sale of ice cream and ice cream products. These are *The Food and Drugs Act 1962* and *Food Standard Regulations 1970.*

To do

- Find out the cost of buying in Crème caramels. Compare the cost to making it in the kitchen.
- Ask your supervisor why copper pans are used for sugar boiling.
- Watch your chef testing cooked egg custard. How do they know that the custard is cooked?
- Find out where gelatine comes from. How is it produced? What vegetarian alternatives are available?
- Ask how to reform a curdled sauce anglaise.
- Watch a fruit bavarois being produced. Check when the fruit purée is added. What faults can occur when making this dish?
- Find out why ice cream with a high sugar content freezes at lower temperatures than water and fruit based ices.
- Watch ice cream being made using a turbine freezer.

Creaming and decorating desserts

- Finishing is very important; each dessert product should be the same size, colour and shape.
- Cream piping should be neat, accurate and equal for each dessert.
- Decorating media should be used to highlight products, so keep it attractive but simple.
- Ensure cream is whipped correctly, in a hygienic manner, and that all cream-filled products are stored according to The *Food Hygiene (Amendments) Regulations 1993.*
- Piping cream is a skill that develops with practice. Finished desserts should always look clean and tidy.
- Check that piping bags have been sterilised: they are a source of cross-contamination.
- Creaming of cooled, baked or chilled egg custard based dessert products should be carried out in a cool area of the kitchen.
- Check that you have used fresh cream following the rotation order and that it is whipped cleanly. All equipment should be hygienically cleaned before and after use, to avoid contamination.
- Be aware of timing when whipping cream: it can easily be over-whipped and formed into butter. Cream is an expensive commodity and should always be used carefully and cleanly.
- Dust any icing or caster sugar *evenly* using a shaker or sieve.
- Handle and store products carefully between finishing and serving them. Chilled charlottes, bavarois, cold soufflés and cream-based desserts should be stored in the refrigerator until required.
- Check when each dessert was made to ensure that it is still safe to serve. *When in doubt, throw it out.*

Egg based desserts

A soufflé

Soufflés

The soufflé is a light and delicate dessert which is popular on the menu and delicious to eat when made well. Soufflés may be sweet or savoury, and served hot or cold. Pudding soufflé can be kept hot and served, but note that most other hot soufflés must be served immediately once cooked.

Pudding soufflé (*Pouding*)

Method

1 Prepare dariole moulds by greasing them with melted clarified butter and dressing with caster sugar.
2 Lightly blend some butter with flour to form a *beurre-manié* (i.e. equal quantities of flour and butter mixed to a paste and used for thickening).
3 Boil the milk and whisk in the beurre-manié until the mixture becomes thick.
4 Cool slightly and then add the beaten egg yolks.
5 Make a meringue with the egg whites and sugar and fold carefully, cutting the meringue into the pastry cream and yolk base.
6 Add the soufflé mix to the lined dariole moulds until they are two-thirds full, then place them in a bain-marie tray with a tray lid.
7 Simmer the tray on top of the stove until the mix has reached the top of the dariole mould.
8 Remove the lid and place the moulds into the oven to cook for 20–25 minutes at 204 °C (400 °F).
9 When cooked, turn out onto a hot plate or dish, garnish with sauce and serve.

This soufflé can be held if not required immediately. Keep it hot in the bain-marie until required. Remember that although pudding soufflé can be held for service *most soufflés must be served immediately.* Other exceptions are cold pre-set varieties, such as Cold lemon soufflé (*Soufflé milanaise*).

Crème pâtissière based soufflés

Confectioners' custard (or pastry cream) is used as a base for the production of soufflés. These are served hot, direct from the oven. Remember the rule: *the customer can wait for the soufflé but the soufflé cannot wait for the customer.* Timing is all important when hot soufflés are on the dessert menu in a restaurant or hotel. Careful timing and communication between chef, waiter and customer will ensure an enjoyable dessert; poor timing and communication will produce poor results and dissatisfied customers.

Method

1 Make the pastry cream and keep it warm.
2 Prepare a soufflé dish by lightly greasing it with melted clarified butter and dressing with caster sugar.
3 Fold the egg yolks, butter and flavouring into the warm pastry cream.
4 Make a meringue with the egg whites and sugar. Do not over-whip the egg whites, as this produces a coarse meringue which may be soft. (This is due to the albumen strands being over-extended, allowing the water molecules to come into direct contact with the sugar grains. This

in turn dissolves the sugar and makes it heavy and wet.)

5 Fold the light meringue into the pastry cream and egg yolk mix, folding the two mixtures into one.
6 Cook in a hot oven at 204 °C (400 °F) for 20–25 minutes or until the soufflé has risen and developed a light brown top. Serve immediately.

Dessert omelettes

Sweet dessert omelettes are not widely used on the menu in this country. There are many variations of filling, such as fruit, jam and liqueurs. Omelettes are always made to order.

Method

1 Make up the omelette mixture using eggs, sugar and a little milk or cream.
2 Cook the mixture in clarified butter, cooking the eggs until just cooked, but not over-cooked (*baveuse*).
3 Add the filling, e.g. jam, by placing a portion of filling in the centre of the omelette, then folding the omelette in half.
4 Dust with fine icing sugar and then either mark with a hot iron, colour under a salamander (grill) or use a blow torch to glaze the omelette.

Soufflé omelettes

A soufflé omelette is made by first beating the egg yolks and flavouring them with 50 per cent of the sugar until light. Then make a meringue ordinaire (see page 96) with the remaining sugar and fold (cut) it carefully into the yolk and sugar mix. Form the mixture into an oval omelette shape on an oval dish that has been buttered and sugared. Dredge the omelette with fine sugar and bake in a hot oven for 15 minutes at 204 °C (400 °F). Serve at once, prepared for 4–6 portions, rather than individually.

Pancakes (*Crêpes*)

The pancake is indicative of Shrove Tuesday when everyone delights in making pancakes to eat and to toss in races for fun and for charity. This egg dessert is made from milk, flour, eggs, salt, sugar and clarified butter.

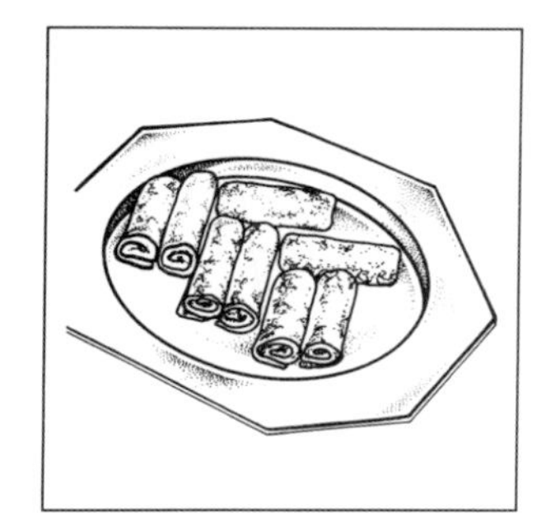

Filled pancakes

Method

1 Mix all of the ingredients except the butter thoroughly by whisking, then strain the mixture through a conical strainer.
2 Pour in the melted butter and stir.
3 Place a little butter in a crêpe pan to melt, then pour off any excess butter. (A crêpe pan is a specialised piece of equipment, used only for the production of pancakes. This small, flat-handled pan should not be washed but wiped with clean paper after cooking and stored hygienically.)
4 Heat the pan well and ladle in a portion of the pancake batter. Swill this until it just covers the base of the pan, setting almost at once. When no moist batter can be seen turn (toss) over the pancake to finish cooking.
5 For production, the crêpes are made then stacked between plates and kept warm in the hot cupboard or used as they are required.

Savoury pancakes are a popular dish for filling with vegetables, cheese and herbs for vegetarian diets, or for wrapping other food items in, such as in Russian pasties (*coulibiac*). Pancakes are also used as a garnish for soups, and small trimmed pancakes called *pannequets* are used for sweet and savoury dishes and hors d'oeuvres.

Egg white based desserts

Meringues

Egg whites are used as a foam aerating agent for many dishes and desserts. Meringues are a popular example of this and they add variety to a whole range of dessert items.

We classify meringue in the following ways:
- cold basic meringue (*Meringue ordinaire*)
- warm Swiss meringue (*Meringue suisse*)
- hot boiled Italian meringue (*Meringue à l'italienne*)
- almond meringue (*Japonaise*).

The pastry kitchen or section always has a stock of fresh egg whites, both from the kitchen and pastry production, where egg yolks are used for recipes. Take care when using whites from the kitchen as these may be tainted by the smell of fish or other pungent foods. Frozen egg whites can be used, as they produce good meringue foams.

In order to produce meringue goods successfully, some basic points need to be considered:
- the egg white to sugar ratio is 1 egg white to 2 teaspoons or cubes of sugar
- make sure that equipment is very clean, by scalding it with boiling water before starting. Mixing machine bowls used for fat based recipes and not cleaned properly after use can prevent the whites from whipping, as the grease does not allow the albumen strands to bond and trap air bubbles
- egg yolk has a fat content and this has the same effect as grease from equipment. Always use clean egg whites, free from yolk particles, shell or blood spots
- a little acid (cream of tartar or lemon juice) added to the whites will strengthen and stabilise the albumen foam, and extend the foam, producing more meringue
- meringue is either dried in a hot cupboard or above an oven range or baked to produce a fawn-coloured, firm meringue product.

Cold basic meringue *(Meringue ordinaire)*

Place the egg white and acid (cream of tartar or lemon juice) in a clean machine bowl, whisk to a light foam and add the sugar over 60 seconds to produce a light firm meringue with a pearl glow.

Technology tip
If the foam is over-whipped, the albumen strands (which hold the water molecules with the sugar suspended on the outside of the bubble) are over-stretched. The water and sugar make contact and the sugar dissolves. This makes the meringue wet and heavy. However, if you whip for a further 10–15 minutes this will firm slightly, and the meringue will not have to be wasted.

Warm Swiss meringue *(Meringue suisse)*

The recipe is the same as for cold meringue (above) but the whites and sugar are whisked to a foam over a bain-marie until very warm: approximately 50 °C (120 °F). Continue to whisk until cool, then pipe into petit four shapes or dry meringues.

Piping meringues

Hot boiled Italian meringue *(Meringue à l'italienne)*

This meringue is used for a wide variety of dessert, pastry and confectionery products: baked Alaska, soufflés, ice cream, marshmallows, nougat de montelimar, mousses, crêpe soufflé, buttercream, vacherins, meringue tranches, shells, swans and figures. Marshmallows are made by adding gelatine to the Italian meringue and piping the mixture while still warm.

Method

1 Dissolve some cube sugar in a little water and boil until it reaches the soft ball stage (118 °C/245 °F).
2 When the boiled sugar is nearly cooked, whisk up the egg white to half volume and pour the boiled sugar onto the egg white while whisking on Speed 3.
3 Take care not to over-whip the egg whites, or the boiled sugar will cook the albumen foam structure and produce a coarse grain finish to the meringue.
4 Continue to whisk until cool. Use a paddle attachment to keep the meringue smooth. Avoid over-whisking.

Almond meringue *(japonaise)*

Japonaise is a versatile mix produced by making a semi-meringue and mixing this with ground almonds and or hazelnuts. A semi-meringue is made with equal quantities of egg white and sugar and is only used for japonaise. Piped and cooked in the oven these meringues are the basis of gateaux, tea fancies, petit fours and biscuits.

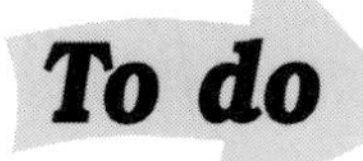

- Check that piping bags are made from nylon and not cotton in your kitchen.
- Find out the cost of the following types of fresh cream: fresh, whipping, double, single and clotted.
- Watch the techniques used by your supervior when piping cream and chocolate.
- Find out where meringues are dried in your kitchen.
- Find out how to clean omelette and crêpe pans after use.
- Ask your supervisor how to clarify butter.

Planning your time

Refer to *Planning your time* on pages 1–4, concerning general points to note when working in a kitchen or pâtisserie. When preparing hot and cold egg based dessert and dessert products the following points are especially important to remember:

- always check the cooking time and temperature for the dessert product. An error here not only wastes valuable time and costs money, but can result in either an undercooked dessert (which is dangerous), or an overcooked dessert (which will not meet customer expectations)
- check oven temperatures very carefully. If the oven is not at the required cooking temperature the dessert will have a poor colour and might be undercooked; so that when turned out of the mould, the custard may collapse. Always ask about the oven setting for each individual dessert product
- check the setting and cooling times carefully: if the dessert is not cooled sufficiently when turned out of the mould the cream will collapse
- make sure that desserts are mixed according to production method/s, with the stated ingredients and then cooked for the required time
- while desserts are cooking, cooling or setting, use the time to clean and clear work surfaces ready for the next batch of work. Use any further time to prepare any service equipment, accompaniment sauces, creams or decorating media used for finishing the dessert items
- check that you know how to test hot egg custard desserts to ensure they are cooked and set before serving. Remember the rule: *when in doubt – check it out*
- when producing a large amount and range of dessert products make sure you have enough moulds, dishes and oven or refrigeration space
- always cool custards, creams and sauces before placing in the refrigerator
- work in a clean and organised manner
- clean up after yourself as you work.

Dealing with unexpected situations

- An unexpected situation might be a rush order, customer request or emergency. Be clear about the priorities in any of these situations, and always follow laid down procedures.
- Check that you know what to do in the case of accidents and emergencies. Note all the points given in Unit G1 of the Core Units book: *Maintaining a safe and secure working environment.*
- Be aware of your personal responsibilities within the kitchen. If in doubt ask your supervisor.

Disposing of waste

- Waste materials should be disposed of cleanly and efficiently, to prevent contamination of uncooked and cooked dessert products and ingredients.
- If you are unsure whether an egg custard based dessert has been cooked correctly, check with your pastry chef. An incorrectly cooked dessert can cause food poisoning. Remember the rule: *if in doubt, throw it out.*
- Wipe down working surfaces between each task or job. Never use glass or oven cloths to wipe down work surfaces: this causes contamination.

What have you learned?

1 What are the main contamination threats when preparing and cooking egg custard based desserts?
2 Why is it important to keep preparation and storage areas and equipment hygienic?
3 Why are time and temperature important when cooking egg custard based desserts?
4 How should you store egg custard based desserts in order to avoid contamination?
5 Why must all cream products be kept in a refrigerator?
6 What should you do in the event of a burn caused by a splash of boiling sugar solution?
7 What are the main differences between liquid egg custard and crème pâtissière?

Extend your knowledge

Eggs, sugar and dairy products are the basic food tools that pastry chefs use to produce artistic dessert, pastry and confectionery dishes to please the customer and make a profit from sales. You can extend your knowledge of these food items by reading, researching and talking to other chefs and crafts people who have developed high levels of technical and artistic skill by dedicating themselves to excellence. Books are also a rich source of information, giving help and tips in the production of quality desserts.The following is a list of ways in which you could extend your knowledge:

1 Find out how to make some classic creams, such as Opera and Edna May.
2 Research ice cream soufflés such as Mylady and Mylord.
3 Find out how to make a range of soufflés, such as Soufflé Irma, Infante, Aida, Pacquita, Grand Success, Volcano and Jubilee.
4 Find out how to make a range of puddings, such as d'Aremburg, Malakoff, Nesselrode and Reine des Fees.
5 Look up recipes for some classic desserts such as Sarah Bernhardt, Parisien and Sabayon.
6 Read books by Hanneman, Barker and the Roux brothers. Read up on recipes from classical and modern experts such as Kollist or Nicolello.
7 Find out about the range of Charlottes and rice based desserts using gelatine and cream such as Riz Maltaise.
8 Extend your knowledge of ice creams by reading up on bombes, parfaits and the biscuit glacé ranges of ice cream desserts. How would you make soufflé glacé, plombiere, granite, fruit or water ices? Investigate coupes and sundaes. What is the difference between these and which are the most classical and widely used dishes?

UNIT 2D4

Preparing and cooking stocks, sauces and soups

ELEMENTS 1, 2 AND 3: Preparing and cooking stocks, sauces and soups

What do you have to do?

- Prepare cooking and preparation areas and equipment ready for use and clean after use, satisfying all health, safety and hygiene regulations.
- Carry out your work in an organised and efficient manner taking account of priorities and laid down procedures.
- Prepare and cook stocks, sauces and soups according to customer and dish requirements.
- Finish and present soups and sauces according to customer and dish requirements.
- Store prepared stocks in accordance with food hygiene regulations.

What do you need to know?

- The type, quality and quantity of ingredients required for each type of stock, sauce and soup.
- How to plan your time efficiently, taking account of priorities and any laid down procedures.
- Why time and temperature are important when cooking stocks, sauces and soups.
- What the main contamination threats are when preparing, cooking and storing stocks, sauces and soups.
- Why it is important to keep preparation, cooking and storage areas and equipment hygienic.

ELEMENT 1: Preparing and cooking stocks

Introduction

Stock is one of the most important ingredients in many wet-cooked dishes: without good quality stock it is impossible to achieve good results in finished dishes. *Stock* is the liquid in which meat, bones, fish or vegetables have simmered. During the cooking process, flavour and soluble nutrients are extracted from the bones and/or vegetables, enriching the liquid. Other flavours can be absorbed into the liquid by the addition of other ingredients, particularly selected herbs.

There are three basic types of stock: fish, vegetable and meat stock.

Preparing stocks

Quality of ingredients

Use only good quality, fresh ingredients for making stocks. Make sure that you are using ingredients in the correct proportions. (See the table on page 102: *Ingredient ratios and cooking times for stocks*).

Meat stock

This can be made from various types of animal bones, including beef, veal, lamb, mutton, chicken and game. Each type produces its own distinctive flavour and taste.

Meat stocks are divided into two basic types:

- white stock
- brown stock.

White stock is made from blanched bones, water, vegetables and herbs simmered together. Brown stock is made from the same ingredients, but the bones and vegetables are browned (see *Method* given below) before simmering. The browned bones give the stock a darker colour and a nuttier taste. The cooking times of meat stocks are dependant on how large the bones are and the type of bones being used.

Meat (i.e. the flesh from animals) can be added to stocks to enhance and concentrate the flavour. For example, boiling fowls can be added to white stock, while shin beef or veal (estouffade) can be added to enrich brown stock.

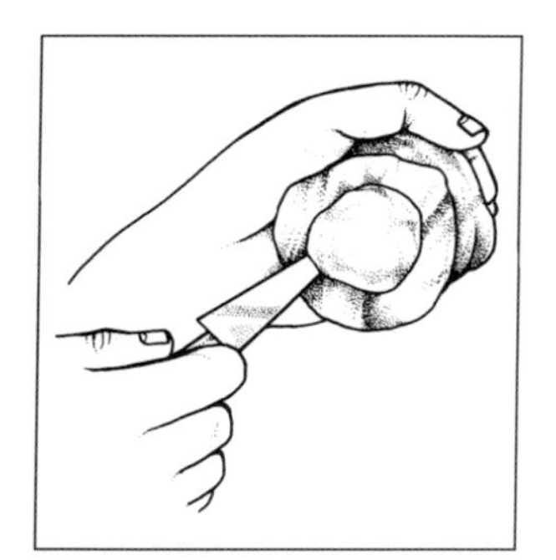

Removing the bone marrow

Method

1. Make sure the bones are chopped into small pieces as this will help to extract as much flavour as possible during the cooking process. The marrow of shin bones should be removed and reserved for other uses.
2. Leave the vegetables whole for white stocks to avoid clouding the stock. Roughly chop the vegetables for a brown stock.
3. *White stock:* go to Step 4.
   Brown stock: brown the bones in a moderately hot oven. Make sure that the bones are browned right through to the centre, not only on the surface. This is necessary because all the flavour from *inside* the bones will be extracted during the long cooking process. The browning of the bones will also give a deep golden colour to the stock.
   When the bones are cooked, drain off the fat, place the bones in a stock pot and *deglaze* the roasting tray. *Deglazing* is the term given to the process of removing any caramelised meat juices from the bottom of the roasting tray. It is done by pouring a little water or stock into the hot tray and bringing it back to the boil. This liquid can then be added to the stock. These residues are rich in flavour and colour, but it is important to make sure that the bottom of the tray is not burnt as this will spoil the stock (by giving it a very bitter flavour).
   Fry the vegetables for brown stock, cooking them separately from the bones. If cooked together, the vegetables would burn before the bones were browned and this would make the stock bitter.
4. Place the vegetables, bones, herbs and water into a stockpot. Always use *cold water.* Gently bring the stock to the boil.
5. As the stock begins to heat up, a scum will start to form on the surface of the liquid. Allow the stock to come to the boil, when the scum will form into solids which can be skimmed off more easily. The scum is

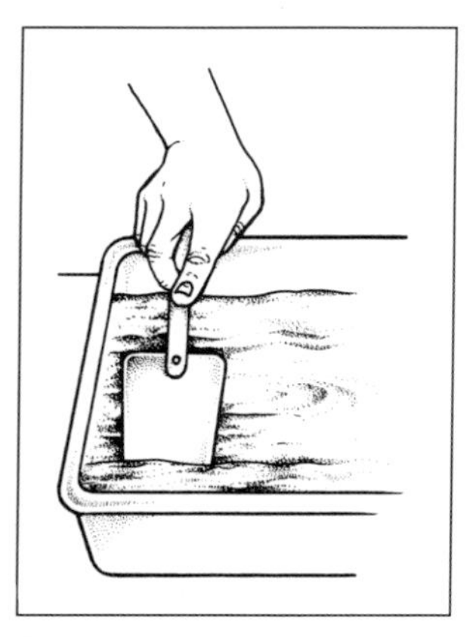

Deglazing the roasting tray

Ingredient ratios and cooking times for stocks

	Beef stock	Veal stock	Mutton/lamb stock	Chicken stock	Game stock	Fish stock	Vegetable stock
Bones	2 kg (4 lb)	2 kg (4 lb)	2 kg (4 lb)	2 kg (4 lb)	2 kg (4 lb)	2 kg (4 lb)	
Vegetables	1/2 kg (1 lb)	1/2 kg (1 lb)	1/2 kg (1 lb)	1/2 kg (1 lb)	1/2 kg (1 lb)	250 g (10 oz) onions: 200g (8 oz)	3 kg (6 lb)
Water	4 litres (8 pt)	4 litres (8 pt)	4 litres (8 pt)	4 litres (8 pt)	4 litres (8 pt)	4 litres (8 pt)	4 litres (8 pt)
Seasoning	Bouquet garni peppercorns	Bouquet garni peppercorns	Bouquet garni peppercorns	Bouquet garni peppercorns	Bouquet garni peppercorns	Peppercorns, bay leaves, parsley stalks, lemon	Herbs and seasonings as required
Cooking time	6–8 hours	6–8 hours	4 hours	3–4 hours	4 hours	20 minutes	10–30 minutes depending on type of vegetable

Skimming the stock

made up of fat and coagulated insoluble protein which rises to the surface and brings with it any impurities from the bones.

6 Keep the stock on a gentle simmer, as this makes sure that all the impurities, fat and scum are constantly being thrown up to the surface, where they can be collected and skimmed off. The culinary term for skimming fat off liquid is degraisser. If you boil stock too rapidly or if you leave the scum on the surface for too long, it boils back into the stock. Once this happens it is difficult to remove it again and the finished stock will be cloudy and dull. This is also why a lid is *never* left on a stockpot during the simmering process.
   Always keep an eye on your stocks while they are cooking, skimming them regularly and adjusting the temperature as necessary.
   Water evaporates from the stock during the cooking process: add water regularly to replenish the stock.
7 Ensure that the stock is cooked for the correct time. *Over-cooking* can result in a deterioration of taste. It will also cause the bones and vegetables to start to break up, and they will then begin to reabsorb some of the flavour from the liquid and turn the stock cloudy. *Under-cooking* will result in a lack of flavour.
   It is not advisable to season the stock, as the seasoning can become excessive if the stock is later reduced. Bear in mind that the stock may also be required for use in dishes already seasoned at a later stage.

Straining the stock

When you are making a clear stock, strain it through a chinois, tapping the side of the chinois with a ladle (as shown below left). If you force the liquid through by placing the ladle inside the strainer the stock will become cloudy. However, if the stock is thickened or does not need to be clear, the liquid can be forced through the strainer by moving the ladle inside the chinois (as shown below right).

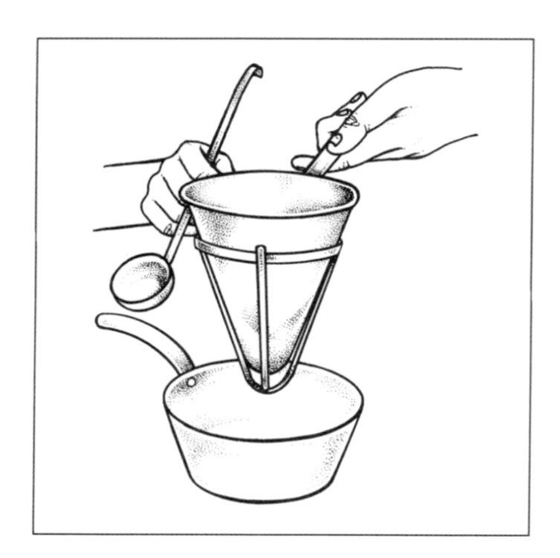

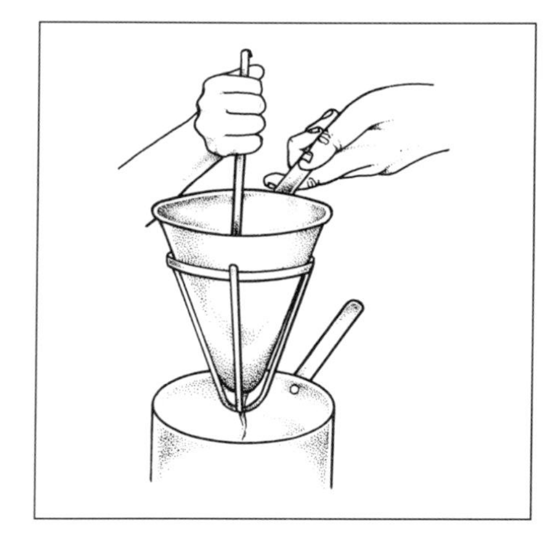

Left: straining a clear stock
Right: straining a stock to be thickened

Fish stock

The basic ingredients for fish stock are as follows:

- washed bones of white fish (preferably whiting, sole or turbot)
- onions
- butter
- bay leaf, parsley stalks, peppercorns, lemon (seasonings)
- water.

Method

1 Sweat the washed bones of white fish with the sliced onions in butter, cooking without developing colour.
2 Add the water and seasonings to the stock and allow to simmer gently for 20 minutes. Note that if the stock cooks for more than 20 minutes it will develop a bitter flavour (due to the release of calcium from the bones) and render the stock unpalatable. (See also Steps 5–7 of *Meat stock* on page 101 regarding skimming off impurities.)
3 Strain the stock ready for use.

Vegetable stock

The basic ingredients for vegetable stock are as follows:

- a selection of chopped vegetables
- butter, vegetable oil or sunflower margarine
- herbs
- water.

Method

1 Sweat the chopped vegetables in the oil or fat.
2 Cover with water, add the herbs and allow to simmer.
3 Cook for 10–30 minutes, depending on the type of vegetables used. (See also Steps 5–7 of *Meat stock* on page 101 regarding skimming off impurities.)
4 Strain ready for use.

Brown vegetable stock may be made following the same procedures, but with the addition of yeast extract and tomatoes when the stock begins to simmer (Step 2). The stock is then cooked and strained ready for use in the same way.

Essential knowledge

Time and temperature are important when cooking stocks in order to:

- achieve the required flavour and consistency
- ensure that the finished product is correctly cooked
- prevent food poisoning.

Convenience products

Convenience products are available for most types of stock and can be purchased in paste, cube, powder or granule form. These products are useful in emergencies and are used by some establishments to eliminate the hygiene and storage risks associated with stocks (see Essential knowledge, page 106).

Some convenience products have one or more of the following disadvantages:

- added colours
- high pre-seasoning, which does not allow for reductions (the concentration of flavour by evaporation)
- monosodium glutamate as an extra ingredient, which produces an unnatural seasoned taste.

To do

- Watch your supervisor carry out general preparation on at least two types of stock.
- Notice the equipment and techniques used when preparing each type of stock.
- Find out the recipes and cooking times for each type of stock prepared in your workplace.
- Find at least three different uses for each type of stock.

Storing stocks, sauces and soups

Stocks, sauces, gravies and soups are known as *high risk foods.* This means that they are the foods most at risk from bacterial infection because they provide the nutrients that bacteria need to grow. If stored incorrectly, these items provide all the optimum conditions for bacteria growth: food, moisture, warmth and time.

If stocks, sauces, gravies and soups are to be stored, rapid cooking and correct temperature control are of vital importance in the prevention of food poisoning. The products required for storage should be placed into a clean pot, bucket or bain marie container. They must then be cooled down as quickly as possible, using one of the following methods:

- place the container in a large blast chiller (see Unit 2D15: *Preparing cook-chill food,* page 270)
- place the container in a sink full of running cold water, stirring frequently
- rest the container on a wooden stand in a cool place so that the air can circulate under it.

If a blast chiller is not available, make sure that the cooling down process does not take more than 90 minutes before refrigeration. Once the product is cold, cover it and attach a clear label stating the contents and date produced. *It is essential to label and date every food product that is stored in the refrigerator.* Keep these products away from uncooked food; store in a separate refrigerator if possible. The refrigerator should be kept at a temperature below 5 °C (41 °F). If the product is to be stored in a freezer, make sure that the freezer maintains a temperature below –18 °C (0 °F).

Key points when storing stocks

- Stocks, sauces gravies and soups are high risk foods and bacteria will grow rapidly in the *danger zone:* 5–63 °C (41–145 °F). However, bacteria cannot grow at temperatures outside of this range, i.e. below 5 °C (41 °F) or above 63 °C (145 °F).
- Stocks, sauces, gravies and soups taken from storage must be boiled again for at least two minutes before use.
- Never reheat stocks, sauces, gravies or soups more than once.

Health, safety and hygiene

Ideally stock should be made from fresh ingredients every day, kept simmering throughout the period of use and any leftovers discarded at the end of the day. If stock is to be stored it must be strained through a conical strainer or chinois into a clean pot, bucket or bain marie container (see above: *Storing stocks sauces and soups*).

Make sure you are familiar with the general points given in Units G1 (*Maintaining a safe and secure working environment*) and G2 (*Maintaining a professional and hygienic appearance*) of the Core Units book. Remember that stocks are *high risk foods*.

Essential knowledge

The main contamination threats when preparing, cooking and storing stocks are as follows:

- food poisoning bacteria may be transferred through dirty surfaces and equipment.
- food poisoning bacteria may be transferred between yourself and food items through inadequate attention to personal hygiene
- contamination can occur due to incorrect storage temperatures. Remember that bacteria will grow most rapidly in the danger zone: 5–63 °C (41–145 °F); bacteria growth will not occur above or below this range of temperatures
- contamination can occur if items undergo prolonged cooling. Stocks must be rapidly cooled (within 90 minutes) before refrigeration
- incorrect waste disposal procedures can lead to contamination
- contamination can occur through items being left uncovered. Cooled stocks prepared for storage must be covered, labelled and dated before being placed in the refrigerator
- cross-contamination can occur between cooked and uncooked food if these are stored together.

Equipment

Stocks are usually made in stockpots, which have a tap at the base to drain off the cooked stock

- Always use *clean* equipment to prevent food from becoming contaminated through harmful bacteria and dirt.
- Use the correct chopping boards for preparing meat, fish or vegetables for soups, stocks, gravies and sauces; check the colour coding system used in your kitchen. Do not use unhygienic wooden boards.
- When chopping vegetables, use the correct knife and correct techniques. Check back to Unit 2D18: *Preparing and cooking vegetables and rice* (page 311–13) and Unit 2D10 of the Core Units book: *Handling and maintaining knives*.
- Do not use unhygienic wooden spoons or aluminium whisks for stirring soups, sauces or gravies. Utensils made from plastic or a similar polythene compound are more hygienic.
- Do not use mechanical bowl choppers or vegetable cutting and dicing machines unless you have received instruction from your supervisor on their use and you are familiar with the safety procedures outlined in Unit 2D12 in the Core Units book: *Cleaning cutting equipment.*

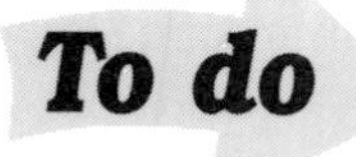

- Check today's menu and identify the types of stock used in the various dishes.
- Find out how frequently stocks are made in your kitchen and ask how each type is stored.
- Check the temperatures of areas in which stocks are stored in your kitchen.
- Ask your supervisor how long each type of stock can be safely stored (including stock stored in deep-freezers).
- Watch your supervisor preparing vegetables for inclusion in a stock. Notice the equipment and techniques used.

Preparation and cooking areas

Make sure you are familiar with the general points given in Unit G1 (*Maintaining a safe and secure working environment*) and Unit G2 (*Maintaining a professional and hygienic appearance*) of the Core Units book. Remember that stocks, soups, gravies and sauces are *high risk foods*.

The points below should all be considered when preparing stocks, soups, gravies and sauces.

- Make sure the vegetables for stocks, soups, gravies and sauces are thoroughly washed, peeled and rewashed before use.
- Only put out the tools and equipment you need; keep your work area uncluttered.
- Do not leave pots or pans on the floor.
- Keep cooking areas clean.
- Do not leave pan handles jutting over the edge of the stove where they could cause accidents.
- Remember that pan handles become hot when the pan is on top of a stove; always use a suitable dry cloth (*rubber*) when lifting.
- Always mark hot pans or trays taken from the oven with something white such as a sprinkling of flour as a warning to others.
- Remember to seek assistance when lifting large, heavy pans and to lift them in the correct manner. Stockpots are especially large and heavy, but these should have an emptying tap at the base to enable the stock to be drawn off with ease and the minimum of disturbance.
- Always mop up spillages immediately.
- When straining or passing stocks, sauces, gravies, or soups into suitable containers, hold your face back and away from the hot products to avoid splashes from the hot liquid.
- When using an electric liquidiser, remember that they should never be more than two-thirds full. If they are filled above this level the initial power surge would send the contents over the top and you could suffer a serious scald or burn from the hot liquid.
- Make sure you are fully aware of what to do in the case of burns or scalds. Check back to Element 3 of Unit G1 in the Core Units book: *Maintaining a safe and secure working environment.*
- When tasting soups, gravies or sauces for seasoning, remember to use a clean spoon *on each occasion.*

Essential knowledge

It is important to keep preparation and storage areas and equipment hygienic in order to:

- prevent the transfer of food poisoning bacteria to the stock
- ensure that standards of cleanliness are maintained
- comply with statutory health and safety regulations.

Planning your time

Be aware of what stocks you will be preparing so that you know what equipment, ingredients and utensils you will need to use. Preparing a production time plan will help you think about priorities. When drawing up your plan, make sure that you have taken account of the daily working schedules within your kitchen.

What have you learned?

1 What are the main contamination threats when preparing, cooking and storing stocks?
2 Why is it important to keep preparation, cooking and storage areas hygienic?
3 Why are time and temperature important when cooking stocks?
4 What faults can occur when making stock and how can they be avoided?
5 What is meant by the terms *high risk foods* and *the danger zone*?
6 What are the three main categories of stock?

ELEMENT 2: Preparing and cooking sauces

Introduction

The French word *sauce* comes from the Latin *saltus* or *salted*, reminding us that sauces were originally liquid seasonings for food. Sauces now have a much more important part to play as they are used to complement the flavour, texture and appearance of foods.

A sauce should never overpower the dish it accompanies but bring harmony and balance. A sauce may also aid digestion: traditionally acidic sauces often accompany fatty foods, we often serve orange sauce with duck and apple sauce with pork. Similarly, fatless foods such as fish and vegetables are often complemented by emollient sauces made with butter or oil. The strong taste of game is often balanced by a sharp, piquant sauce.

Today, due to increased public awareness of the relationship between health and diet, and our changing tastes and attitudes towards food, sauces that are natural, light and quick to prepare have grown in popularity.

Traditional sauces are often categorised or put into main classifications according to their method of preparation. During this element we will look at the following types of sauces: roux sauces, starch thickened sauces, egg based sauces, and meat, poultry and vegetable gravies.

Roux-based sauces

A *roux* is generally defined as *a type of thickening for sauces composed of cooked fat and flour to which liquid is added.* The type of fat and liquid used, and the degree to which the roux is cooked depend on the type of sauce being made.

Roux sauces are classified as *basic sauces.* This means that they are used as a foundation for other dishes or as a base for derivative sauces.

White roux

This is made from equal quantities of fat and flour. The fat used for the preparation of a white roux is usually butter or margarine, although a white roux can be made using vegetable oil.

Method: white roux

1 Melt the fat over a gentle heat.
2 Add the flour and mix well.
3 Gently heat the roux, cooking without developing colour.

Making a béchamel sauce using white roux:

4 Allow the roux to cool.
5 Gradually pour in some warmed, infused milk; i.e. milk that has been flavoured by being gently heated with an *onion clouté* or *studded onion.* (To make an onion clouté, press a number of cloves into a peeled onion, using one clove to staple a bay leaf to the onion.)
6 Cook the sauce for approximately 20–30 minutes.
7 Remove the onion and strain the sauce ready for use.

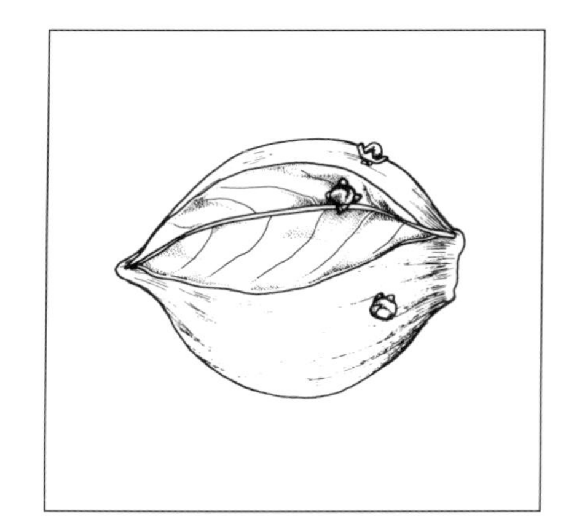

Left: preparing a roux
Right: an onion clouté

Ingredient ratios for basic white sauce

Butter/margarine	100 g (4 oz)
Flour	100 g (4 oz)
Milk	1 litre (2 pt)
Small onion clouté	

Blond roux or yellow roux

A blond roux is prepared as for a white roux (above), but the roux is cooked for a little longer than for a white roux, until it develops a sandy texture and colour.

It may be used for velouté or *velvet* sauce and in the preparation of tomato sauce.

Method: blond roux

1 Melt the fat over a gentle heat.
2 Add the flour and mix well.
3 Gently heat the roux, cooking until it is light fawn in colour and has developed a sandy texture.

Making a velouté sauce using blond roux:

4 Allow the roux to cool.
5 Gradually add boiling white stock to the roux.
6 Cook the sauce for approximately 45 minutes–1 hour.
7 Strain the sauce through a chinois.
8 If a liaison is to be added (optional): mix a liaison of egg yolks and cream with a little of the sauce, then add the liaison to the main portion of sauce. Do not re-boil the sauce or the egg yolk will cook and scramble (*coagulate*), causing the sauce to develop a curdled texture and appearance.

Ingredient ratios for basic velouté sauce

Butter/margarine/oil	100 g (4 oz)
Flour	100 g (4 oz)
White stock	1 litre (2 pt)
Liaison (optional):	
Cream	120 ml ($^{1}/_{4}$ pt)
Egg yolks	2

Brown roux

A brown roux is prepared as for a white roux (page 109), with the following differences:

- use lard or dripping as the fat
- add a little more flour than for a white roux, as the browning of the roux diminishes its thickening properties
- cook the roux over a gentle heat until it turns golden brown and smells of biscuits or lightly roasted hazelnuts.

Making a sauce espagnole using brown roux:

1 Cook the roux until it turns golden brown. Allow it to cool.
2 Gradually add boiling brown stock to the roux over a gentle heat.
3 Add some browned vegetables, tomato purée and mushroom trimmings.
4 Cook the sauce for approximately 4–6 hours.
5 Strain ready for use.

Sauce espagnole can be refined to produce a *demi-glace* or *half-glaze* sauce. This is prepared by adding brown stock to the espagnole sauce, in equal quantities. The thinner sauce is then brought to the boil, simmered until it has reduced by half, and strained ready for use.

Ingredient ratios for basic espagnole sauce

Beef dripping	50 g (2 oz)
Flour	60 g (2$^{1}/_{2}$ oz)
Tomato purée	25 g (1 oz)
Brown stock	1 litre (2 pt)
Mirepoix (vegetables):	
Onions	100 g (4 oz)
Carrots	100 g (4 oz)
Mushroom trimmings	25 g (1 oz)

Beurre manié

This is an uncooked roux made from equal quantities of fresh butter and flour kneaded together into a paste. Small lumps are added to boiling liquids and beaten in with a spoon or whisk until the required consistency is achieved.

Preparing roux sauces

Keep the following points in mind when preparing roux sauces:

- the principle of the roux is that the flour (the thickening agent) is coated in fat and will not therefore clump together when mixed with a liquid
- always stir the roux when cooking so that it cooks evenly; do not allow it to catch in the corners of the pan where it might burn
- when preparing a sauce, it is important to have a cold roux and a hot liquid or a hot roux and a cold liquid. If the roux and liquid are mixed together when both are hot, the flour will cook and thicken immediately into lumps before being evenly distributed throughout the sauce. As the hot liquid meets the hot pan it will also cause some of the liquid to rapidly rise as steam, which can scald the face
- a smooth sauce is achieved by *gradually* adding the liquid, stirring thoroughly and continuously as the liquid comes to the boil. If you add the liquid too fast the sauce will become lumpy. Never use aluminium whisks for stirring sauces, as these are unhygienic and may discolour the sauce
- remember to cook the sauce for the correct period of time. Under-cooking will cause the sauce to be dull, lacking gloss and shine. The sauce may also have a raw flavour and grainy texture and it will not be the correct consistency
- remember that the sauce will only begin to thicken as the flour grains begin to swell up with moisture and eventually burst releasing starch. The sauce then needs to simmer for a further period to avoid a raw starch flavour
- when cooking the sauces, simmer gently and stir frequently to prevent the sauce settling on the bottom of the pan, developing a burnt taste. Sauces requiring long cooking times such as sauce espagnole can be gently cooked in the oven with a tight fitting lid to avoid the constant attention the sauce requires if it is cooked on top of the stove
- when the sauce is cooked it needs to be passed through a conical strainer or chinois into a clean container and kept at a temperature above 75 °C (167 °F)
- as basic sauces have many different uses it is best to keep the seasoning to a minimum. This can always be adjusted later. If the sauce is to be used immediately you will need to adjust the seasoning and consistency to suit the finished dish
- when preparing roux sauces, place some knobs of butter or a buttered greaseproof paper cartouche on the surface to prevent a skin forming.

Essential knowledge

Time and temperature are important when cooking sauces in order to:

- achieve the required flavour and consistency
- ensure that the finished product is correctly cooked
- prevent food poisoning.

- Watch your supervisor carry out general preparation on at least two types of roux based sauces. Notice the equipment and techniques used.
- Find out the recipes and cooking times for each type of roux based sauce prepared in your workplace.
- Check today's menu and identify any roux based sauces used in or with the various dishes.
- Find at least three different uses or derivatives for each type of basic roux sauce.
- Check the temperatures of areas in which basic roux based sauces are stored in your kitchen.
- Find out if there are any convenience basic sauces in your foodstores and ask your supervisor if you could arrange to make comparisons.

Starch-thickened sauces

Sauces may be thickened by the addition of a starch such as rice flour, potato flour or arrowroot. Arrowroot is a very pure starch which has no taste and cooks very quickly. It is the refined starch from the root of the West Indian Maranta plant.

To thicken a sauce with a refined starch, first dissolve the starch in cold water. Take the liquid to be thickened off the boil and pour in the dissolved starch, stirring all the time. Return to the boil and simmer gently for 10–15 minutes. Pass the finished sauce through a conical strainer or chinois. Starch-thickened gravy (*jus lié*) made with gelatinous reduced stock is now widely used as a brown base sauce and is a modern alternative to traditional demi-glace sauce.

Egg-based sauces

Egg-based sauces are divided into two types.

1. Egg sauces that are served hot, made from emulsions of egg yolks and butter.
2. Egg sauces that are served cold, made from emulsions of egg yolks and oil.

Emulsion is the scientific name for a mixture of two substances that do not normally bind together. In cookery, an *emulsion* normally refers to fat and liquid bound together in a creamy mixture.

Preparing basic hot egg-based sauce

Sauce hollandaise is a hot emulsion sauce of egg yolks and butter and is prepared using the method below.

Method

1. Make a reduction of vinegar and peppercorns in a sauteuse pan.
2. When completely reduced, add a little cold water.
3. Place some egg yolks and strained reduction into a clean mixing bowl.
4. Whisk up the yolks and reduction with a balloon whisk over a pan of hot water or over a gentle heat until the mixture becomes light and fluffy and thickens slightly. The cooked mixture, which should form

Gradually adding butter to hollandaise sauce

stable peaks with the whisk is called a *sabayon.*

5 Remove from the heat and gradually whisk in warm, melted, unsalted butter a little at a time until the sauce thickens.
6 Season and strain through a fine chinois or through muslin into a clean bowl.

Ingredient ratios for hot egg-based sauce

Vinegar	25 ml (1 fl oz)
Crushed peppercorns	8–10
Cold water	1 tbsp
Unsalted butter	400 g (1 lb)
Egg yolks	4

Key points in preparing hot egg-based emulsion sauces

- Always use good quality ingredients.
- Take care not to overcook the sabayon mixture or the eggs will scramble.
- Make sure the butter is not too hot.
- Remember not to add the butter too quickly and always make sure all the butter has been incorporated into the sauce before adding any more.
- If your sauce should separate or split: put a tablespoon of hot water into a clean bowl and gradually add the separated sauce to it, whisking vigorously. Should this fail the yolks are probably slightly overcooked and cannot hold the butter, so you need to whisk up a fresh egg with a spoonful of water. Then repeat the procedure, adding the separated sauce.

Making mayonnaise sauce by hand

Preparing basic cold egg-based sauces

Sauce mayonnaise is a cold emulsion of egg yolks and oil.

Method

1 Place the egg yolks, vinegar and mustard into a bowl and whisk together*.
2 Gradually whisk in the oil a little at a time.
3 Season to taste.

*Some chefs like to add half the vinegar with the egg yolks and the remaining half when all the oil has been added.

Key points in preparing cold egg-based emulsion sauces

- In order to reduce the danger of salmonella infection from raw eggs, use pasteurised egg. Remember that pasteurised egg must still be handled and stored with care.
- Keep all ingredients at room temperature. Gently warm the oil if it has been chilled.
- Add the oil very gradually at first, adding it in small drops until the egg yolks have had time to absorb the oil. The oil can be added more quickly as the sauce starts to thicken.

- If the oil is added too quickly or if the recipe balance is incorrect the sauce will curdle. If your sauce should separate or curdle: put a table-spoon of hot water into a clean bowl and gradually add the curdled sauce to it, whisking vigorously. Should this fail, whisk up a fresh egg yolk with a little cold water. Then proceed as before, gradually adding the curdled sauce.

Ingredient ratios for cold egg-based sauce

Vinegar	1 tbsp
Dry mustard	$^1/_2$ tbsp
Oil	500 ml (1 pt)
Egg yolks	4

Storing hot egg emulsion sauces

Hollandaise sauce is usually held at a warm temperature rather than hot. If the sauce is allowed to become cold the butter will set and the sauce will split when reheated. Alternatively, if the sauce becomes too hot it will also split. The sauce cannot therefore be effectively stored outside the danger zone, i.e. the temperature range within which bacteria grow most quickly 5–65 °C (41–149 °F). The solution is to avoid storing the sauce: once prepared and it can only be kept warm for a *maximum* of two hours. Any remains must then be thrown away. However, it is strongly recommended that the sauce is prepared as close to the service time as possible and any surplus discarded.

Storing cold egg emulsion sauces

Mayonnaise should be stored in a cool place away from extreme temperatures in airtight containers or screw top jars. If mayonnaise is stored in a refrigerator it must be allowed to come to room temperature slowly before it is stirred or it will separate.

- Watch your supervisor carry out general preparation on a hot and cold egg-based sauce. Notice the equipment and techniques used.
- Find at least three different uses or derivatives for each type of egg-based sauce.
- Check today's menu and identify any starch-thickened sauces or gravies used in or with the various dishes.
- Ask your supervisor for the recipes and methods for egg-based sauces and gravies prepared in your workplace.
- Check the temperatures of areas in which egg-based sauces and gravies are stored in your kitchen.

Finishing and presenting sauces

If you have prepared your sauce correctly it should only require minor adjustments to finish and present. You will need to check the consistency and correct this if necessary, as shown in the table on the next page.

Correcting sauce consistencies

Type of sauce	If the sauce is too thick	If the sauce is too thin
Roux based	Gradually add more liquid to required consistency	Leave to simmer gently and reduce to required consistency *or:* thicken with beurre manie or refined starch (arrowroot)
Hollandaise	Add a little hot water	Gradually add sauce to fresh egg yolk which has been whisked over heat with a little water
Mayonnaise	Add a little hot water	Gradually add sauce to fresh egg yolk/s

You will also need to check whether the sauce is sufficiently seasoned; i.e. seasoned to capture the full taste. This comes with experience and you may need to check the seasoning with your supervisor when you first start to prepare sauces. Take care not to over-season, as it is easy to add more seasoning but virtually impossible to remove.

Make sure your sauce is served at the correct temperature and that any garnish is of the correct size and quantity and correctly cooked.

Serving sauces
Do not over-fill sauce boats. When serving hot sauce for plate service, do not over-fill the plate. Use a suitable jug with a lip for pouring the sauce to prevent unsightly drips on the plate.

Meat, poultry and vegetable gravies

The French term for gravy is *jus rôti*, which translates to *the juices of the roast.* The Oxford dictionary also defines gravy as *the juice that comes out of meat while it is cooking and the sauce that is made from this.* These definitions are loosely correct but obviously do not take account of vegetable gravy preparations. Vegetable gravies provide liquid flavour and complement the vegetable dishes in the same way as traditional sauces and gravies accompany and complement meat dishes.

Preparing meat and poultry gravies

1. Remove the cooked meat or poultry from the roasting tin.
2. Allow the sediment to settle on the bottom of the roasting tray then carefully decant off the fat.
3. Gently heat the sediment on top of the stove to cook brown.
4. Swill out the roasting tin with the appropriate brown stock (*deglazer*).
5. Simmer, season to taste, strain and serve.

Preparing vegetable gravies

Vegetable gravies are not classified in the same descriptive way as traditional gravies and are more open to the individual flair, interpretation and imagination of the cook (provided of course that the resultant liquor is derived from vegetables). A simple vegetable gravy can be prepared by lightly thickening a good vegetable stock with arrowroot, following the principles described under *Starch-thickened sauces* (page 112)

Essential knowledge

The main contamination threats when preparing, cooking and storing sauces are as follows:

- food poisoning bacteria may be transferred through dirty surfaces and equipment
- food poisoning bacteria may be transferred between yourself and food items through inadequate attention to personal hygiene
- contamination can occur due to incorrect storage temperatures. Remember that bacteria will grow most rapidly in the danger zone: 5–63 °C (41–145 °F); bacteria growth will not occur above or below this range of temperatures
- contamination can occur if items undergo prolonged cooling. Sauces must be rapidly cooled (within 90 minutes) before refrigeration
- incorrect waste disposal procedures can lead to contamination
- contamination can occur through items being left uncovered. Cooled sauces prepared for storage must be covered, labelled and dated before being placed in the refrigerator
- cross-contamination can occur between cooked and uncooked food if these are stored together.

Planning your time

Be aware of which sauces you will be preparing so that you know what equipment, ingredients and utensils you will need to use. Preparing a production time plan will help you think about priorities; make sure that you take account of the daily working schedules within your kitchen.

Health, safety and hygiene

Make sure you are familiar with the general points given in Units G1 (*Maintaining a safe and secure working environment*) and G2 (*Maintaining a professional and hygienic appearance*) of the Core Units book. Remember that sauces are high risk foods.

Preparation of equipment, preparation and cooking areas

Refer to *Equipment* (page 106) and *Preparation and cooking areas* (page 107) for stocks, sauces, gravies and soups.

When making hollandaise sauce, use a copper or stainless steel pan (sauteuse) if possible, as this will conduct the heat evenly and retain the sauce's colour. It is important to ensure that any copper pans are well tinned to prevent the risk of contamination from the copper.

The wide surface area of a sauteuse makes it ideal for preparing sauces, especially those which need reducing

Use a thin, springy balloon whisk when making either hollandaise or mayonnaise sauce, as this will effectively combine the mixture. If mayonnaise is normally prepared in your workplace using a mechanical mixing machine, make sure that you have received instruction in its use and that you are familiar with the safety procedures before attempting to use the machine.

Essential knowledge

It is important to keep preparation and storage areas and equipment hygienic in order to:

- prevent the transfer of food poisoning bacteria to the stock
- ensure that standards of cleanliness are maintained
- comply with statutory health and safety regulations.

What have you learned?

1. What are the main contamination threats when preparing, cooking, and storing sauces?
2. Why is it important to keep preparation, cooking and storage areas hygienic?
3. Why are time and temperature important when preparing and cooking sauces?
4. What faults can occur when preparing sauces? How can they be avoided and, in some cases, rectified?
5. What are the main classifications of sauces and how are they prepared?
6. Why are sauces served with food?

ELEMENT 3: Preparing and cooking soups

Introduction

Soups, unlike sauces, are a dish in their own right. Soups may be served as a snack meal, a meal in themselves, or more often, as a first course on a menu to stimulate the appetite. The place that soups have on the menu means that they are one of the first dishes to be eaten by a customer; this is important as they can influence the whole meal.

A good soup should be light and flavoursome, capturing the various flavours and aromas of the chosen ingredients. Soups can be classified according to their method of preparation. In this element we will look at the following types of soup: purée soups, cream soups, broths.

You will need to be familiar with the following culinary terms:

- *croûtons or sippets*: small cubes of shallow-fried bread served with soup. Additional flavours may be added to croûtons such as garlic or celery seeds
- *flûtes:* small circles of toasted French bread served with soup, e.g. minestrone
- *mirepoix:* roughly cut vegetables used for flavouring
- *bouquet garni:* herbs bundled together in leek, celery or muslin, usually parsley stalks, thyme and bay leaf.

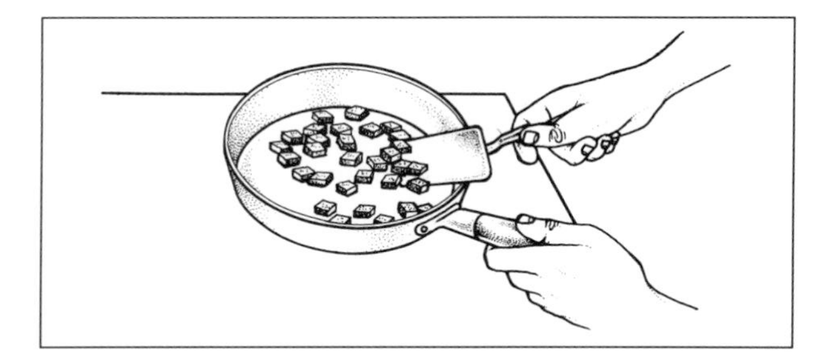

Frying croûtons

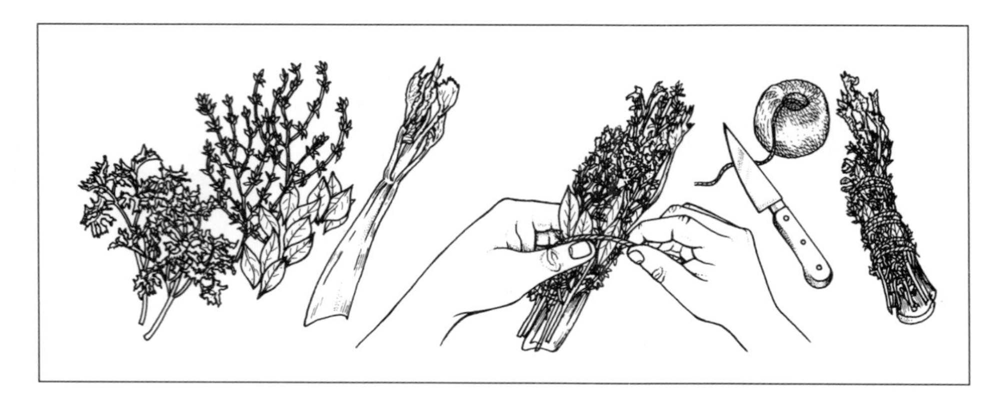

Tying a bouquet garni

Purée soups

These are soups where the main ingredients (usually vegetables, dried vegetables or pulses) are cooked with a stock and, as the name implies, puréed and strained. Where starchy vegetables and pulses are included as ingredients, these act as self-thickeners to the soup. Soups made using other types of vegetables need to be thickened by an extra ingredient: a starch-based thickening agent.

Purée soups are usually served accompanied by croûtons.

Purée soups using starchy vegetables or pulses

Method

1 Place the main vegetables or pulses into a saucepan, add the stock and mirepoix.
2 Bring to the boil, reduce the heat and allow to simmer.
3 Add the bouquet garni and season as necessary. Skim off any impurities that rise to the surface.
4 Simmer for approximately 45 minutes, depending on the type of vegetable used.
5 Remove the bouquet garni and liquidise the soup.
6 Pass the liquidised soup through a chinois or conical strainer into a clean pan.
7 Adjust the consistency and seasoning, and add garnish required.

Recipe examples: Purée of haricot bean soup, Purée of lentil soup.

Purée soups requiring a thickening agent

Method

1 Sweat the main vegetable and mirepoix in fat.
2 Add flour to form a roux, and cook over a gentle heat without colouring.
3 Cool slightly and gradually add the hot stock .
4 Bring to the boil, reduce the heat and allow to simmer.
5 Add the bouquet garni and season as necessary. Skim off any impurities that rise to the surface.
6 Simmer for approximately 45 minutes, depending on the type of vegetable used.
7 Remove the bouquet garni and liquidise the soup.
8 Pass the liquidised soup through a chinois or conical strainer into a clean pan.
9 Adjust the consistency and seasoning, and add garnish required.

The soup can also be thickened by the addition of rice or potatoes (at Step 5) instead of flour. The roux is cooked a little longer to a sandy texture for some soups, such as mushroom and tomato soup.

Recipe examples: Purée of carrot soup, Purée of leek soup.

Velouté soups

Velouté soups are prepared from an appropriately flavoured white stock thickened by a blond roux (see page 109). The stock may be derived from meat, fish or vegetables according to recipe requirements.

Velouté soups without a vegetable content

Method

Gradually add the white stock to the blond roux and then gently simmer the soup until cooked, following the same procedures as for the preparation of velouté sauces (see page 109).

Recipe examples: Chicken velouté soup, Fish velouté soup.

Velouté soups with a vegetable content

Vegetable veloutés may also contain a vegetable content appropriate to the main character of the soup which is cooked separately and puréed into the soup. The vegetable content should not usually account for more than 25 per cent of the volume of the finished soup. A velouté soup can be thickened by the addition of the appropriately flavoured velouté sauce or thinned by the further addition of stock.

Method

1. Sweat the vegetable in butter, then cover with appropriately flavoured stock.
2. Add the bouquet garni and simmer until the vegetables are tender.
3. Remove the bouquet garni, add the prepared velouté to the vegetables and bring the soup back to a boil.
4. Liquidise the soup and pass it through a chinois or conical strainer into a clean allen pan.
5. Adjust the consistency and seasoning and add garnish if required.

Recipe examples: Asparagus velouté soup, Mushroom velouté soup.

Finishing a velouté soup

Velouté soups are finished by taking the soup off the boil and adding a liaison of egg yolks and cream which has been mixed with a little of the soup. The finished soup should have a light, smooth, velvet-like texture and a delicate flavour.

Essential knowledge

It is important to keep preparation and storage areas and equipment hygienic when preparing soups in order to:
- prevent the transfer of food poisoning bacteria to the stock
- ensure that standards of cleanliness are maintained
- comply with statutory health and safety regulations.

Cream soups

The classical cream soup is a purée soup made with the addition of béchamel sauce and stock, where the béchamel sauce acts as the main thickening agent. However, purée soups finished with the addition of cream or velouté soups finished with the addition of cream (rather than cream and egg yolks) can also be identified as cream soups on a menu.

Method

1 Place the main vegetables or pulses into a saucepan, add the stock and mirepoix.
2 Bring to the boil, reduce the heat and allow to simmer.
3 Add the bouquet garni and season as necessary. Skim off any impurities that rise to the surface.
4 Simmer for approximately 45 minutes, depending on the type of vegetable used.
5 Remove the bouquet garni, add the béchamel sauce, mix well.
6 Bring back to the boil.
7 Liquidise the soup.
8 Pass the liquidised soup through a chinois or conical strainer into a clean pan.
9 Adjust the consistency and seasoning, and add garnish required.

Recipe examples: Cream of cauliflower soup, Cream of celery soup.

Key points

- If the soup is too thin, you can thicken it with a little refined starch; if it is too thick, you can add more stock. A thickened soup should have the consistency of single cream.
- Take care when seasoning: salt is easy to add but difficult to remove.
- If using convenience stock, remember that these products often have a high salt content.
- Always closely follow the recipes given to you by your supervisor.
- Remember: if you add cream to a soup, you must not re-boil the soup or it will separate. The same rule applies if you add a liaison to a velouté soup: do not re-boil the soup or the egg yolk will cook and scramble (*coagulate*), causing the soup to develop a curdled texture and appearance.

Broths

Broths are made from unthickened stock, finely cut vegetables and either meat or fish. They often contain a cereal such as pearl barley or rice which, due to the release of starch during cooking, slightly thickens and clouds the soup. A distinguishing feature of broths is that they are not strained or passed.

Broths can be divided into two main types of preparation:

- sweated broth type
- unsweated broth type.

Sweated broths containing fish or shellfish are usually referred to as *chowders* (see page 121).

Unsweated broths

Method

1 If using meat, blanch and refresh it. This is not necessary with chicken.
2 Place the meat in a clean pan and add appropriately flavoured stock or water. Bring to the boil and skim any impurities rising to the surface.
3 Add the bouquet garni, seasoning and cereal. Simmer until the meat is nearly cooked.
4 Add the washed and cut vegetables, then simmer until cooked.
5 Remove the meat, cut it into neat dice and return it to the soup.
6 Remove the bouquet garni, adjust the seasoning and skim again if necessary. Add chopped parsley.

Recipe examples: Chicken broth, Mutton broth.

Sweated broths

Method

1 Sweat the washed and cut vegetables in fat without developing colour.
2 Add appropriately flavoured stock then bring to the boil, skimming if necessary.
3 Add the bouquet garni, season and simmer until the vegetables are nearly cooked.
4 Add any garnish and simmer until cooked.
5 Remove the bouquet garni and adjust the consistency and seasoning.

Some recipes for sweated broths contain vegetables (e.g. potatoes, peas or beans) that would overcook if they were sweated with the other vegetables. These vegetables are therefore added to the simmering broth at Step 3.

Recipe examples: Minestrone, Leek and potato soup (Potage bonne femme), Cock-a-leekie soup.

Key points

- Broths should not be too thick; the thickness can be adjusted by the addition of stock.
- Cream may be added to finish a broth. The soup must not then be re-boiled or the cream will separate.
- It is possible to omit meat from most broth recipes for vegetarian preparations.

Essential knowledge

Time and temperature are important when cooking soups in order to:
- achieve the required flavour and consistency
- ensure that the finished product is correctly cooked
- prevent food poisoning.

Chowders

These are sweated broths containing fish or shellfish. Chowders may also contain diced salted pork which needs to be blanched and refreshed before use.

Method

1 Sweat any diced blanched pork in fat, then add the vegetables and continue to sweat without developing colour.
2 Add the fish stock then bring to the boil, skimming as necessary.
3 Add the bouquet garni, season and simmer until the vegetables are nearly cooked.
4 Add the cut fish or shellfish then simmer until cooked*.
5 Remove the bouquet garni and adjust the consistency and seasoning.
6 Finish with chopped parsley and cream.

**If using shellfish such as clams, cook them until the shells open. Remove and dice the meat. Add the diced meat and retained liquor from the shell, at Step 4.*

Optional ingredients: Some recipes for chowder also include tomato concassée, which would be added at Step 4. Water biscuits (whole or broken into small pieces) may be served separately.

Finishing and presenting soups

If you have prepared your soup correctly it should only require minor adjustments to finish and present.

Check the consistency and adjust if necessary following the guidlines given in the table below.

Correcting sauce consistencies

Type of soup	If the soup is too thick	If the soup is too thin
Cream/purée	Add more stock or milk	Leave soup to simmer gently and reduce to required consistency *or:* thicken with a refined starch
Broths	Add more stock	Strain some stock from the soup

Check the seasoning: soups should be seasoned to capture the full taste. You will recognise this with experience; check the seasoning with your supervisor when you first begin to prepare soups. Take care not to over-season the soup, as it is easy to add more salt or pepper but virtually impossible to remove them. Ensure that any garnish in your soup is of the corret size and quantity and is correctly cooked.

Serving soup

Soup may be served using a ladle, which also provides a method of portion control. Where a large number of customers require serving at the same time (e.g. at banquets and functions) it is often easier to use a large jug with a lip for pouring the soup.

Garnishes

The main garnish may be added to the finished soup or kept separate and placed in each bowl before the liquid is added. This method has the advantage of ensuring that each customer receives an equal amount of garnish and so provides an effective method of portion control.

It is important to ensure that the garnish is kept at the correct temperature prior to service.

A purée soup may be finished with a swirl of cream

Purée soups may be finished with a small swirl of cream which is added when the soup is in the bowl. Croûtons and flûtes on traditional menus are served separately in sauce boats and offered to the customer. However, for bistro-style service, they may be sprinkled on top of the soup prior to service.

Storing soups

See *Storing of stocks, sauces, gravies and soups* on page 105. If the soup is to be used immediately it should be kept at a temperature above 75 °C (167 °F). Soups should be served very hot, at 90 °C (194 °F) in heated cups or bowls. Cold soups should be served at 3 °C (37 °F) in ice-cold bowls.

If a cooked garnish is added separately to a soup, ensure that the correct temperature control is maintained for both the garnish ingredient and the soup. Take care to avoid cross-contamination occurring between raw garnishes and cooked soup. Refrigerate raw garnishes which are cooked at the last minute (such as quenelles) until they are required.

Essential knowledge

The main contamination threats when preparing, cooking and storing soups are as follows:

- food poisoning bacteria may be transferred through dirty surfaces and equipment
- food poisoning bacteria may be transferred between yourself and food items through inadequate attention to personal hygiene
- contamination can occur due to incorrect storage temperatures. Remember that bacteria will grow most rapidly in the danger zone: 5–63 °C (41–145 °F); bacteria growth will not occur above or below this range of temperatures
- contamination can occur if items undergo prolonged cooling. Soups must be rapidly cooled (within 90 minutes) before refrigeration
- incorrect waste disposal procedures can lead to contamination
- contamination can occur through items being left uncovered. Cooled soups prepared for storage must be covered, labelled and dated before being placed in the refrigerator
- cross-contamination can occur between cooked and uncooked food if these are stored together.

Planning your time

Be aware of which soups you will be preparing so that you know what equipment, ingredients and utensils you will need to use. Preparing a production time plan will help you think about priorities; make sure that you take account of the daily working schedules within your kitchen.

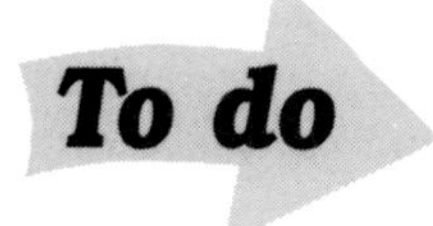

- Watch your supervisor carry out general preparation on a purée soup, a cream soup and a broth. Notice the equipment and techniques used.
- Check today's menu and see if any purée, cream or broth type soups are featured. Ask your supervisor for the recipes. What methods of preparation are used?
- Check the temperatures of areas in which soups are stored in your kitchen.
- Find out if there are any convenience soups used in your work place and ask your supervisor if you may make comparisons.
- Find a suitable soup garnish where the main ingredient of the soup is:
  - vegetables
  - pasta
  - meat
  - fish
  - cheese.

Health, safety and hygiene

Make sure you are familiar with the general points given in Unit G1 (*Maintaining a safe and secure working environment*) and Unit G2 (*Maintaining a professional and hygienic appearance*) of the Core Units book.

Remember that soups are *high risk foods.*

Preparation of equipment, preparation and cooking areas

Refer to *Equipment* (page 106) and *Preparation and cooking areas* (page 107) for stocks, sauces, gravies and soups.

Dealing with unexpected situations

Unexpected situations usually fall under two categories: those that affect the safety of yourself or your colleagues and those that affect the product you are preparing. Kitchens are busy environments where people are often working to deadlines; if an unexpected situation arises, remember not to panic. Tackle each situation in a professional manner using the knowledge you have learnt in this unit.

An example of an unexpected situation might be a stockpot spillage. If this happens:
- mop and dry the area where the spillage has occurred immediately
- warn your colleagues
- make sure that you are familiar with the procedure for scalds (see Unit G1: *Maintaining a safe and secure working environment* in the Core Units book).

What have you learned?

1. What are the main contamination threats when preparing cooking and storing soups?
2. Why is it important to keep preparation cooking areas and storage areas hygienic when preparing soups?
3. Why are time and temperature important when cooking soups?
4. How should you prepare and serve cream broth and purée types of soup?
5. What faults can occur when finishing soups and how they can be rectified?

Extend your knowledge

1. Find out how stocks can be reduced to form *essences* and *glazes*.
2. Carry out some research to find out why veal bones are considered to be good for making reduced stocks.
3. Find out how stock can be clarified to make consommés or aspic jelly.
4. Certain soups are unclassified. Find out what they are and how you might prepare and cook them.
5. Use recipe books to find examples of hard butter sauces. How are these made? What dishes are they usually served with?
6. Ask your supervisor if you have any coulis type sauces on your menu. Find out how they are prepared and cooked.
7. Small pieces of butter are sometimes shaken into finished sauces (*monter au beurre*). What effect does this have on the sauce?
8. Find out about classifications of soup not included in this unit, such as consommés and shellfish bisques. How might you prepare and cook them?
9. Find out what types of soups are made using a brown roux as a thickening agent.

UNIT 2D5

Preparing and cooking pulses

ELEMENT 1: Preparing and cooking pulses

What do you have to do?

- Prepare cooking and preparation areas and equipment ready for use and clean after use.
- Combine pulses correctly with other ingredients according to customer and dish requirements.
- Prepare, cook, finish and present pulses according to customer and dish requirements.
- Work in an organised and efficient manner taking account of priorities and laid down procedures.

What do you need to know?

- The type, quality and quantity of pulses required for each dish.
- What the main contamination threats are when preparing and cooking pulse dishes.
- Why it is important to keep preparation, cooking and storage areas and equipment hygienic.
- Why time and temperature are important when cooking pulses.
- How to satisfy health, safety and hygiene regulations concerning preparation and cooking areas and equipment.
- How to plan your time efficiently, allocating time appropriately to meet daily schedules.
- How to deal with unexpected situations.

Introduction

Pulses is the name given to the edible seeds of the legume family of plants. Peas, beans, chickpeas and lentils are all types of pulses. They are an important part of our diet as they are high in fibre, protein, iron and Vitamin B and contain no saturated fat. This makes them a healthy alternative to meat and especially useful for vegetarians, who need to rely on non-meat sources for protein. Our bodies can make the best use of the protein in pulses if they are served either with a cereal like wheat (bread) or rice, nuts or dairy products (milk, yoghurt, cheese or eggs).

There are many hundreds of varieties of pulses used throughout the world. They are harvested in warm temperate climates, so most pulses used in the UK are imported. They can be purchased in a variety of forms: fresh, dried or canned. Fresh pulses, like fresh peas and beans can be treated like vegetables (see Unit 2D18: *Preparing and cooking vegetables and rice*, pages 305–28). Canned pulses have already been cooked and can be used immediately as an ingredient, while the dried varieties need to be prepared before they can be eaten.

Pulses are very flexible and can easily be used to replace meat or fish on the menu. Some suggestions for using pulses on the menu are given below:

Starter:	Lentil and mushroom pâté
Soup:	Split green pea soup
Pasta:	Lentil lasagne
Main course:	Kidney bean moussaka
Vegetables:	Stir fried butter beans
Salad:	Chickpea bean sprout and apple salad
Dessert:	Adzuki bean dumplings.

Types of pulses

Pulses can be divided into three types: peas, beans and lentils.

Peas

Dried peas contain a large amount of starch and they disintegrate easily when cooked. This makes them especially useful when making soups, and they are often used as an ingredient in soups and broths. Chickpeas are an exception to the rule: they keep their shape even after 1 1/2 hours of cooking.

The most commonly used peas are listed below:

- *Marrowfat (blue) peas:* these have a sweet flavour. Their skin is tough, but they have a floury texture when cooked.
- *Split green peas:* these are used to make traditional English pease pudding, served with boiled pork. They disintegrate when cooked and are used to make thick purée soups.
- *Split yellow peas:* these are cooked to a purée and served as a vegetable. They are also used to make thick purée soups.
- *Chickpeas:* these have a nutty taste and will absorb other flavours easily. They are widely used as a vegetable in Middle Eastern and Asian cookery. Chickpeas are also milled into gram flour.

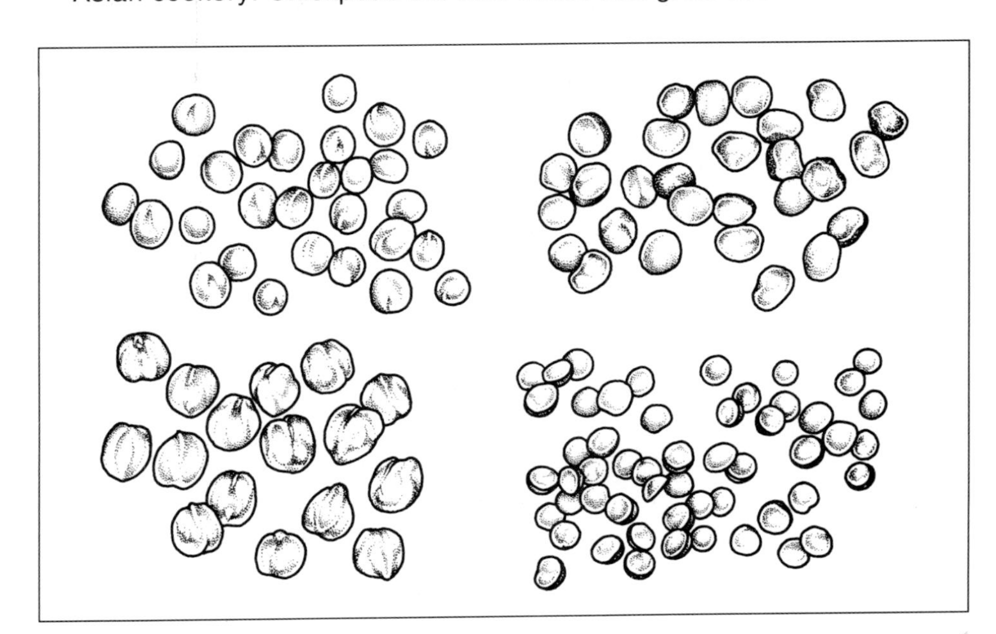

Clockwise from top left: Marrowfat (blue) peas; split green peas; split yellow peas; chickpeas

Beans

Beans are the largest group of pulses: in some Mexican markets you can find over 25 different varieties, each suited to a different local dish. Examples of some beans now commonly available in this country are listed below.

- *Red kidney beans:* these have a floury texture and sweetish flavour. They contain an enzyme which needs to be destroyed by rapid boiling for the first ten minutes of cooking time. This is essential: failure to do this will cause food poisoning.
- *Mung beans:* these are mostly used in stews. They are very popular in their sprouted form, which can be used as a vegetable.
- *White haricot beans:* these are by far the most popular bean in the UK. Most of us will have eaten them as baked beans, where they are cooked with a tomato sauce. They are also used as an ingredient in soups.
- *Green flageolet beans:* these have a delicate taste and can be served hot as a vegetable or cold in salads. They are actually haricot beans which have been picked when still young and tender.
- *Lima beans* (butter beans): these come from Peru in South America and have a sweet buttery flavour which is why they are sometimes called butter beans.
- *Black-eyed beans:* these have a savoury flavour and succulent texture, and are popular in traditional North American cookery.
- *Adzuki beans:* these have a strong, nutty, sweet flavour and can be served as a savoury side dish. In the Far East they are sweetened with sugar and used in cakes and desserts.
- *Borlotto beans:* these are a member of the kidney bean family and can be substituted in any dish that requires kidney or haricot beans. They have a pleasant flavour and a moist and tender texture.
- *Broad beans:* these are most often eaten when fresh, but are also available dried. The tough skin of the dried beans needs to be removed before serving. They have a fine texture and can be used as a vegetable, in salads, or as a thickening agent for sauces in some modern recipes.

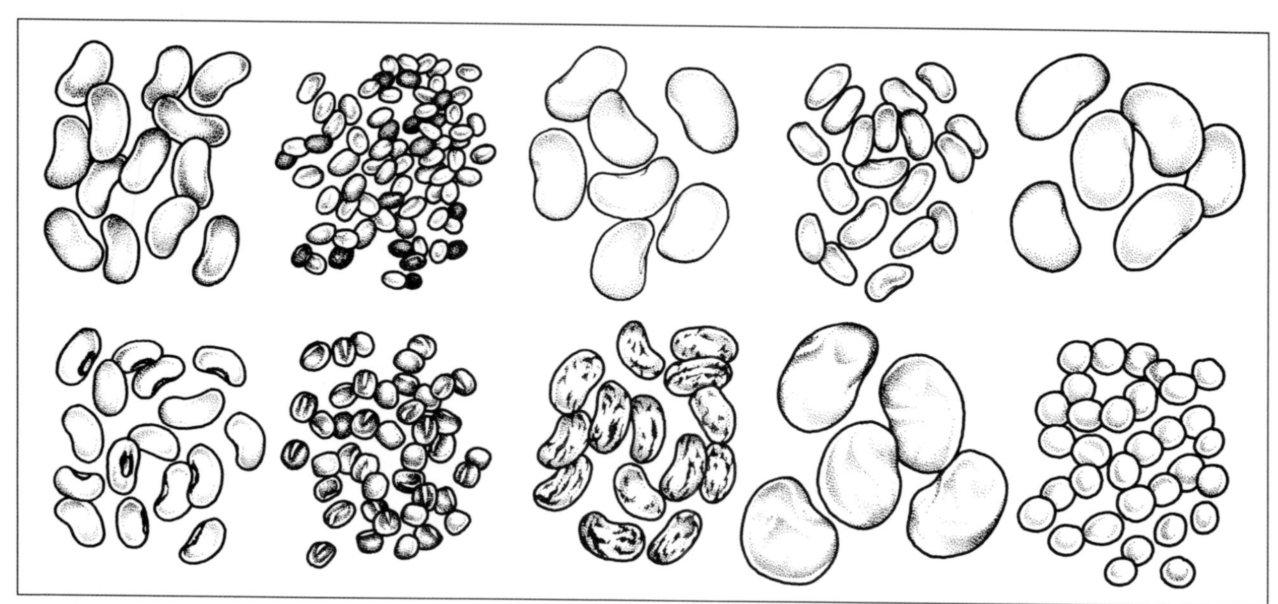

Top row: red kidney beans, mung beans, white haricot beans, green flageolet beans, lima beans
Bottom row: black-eyed beans, adzuki beans, borlotto beans, broad beans, soybeans

- *Soybeans:* these are very high in nutrients, especially protein and fat. They are available in many forms: ground to make soya flour; processed into chunks and granules for use as a meat substitute (TVP); processed into bean curd (Tofu); pressed and used to make soya oils and margarines; processed to form soya milk; and processed to make condiments such as soy sauce.

Lentils

Lentils are rich in protein and do not have to be soaked before cooking, the different types are identified and named by their colour. They are available whole or split.

- *Red and yellow lentils:* like peas, these disintegrate when cooked and are used to make soups, stews and vegetable loaves.
- *Green and brown lentils:* these stay whole when cooked and can be served as a vegetable or used in salads.

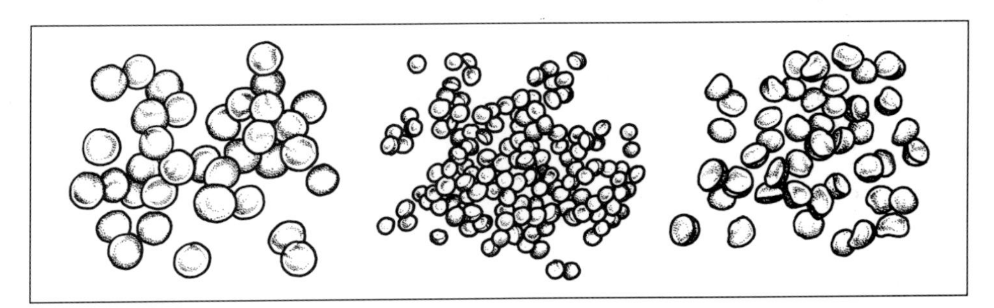

From left: green lentils, red lentils, yellow lentils

Storing pulses

When storing pulses, make sure that wherever they are stored, they are part of a stock control system so that stocks will be rotated. Check the 'sell by' dates regularly and remember the rule: first in, first out. If pulses are stored for too long they become hard and inedible. Check the conditioning of their packaging regularly: any pulses in damaged packaging should be discarded as they may have become contaminated by food pests. Any cans showing signs of rust, dents or bulging should also be discarded.

- Fresh pulses should be stored in a refrigerator at a temperature below 5 °C (41 °F).
- Frozen pulses should be stored in a freezer at a temperature below −18 °C (0 °F).
- Dried or canned pulses should be stored in a dark, cool, dry and well ventilated place (the dry store). They should be kept in airtight containers placed off the floor and out of direct sunlight.

Planning your time when preparing pulses

In order to work efficiently and effectively you need to plan ahead. Take the time to work out a schedule before you start: this will save you considerable time and trouble during preparation and cooking.

Use the following guidelines when planning to prepare pulses:

- identify the dishes you will be required to prepare. How many portions are needed? What time do they need to be ready by?

- make sure that all your equipment is ready, including your utensils and pans
- make sure that space is available on the stove or in the oven
- check the soaking time and the cooking time for the pulses that you will be preparing
- make sure that you have all the ingredients ready for the dishes you have to prepare
- check whether you will be cooking any pulses that need longer cooking times. This time could be used to prepare other ingredients for your dish.

Health, safety and hygiene

Make sure that you are familiar with the general points given in Unit G1: *Maintaining a safe and secure working environment* in the Core Units book. The following points are especially important to remember when preparing pulses:

- pulses can be very dusty and need to be checked for food pests, like flour moths. Always wash the pulses in plenty of cold water and check for any foreign matter such as bits of wood or little stones
- cooked pulses contain both moisture and protein and can easily become contaminated by bacteria. Always keep them covered and at temperatures either below 5 °C (41 °F) or above 63 °C (145 °F)
- cooked pulse dishes must be stored separately from raw foods to prevent cross-contamination
- high standards of personal hygiene must be maintained at all times. Wash your hands frequently, avoid touching your hair, nose and mouth, and cover any open cuts or sores with visible waterproof dressing
- use of unhygienic preparation areas, utensils and equipment can cause contamination
- all waste must be disposed of in a hygienic manner.

Essential knowledge

The main contamination threats when preparing and cooking pulse dishes are as follows:

- cross-contamination between cooked and uncooked food during storage
- contamination of food through inadequate personal hygiene, where bacteria are transferred from the nose, mouth, open cuts or sores or unclean hands to the food
- cross-contamination from unhygienic equipment, utensils and preparation areas
- contamination through products being stored at incorrect temperatures
- contamination through incorrect waste disposal
- contamination through products being left uncovered and foreign objects (e.g. pieces of string, ash, etc.) falling into them.

Preparing cooking areas and equipment

Make sure that your preparation area is clean and tidy. Only put out the tools and equipment you need: large bowls for soaking, a colander for draining and pans for cooking. You may also need special equipment, such as a food processor to blend or purée pulses, or a loaf tin for baking certain dishes.

To prevent cross-contamination, check that all your equipment is clean and in good working order (e.g. none of the handles are loose). Report any damage or malfunction to your supervisor.

Keep your cooking area clean and always consider the safety of both yourself and your colleagues. Make sure that pan handles do not jut out over the edge of the stove, as this may cause accidents. Take care when carrying pans of boiling water; make sure that your path is clear. Lift any heavy loads correctly to avoid straining your back.

When you have completed each task wipe down and clean your preparation area and equipment.

Preparation methods

Soaking

All dried pulses benefit from an initial soaking before cooking. This makes them more digestible and shortens the cooking time.

Cover the pulses in twice their volume of cold water and leave for 4–8 hours. Pulses that are not adequately soaked will take longer to cook.The best place to do this is in the refrigerator, so that you keep them below 5 °C (41 °F) and therefore out of the danger zone.

Always rinse the pulses after soaking as this removes some of the sugars which make them difficult to digest.

Preparing dried pulses

Most pulses need to be thoroughly cooked before they can be used either as a vegetable or as an ingredient in a dish. Dried pulses need to reabsorb moisture, and the high percentage of starch they contain needs to be cooked before our bodies can digest it. As many beans take over an hour to cook (see below), they are normally cooked separately so that the other ingredients in the dish are not over-cooked.

1. Put the pulses into a large pan and cover them with double their volume of water or stock. You can add some flavourings (e.g. roughly cut carrots, onions and celery and a bouquet garni of parsley stalks, bay leaf and thyme).
   Do not:
   - add any salt; this toughens the outside of the pulses and prevents them from cooking properly. Avoid using salty convenience stocks as a cooking liquid
   - add any acids; such as lemon juice, vinegar or tomatoes. This will also toughen the pulses
   - add any bicarbonate of soda; although it may speed up the cooking time it destroys the nutritional value and flavour.
2. Cook the pulses on top of the stove or in the oven. Check the amount of liquid in the pan regularly: there should always be enough liquid to cover the pulses. This will prevent them from burning.
3. When the pulses have come to the boil, reduce the temperature to a simmer. Prolonged rapid boiling will cause them to break up.
4. Test the pulses as they are cooking to ensure that they do not over-cook; some varieties will become very soft and difficult to use.

Important: remember that red kidney beans contain an enzyme that must be destroyed by rapid boiling for the first ten minutes of the cooking time, otherwise food poisoning may result.

Cooking times for pulses

The cooking times for pulses will vary from batch to batch. The times given below can be used as a guide.

Pulse cooking times

Pulse	Cooking time	Pulse	Cooking time
Adzuki beans	30 minutes	Black-eye beans	30–45 minutes
Borlotto beans	1 hour	Broad beans	$1\frac{1}{2}$ hours
Chickpeas	$1–1\frac{1}{2}$ hours	Green/brown lentils	
		soaked	30–40 minutes
		unsoaked	$1–1\frac{1}{2}$ hours
Haricot beans	$1–1\frac{1}{2}$ hours	Kidney beans	1 hour
Lima beans	45–60 minutes	Mung beans	
		soaked	20–30 minutes
		unsoaked	30–40 minutes
Peas	45 minutes	Red split lentils	
		soaked	15–20 minutes
		unsoaked	20–30 minutes
Soybeans	3–4 hours	Split peas	
		soaked	30 minutes
		unsoaked	40–45 minutes

- Find examples of three different pulses used in your place of work.
- Note down how they are stored; how long they need to be soaked; and any other preparation needed.
- Find a recipe for each of three different types of pulses. Write down which utensils you would need to prepare them, and how the pulses are cooked.

Cooking processes and example dishes

Stewing or casseroling: casseroles and curries

Red kidney, black-eye, mung and white haricot beans become very tender when cooked and are suitable for these long, slow methods of cooking because they can easily absorb the flavours of the other ingredients.

Stewing or casseroling is a moist method of cooking, where the pulses are gently simmered in a small amount of liquid which becomes part of the dish. They can be cooked with other vegetables (carrots, onions, leeks, garlic, tomatoes, celery, peppers, mushrooms); fruits (lemon, apple, dried apricots, sultanas, raisins); herbs (thyme, bay, parsley, coriander); and spices (cummin, caraway, ginger, chilli, turmeric, paprika).

Always check that there is enough liquid in the pan to prevent sticking or burning. Control the temperature of the stove or oven during cooking, so that it is not too hot or too cold.

***Example recipe:* Mung bean and chickpea curry**

Mung bean and chickpea curry

Pulses:		
	200 g (7 oz)	green mung beans
	200 g (7 oz)	chickpeas
Cooking ingredients:		
	400 g (14 oz)	tinned tomatoes
	400 ml (14 fl oz)	cooking liquid from the chickpeas
	2 tbsp	vegetable oil
Flavouring:		
	1 bunch	fresh coriander
	100 g (4 oz)	fresh ginger
	3 cloves	garlic
	2	fresh green chillis
	1 medium sized	onion (roughly chopped)
	1 tsp	turmeric
	1 tsp	ground coriander
	1 tsp	ground Cummin
	1/2 tsp	chilli powder
To finish:		
	50 g (2 oz)	fresh yoghurt
	25 g (1 oz)	creamed coconut
To accompany:		
	200 g (7 oz)	Basmati rice

Method

1. Cook the chickpeas and mung beans separately in water flavoured with the coriander stalks and some roughly chopped onion, using the guide on page 132.
2. Make the curry mix, by placing all the remaining flavouring ingredients into a food processor and blending them into a paste.
3. Cook the paste in the oil with a little salt and pepper to help extract the flavour of all the spices. Cook the paste out until it is almost dry.
4. When the paste is dry, add the tinned tomatoes, the creamed coconut and the cooking liquid. Then add the cooked pulses.
5. Gently simmer the curry for approximately 30 minutes, stirring occasionally to prevent sticking. Cook until the beans are tender, removing them from the heat before they begin to break up and become mushy.

This example is a very basic curry. By altering the amounts of the spices and the types of pulses you can make infinite variations on this theme. Whichever type of pulses you use, it is always best to choose a variety that will not break up during the cooking.

You can add vegetables to this recipe. Cauliflower, okra, courgettes or sweet capsicum peppers blend well into this type of dish. For authenticity, you can finish the curry with a couple of tablespoons of fresh yoghurt.

Grilling or barbecuing: rissoles and burgers

Grilling is a dry cooking process, using radiated heat generated by infra red waves which come from gas fired or electrically heated elements. It is a very quick method of cooking because of the high heat involved. You can adjust the heat by turning down the gas or electricity, or you can lower the tray so that the food is further away from the heat. However, the burgers and rissoles need to be of an even size and not too thick, so they will cook evenly and not burn on the outside before cooking through.

A barbecue has hot and cool zones: some positions on the grill bars are hotter than others. This allows you to start the food off in a hot zone and then place it in a cooler position to cook through.

Split peas and lentils are commonly used to make rissoles and burgers. They need to be cooked in just enough liquid to be absorbed in the cooking time. It is important not to make the burgers or rissoles too wet or mushy, or they will lose their shape and fall apart during frying or grilling. Other pulses can be used; they should be cooked and drained as normal then dried on a cloth before being blended with other ingredients. During grilling you will need to brush the food with oil to prevent it drying out.

Baking: loaf or bake

Baking is a dry cooking process, where food is cooked by convected heat in an oven. As for rissoles and burgers, pulses should be cooked in as little water as possible during their initial cooking. The cooked pulses can then be mixed with other ingredients and flavourings. The pulses' natural starchy texture helps the mixture bind together, although some recipes use eggs as an additional binding agent.

Loaves and bakes are covered with a lid or aluminium foil for most of the cooking time to prevent the food drying out. The covers can be removed towards the end of the cooking to allow a crisp browned top to develop, adding flavour to the dish.

Example recipe: **Lentil and cider loaf**

Lentil and cider loaf

Ingredients:

150 g (5 oz)	dry red split lentils
250 ml (9 fl oz)	dry cider
100 g (4 oz)	sunflower margarine
50 g (2 oz)	dried breadcrumbs
100 g (4 oz)	chopped onion
100 g (4 oz)	carrots
1 stick	celery
1 clove	garlic (chopped)
3 g	dried thyme
50 g (2 oz)	roasted ground hazelnuts
50 g (2 oz)	Parmesan cheese
1/2 tsp	chopped parsley
1	egg (for binding)

Method

1. Wash and drain the lentils, then put them in a saucepan with the cider and just enough water to cover them. Bring the pan to the boil and cook the lentils until they are almost tender and all the liquid has been absorbed. (If you are using pre-soaked lentils, they will cook more quickly and need less liquid.)
2. Line a 400 g (14 oz) loaf tin with silicone paper, brush the paper with melted sunflower margarine and sprinkle with breadcrumbs.
3. Melt the rest of the margarine and use it to cook the onions, carrots and celery which should be cut into large brunoise or dice.
4. When the vegetables are soft and just beginning to brown, add the garlic and lentils and mix them together well.
5. Mix in the thyme, nuts, cheese, parsley and beaten egg, then taste for seasoning.
6. Add salt and pepper to taste, then put the loaf mixture in the prepared tin and cover with greased aluminium baking foil. Bake it in a preheated oven at 180 °C (350 °F) for approximately one hour.
7. Ten minutes before the end of cooking, remove the foil to brown the top.
8. To serve the lentil and cider loaf, turn it out onto a warm plate and garnish with picked parsley.

This loaf can be served with apple sauce and is perfect for vegetarians.

Shallow, deep and stir-frying: rissoles, burgers, falafel

Shallow-frying

This is a dry cooking process, where the food is heated through contact with a hot cooking surface, using hot fat or oil.

Heat the pan until the fat is at the correct temperature: 175–195 °C (325–380 °F). This is important because if the fat is too cool the food items will absorb too much fat and become greasy, while if it is too hot they will burn.

When frying burgers or rissoles, lower the temperature when they are browned on the outside to allow them to cook thoroughly. Take care when turning them to prevent them breaking up.

Essential knowledge

Time and temperature are important when cooking pulse dishes in order to:
- prevent food poisoning
- prevent toughening
- ensure that the dishes are cooked correctly
- maintain customer satisfaction.

***Example recipe:* Falafel**

Falafel

Falafel are Mediterranean fritters, made from broad beans or chickpeas which have been formed into flat little cakes and shallow-fried, or small balls and deep-fried.

Ingredients:

200 g (7 oz)	dried broad beans (skinless)
6	spring onions (finely chopped)
1 tbsp	fresh parsley (chopped)
1 tbsp	fresh coriander (chopped)
1 clove	garlic (crushed)
pinch	cummin, salt, pepper
small pinch	bicarbonate of soda (optional)

Method

1. Soak the skinless broad beans for at least 8 hours.
2. Drain and dry the beans off carefully by patting them dry with a clean cloth, then put them through a mincing machine or processor to produce a grainy paste.
3. Add the spring onions, parsley, coriander, garlic, cummin, salt, pepper and bicarbonate of soda. Mix all these ingredients together to make a fine paste, then put it in the refrigerator to chill.
4. When the paste has hardened, shape it into mini-burgers or small balls and put them back in the refrigerator to set.
5. The shapes can then be shallow-fried until they are golden brown on the outside (approximately 5 minutes each side). As always when frying, take care not to have the fat too hot or the outside will be burnt before the inside is cooked.

Serve the falafel in the Middle Eastern way, with fresh natural yoghurt, a mixed salad and pitta bread.

Deep-frying

Deep-frying is also a dry process, where food is cooked by being completely submerged in very hot oil or fat. Deep-fryers should never be filled over half-full with fat or oil, as the oil can easily boil over. Take care not to put too much food into the deep-fryer, as this can lower the temperature of the oil and/or cause the fat to boil over. Make sure your oil to food ratio is correct.

Always use clean, fresh oil. Oil deteriorates each time you heat it; any that has started to turn dark or frothy should be discarded. Follow the instructions for your particular fryer, draining and straining it regularly to remove any debris. Food particles left from a previous batch of frying will burn if they are left in the fryer, giving an unpleasant taste to the new batch.

Take all reasonable safety precautions whilst deep-frying:

- keep your sleeves rolled down to protect against any splashes of fat
- make sure all the necessary cooking utensils are to hand before you start
- use a clean, dry, thick cloth when handling the fryer
- make sure that the correct fire prevention equipment is to hand and that you are familiar with fire drill procedures
- keep a close eye on the temperature, and allow the fryer to regain heat before adding a new batch of food
- report any fault in your fryer immediately, as it is potentially a very dangerous piece of equipment.

Deep-frying the falafel

If you are deep-frying your falafel, use a spider or basket to place them gently into the hot oil, which should be preheated to a temperature of 175 °C (330 °F). The falafel are cooked when the outside turns golden brown and they begin to float to the surface of the oil.

When they are cooked, drain them well and put them onto a tray with kitchen paper underneath to absorb any excess fat. If you need to store them for a few minutes, store them onto an open tray. Do not enclose them in any way, e.g. by placing them in a covered container, as they will turn soggy from their own steam.

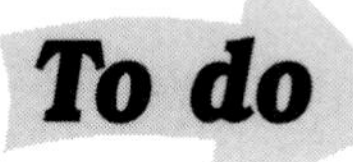

- Find an example of a vegetarian casserole and note the pulses used.
- Find an example of a vegetarian loaf and note which pulses and used and how they are prepared.
- Find another example of a baked pulse dish and note how it is finished.
- Ask your supervisor what sort of foods usually accompany pulse dishes so that their nutritional value is maximised.

Stir-frying

This is a dry cooking process very similar to shallow-frying, where the food is heated through contact with a hot cooking surface (usually a wok), using hot fat or oil.

When stir-frying pre-cooked pulses, make sure they are well drained and dried. Excess moisture will make them stick to each other and to the pan and may cause them to go mushy. Stir-fry in small batches: if you add too many pulses at once the pan will cool down and not fry properly. Only use the pulses that keep their shape well after cooking for this method of cooking (e.g. green and brown lentils; lima; flageolet; black-eye beans and chickpeas).

Storing cooked pulses

Once the pulses are cooked it is always best to use them immediately. If this is not possible, cool them rapidly. All cooked food needs to be cold within 90 minutes to avoid the potential growth of bacteria and the risk of contamination from other foods.

Cool the pulses by spreading them out on a large shallow tray and covering them with a sheet of greasproof paper. You can also use special equipment like a blast chiller which cools food very quickly.

When the pulses are cold, place them in a covered container in the refrigerator or freezer, to avoid cross-contamination. Always check that the refrigerator is at the right temperature (below 5 °C/41 °F).

Essential knowledge

It is important to keep preparation and storage areas and equipment hygienic in order to:

- prevent contamination of food
- prevent pest infestation in storage areas
- ensure standards of cleanliness are maintained
- comply with the law.

What have you learned?

1 What are pulses?
2 Why are pulses important in the diet of vegetarians?
3 How are pulses sold?
4 How should the different types of pulses be stored?
5 Why do you need to cook dried pulses before you can serve them?
6 What danger should you be aware of when cooking red kidney beans?
7 What are the main contamination threats when preparing and cooking pulses?
8 Why is it important to keep your preparation, storage areas and equipment hygienic?
9 Why are time and temperature important when cooking pulse dishes?

Extend your knowledge

1 Find out about pulses not listed on pages 127–9. How long do they need to cook? What type of dishes are they most suitable for?
2 Find out which pulses are most suitable for soups. Find recipes for two soups thickened with pulses.
3 Certain types of flour are made from pulses. Find out what these are and how they are used.
4 Visit some vegetarian restaurants and study the menu. Notice how pulses are used as a fundamental ingredient.

UNIT 2D6

Preparing and cooking fresh pasta

ELEMENT 1: Preparing and cooking fresh pasta dishes

What do you have to do?

- Prepare appropriate pasta equipment ready for use.
- Prepare the ingredients correctly for each individual type of pasta.
- Form the prepared pasta and ingredients according to the dish requirements.
- Cut, mould, shape and cook pasta product according to dish requirements.
- Boil, shallow-fry or bake pasta products for the required time and at the correct temperature.
- Finish, present and store cooked unused pasta dishes according to the product requirements, meeting food hygiene regulations.
- Dispose of waste in the correct manner according to laid down procedures to ensure standards of hygiene.
- Carry out your work in an organised and efficient manner taking account of priorities and laid down procedures.

What do you need to know?

- The type, quality and quantity of ingredients required for each pasta and how to prepare it.
- What equipment you will need to use.
- How to prepare pasta for piping, cutting and shaping for boiling, shallow-frying or baking.
- How to thaw and cook frozen pasta products correctly, i.e. preventing contamination.
- What the main contamination threats are when preparing and cooking pasta products.
- Why hygiene is important for preparation and storage areas and equipment.
- Why time and temperature are important in the preparation of dough products.
- How to satisfy health, safety and hygiene regulations concerning preparation areas and equipment both before and after use.
- How to plan your time to meet daily schedules.
- How to deal with unexpected situations.

Introduction

Pasta (the Italian for *paste*) has been produced since the Middle Ages. Today it is available in fresh or dried form, both of which may be used in catering operations. Fresh pasta has had a resurgence of popularity in recent years, as it is easy and quick to make, costs little to produce and provides a nutritious meal. These qualities, coupled with a growing awareness of the benefits of quality fresh foods, have prompted many chefs to make their own pasta, serving it with traditional fillings or newly invented ones.

Refreshing pasta

Classical cookery uses pasta as a farinaceous course served just before the main course. However, we eat pasta in many ways and more often today, largely because of the range of convenience products available from the supermarket. These shops have used the classical craft of cookery to produce an array of traditional and contemporary dishes to tempt the customer. Pasta is now eaten as a day-time snack, a light lunch or evening meal, a garnish for soups, in broths, meat, fish and shellfish dishes, or as the essential main ingredient for many vegetarian recipes.

There are two main types of pasta: fresh home-made pasta *(pasta fatta in casa)* and dried pasta *(pasta asciutta)*. The best quality dried pasta comes from Naples. Fresh pasta is now more readily available in many shapes, sizes, colours and flavours; both from specialist producers and from supermarkets. Pasta machines are also available which allow you to make spaghetti and noodles easily. It takes less time to cook fresh pasta than the dried variety and it is also superior in both texture and taste; fresh flavoured varieties (such as spinach pasta) have the finish of rich marble and a distinct flavour which the dried varieties lack. The dried varieties are, however, widely used for the convenience market and have the advantage of a two-year shelf life if stored in the correct conditions.

The basic ingredients of pasta are wheat flour (durum flour) and water, with small quantities of other commodities such as egg, oil, tomato, spinach and seasoning. Strong (hard) wheat is used because weaker flours become soft and sticky when cooked and have a floury taste. Durum flour, a yellow flour produced in Italy, the Mediterranean, the Middle East, the former Soviet Union and North and South America, is one of the strongest wheat flours and ideal for use in pasta. The durum wheat is made into a semolina flour which has a high protein content (15 per cent), making pasta a good alternative to rice or potatoes, especially for vegetarians. Pasta also contains starch, which is a carbohydrate that the body can convert into energy.

Types of pasta

Types of popular pasta

- Cannelloni
- Fettuccine
- Gnocchi
- Lasagne
- Macaroni
- Nouilles
- Ravioli
- Spaghetti
- Tagliatelle
- Tortellini
- Vermicelli

Basic white pasta

Although durum wheat is required for true Italian pasta, a good strong bread flour can be used successfully. Mix the flour with water (2–2.5 litres per 10 kg/4–5 pints per 20 lb) and add any other ingredients: e.g. salt and sometimes eggs.

Fresh noodle paste *(Pâte à nouilles fraiches)*

This is also known as *Tagliatelle* or *Fettuccine*, depending on the width of the pasta.

For every 500 g (1 lb) of strong flour you will need 2 whole eggs, 8 additional egg yolks, 2 tablespoons of oil and 2 tablespoons of water or milk with salt to taste. Note that some recipes do not use water.

1 Sieve the flour and salt onto a marble slab. Make a bay in the mixture

(i.e. form a well or hole in the flour and salt).
2 Pour in the mixed egg and/or water and oil and mix until you obtain a clear, smooth dough.
3 Cover with a damp cloth or wrap in cling film and rest for 1–2 hours in the refrigerator.
4 Use a rolling pin to roll the pasta into a thin layer, forming it into the required shape and size. Dust with flour both before and after pinning out to prevent the pasta from sticking.

The ingredients for spinach paste

Spinach noodle paste *(Pâte nouille alla Spinaci)*
This recipe can be made from the noodle paste (above) using the following ingredients as a guide:
500 g (1 lb) of strong flour, 2 whole eggs and 8 additional yolks, 5 g (1/4 oz) of salt and 75 g (3 oz) of cooked drained spinach squeezed well. Add the spinach with the eggs as for noodle paste and mix to a clear, smooth paste. Knead the paste well, then wrap and use as required, according to the dish requirements.

Tomato noodle paste *(Pâte à nouille de tomates)*
Add 50 g (2 oz) of tomato purée to the basic noodle paste (page 140) at Step 2, i.e. when the eggs are added to the bay. Clear to a smooth paste, knead the paste well then cover to rest. The tomato noodles can be mixed with spinach and plain noodles to produce an attractive pasta dish.

Noodle paste with saffron *(Nouille allo zafferano)*
Add 2–3 small sachets of powdered saffron to the basic noodle paste at Step 2, i.e. when the eggs are added to the bay. Work the paste until a rich golden yellow colour results, then wrap and rest for one hour. Pin out, dust with flour and shape as required.

Wholewheat noodle paste *(Pâte à nouille)*
Follow the method for fresh noodle paste, but use wholewheat flour instead of strong flour. Wholewheat pasta has a higher fibre content than plain pasta and may require a slightly longer cooking time.

Essential knowledge

It is important to keep preparation and storage areas and equipment hygienic, in order to:
- comply with food hygiene regulations
- prevent the transfer of food poisoning bacteria to food
- prevent pest infestation in preparation and storage areas
- prevent contamination of food commodity items by foreign bodies
- ensure work is carried out efficiently and effectively.

Lasagne paste *(Pâte à lasagnes)*

The ratio of ingredients used here can also be used for wholewheat and saffron pastes.

Ingredients: 500 g (1 lb) strong flour, 3 eggs, 2 tablespoons of oil (olive or vegetable), salt to taste and 56 ml (2 fl oz) of water.

1 Mix the dough as for noodle paste (above), then wrap and rest in the refrigerator for one hour.
2 Dust the rested lasagne paste with flour and pin out thinly.
3 Cut into rectangle lengths approximately 10 cm × 15 cm (4 in × 6 in) wide.
4 Crimp the edges to produce a frill along the length (if required).

Filled pasta

Ravioli *(Pâte à raviolis)*

For every 500 g (1 lb) of strong flour use 35 ml (1 1/2 fl oz) of oil, 95 ml (4 fl oz) water and salt to taste (approximately 5 g or 1/4 oz).

1 Sieve the flour and salt.
2 Add the oil and water and mix to a smooth paste.
3 Wrap as for noodle paste and rest for 30 mins –1 hour in the refrigerator.
4 Divide the paste into 2 equal portions and pin out each portion thinly.
5 Pipe a suitable filling (spinach, meat or as dictated by the dish), onto one of the pinned portions at 2.5 cm (1 in) intervals.
6 Lightly wash the spaces between the fillings with water and place the other pinned portion of paste on top. Working from the middle, carefully seal down each ravioli removing as much air as possible.
7 When sealed well, cut between each ravioli with a *jigger wheel* (serrated cutting wheel). Dry the squares for approximately 1 hour.
8 To cook: poach in boiling salted water for approximately 12 minutes, drain and finish according to the dish requirements.

Left: piping filling onto the pinned portion
Right: Sealing the ravioli

Cutting the squares using a jigger wheel

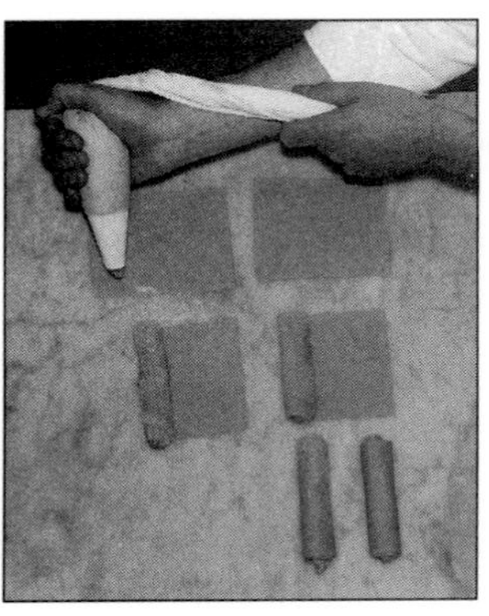

Filling cannelloni

Cannelloni *(Pâte à cannelloni)*

1. Make up a lasagne paste (page 142).
2. Pin out thinly, dust with flour and then cut into squares approximately 6 × 6 cm (3 × 3 in).
3. Cook for approximately 12 minutes. Drain well.
4. Pipe on a suitable filling (see *Typical fillings and garnishes* on page 145) and roll up to form a cylinder. Finish the dish according to recipe requirements.

Tortellini

This is a famous regional dish from Bologna. Here pasta is shaped by filling with veal, chicken, pork, ham or cheese which has been seasoned with salt, pepper and nutmeg, and moistened with melted butter.

1. Thinly pin out the paste. (Any fresh paste can be used to form tortellini, although a fresh egg paste is generally used.)
2. Cut the paste into 2.5 cm (1 in) squares.
3. Place a teaspoonful of filling into the centre of the square.
4. Fold each square into a triangle and seal well.
5. Fold the opposite ends of the triangle together and seal to form the unmistakable tortellini shape. Cook according to the dish and customer requirements.

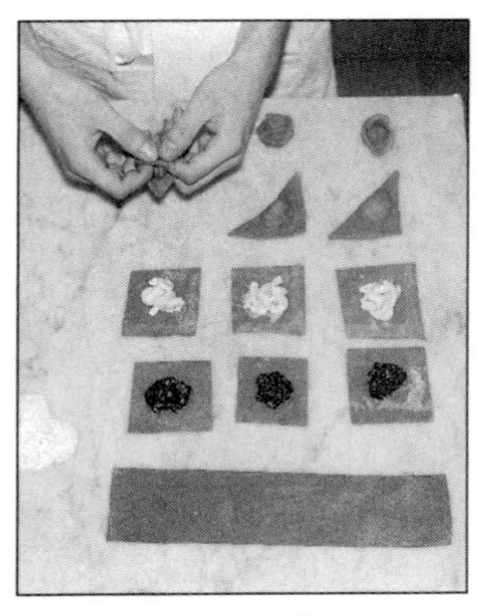

Producing tortellini

Gnocchi

The gnocchi is a small dumpling made from a variety of different ingredients.

Potato gnocchi *(Gnocchi di Patate, Gnocchi Piedmontaise)*
Potato gnocchi are a speciality from Piedmont.

1. Prepare 500 g (1 lb) of mashed potato. This can be creamed potato or (as in Italy) scooped out baked or steamed jacket potato.
2. Mix the hot potato with 25 g (1 oz) of butter, 5 egg yolks and 120 g (5 oz) of flour.
3. Knead gently to form a dough, then roll into a cylinder shape about 1 cm ($^1/_2$ in) thick.
4. Cut the gnocchi into small pieces about 1.5 cm long ($^3/_4$ in long).
5. Mould in the hand into small crescent shapes and finish with a fork to mark with a corrugated pattern.
6. Place onto a floured tray. Cook a few at a time in boiling salted water, then finish according to the dish requirements.

Semolina gnocchi *(Gnocchi alla Romana, Gnocchi Roman style)*
This type of gnocchi is made from semolina, milk, eggs, butter, salt, pepper and nutmeg.

1. Pour the semolina into boiling milk and stir until a thick paste forms.
2. Add the salt, pepper and nutmeg.
3. Allow to cool slightly, then mix in the beaten egg.
4. Mould the gnocchi into small balls, approximately 2.5 cm (1 in) thick.
5. Turn onto a buttered tray and leave to cool and set.
6. When cool, cut into crescent shapes using a plain 5 cm (3 in) cutter. Finish according to the dish and recipe requirements. Plenty of butter is required, as you will need it both for greasing the tray and for brushing the top of the gnocchi once trayed to prevent them from forming a skin or becoming dry.

Gnocchi Parisian style *(Chou paste gnocchi, Gnocchi à la parisienne)*
Pipe some cool choux paste (see page 186) into boiling salted water, cutting the choux into 3 cm ($1^1/_4$ in) pieces as it is piped out, using a damp knife. Poach the choux pieces gently for 10 minutes, then drain and sauté lightly in butter. Finish according to the dish and recipe requirements.

Polenta

Made from yellow or white cornmeal (finely ground maize), this resembles gnocchi but is made with water rather than milk. It is one of the basic foods of northern Italy, where it is often used as a bread substitute.

The polenta may be either fine or coarse, depending on the milling of the maize. The finer variety is generally preferred, and this type benefits from the addition of fresh Parmesan cheese, which can be stirred in when the mixture is first made.

1. Bring 1.5 litres ($2^1/_2$ pt) of water to the boil.
2. Sprinkle approximately 500 g (1 lb) of fine yellow maize into the boiling liquid, stirring constantly with a spoon.
3. Cook over a gentle heat for 20 minutes.
4. Use immediately, or leave to cool before cutting into shapes and finishing by shallow-frying, grilled or baking.

Polenta can also be made from buckwheat flour grown in America and Europe.

Convenience pasta products

The range of convenience pasta products, available frozen, tinned, vacuum packed and dehydrated provides the consumer with a good and varied selection. Prepared fresh pasta can be purchased in large and small quantities, as can true Italian sauces. Multi-pack dishes for commercial and industrial catering are widely available from frozen food wholesalers.

New products are appearing all the time as research kitchens compete to market a different selection of classical and contemporary dishes. Oven-bake pasta dishes which can be cooked from frozen service the ready meal market well, with pasta products across the range available.

Pasta asciutta (dried pasta) is produced in many shapes, including the more unusual ones. Different types of pasta, such as tomato, spinach and wholewheat are readily available from wholesale and retail suppliers.

Dried pasta

Dried pasta shapes *(Pasta asciutta)*

Small pasta (generally used as soup garnishes)
Acine di pepe (peppercorns), Alfabeti (letters and numbers), Anellini (small rings), Capellini (fine hair vermicelli), Cappelletti (small caps or hats), Coralline (small sea shells), Cravatini (small bow ties), Denti di cavallo (tiny horses' teeth), Diavolini (tiny devils' teeth), Ditalin (small thimbles), Lumachini (small snails), Nociette (hazelnuts), Occhi de Pernice (partridge eyes), Pisellini (small pea shapes), Rison (grains of rice), Semi di Melo (apple seeds), Stelline (tiny stars), Vermicelli (fine spaghetti)

Long pasta
Bavette fine (fine noodles), Fettuccine (noodles), Fettuccine verdi (green noodles), Fideline (very fine spaghetti), Fusilli (twisted spaghetti), Lasagne (standard wide lasagne), Lasagne verdi (green spinach lasagne), Lasagnette (small lasagne), Linguine (narrow tongue noodles), Mafaldine (twisted noodles /small lasagne), Mezzani (macaroni (medium), Nastri (medium sized lasagne ribbons), Perciatelli (fine macaroni rods), Perciatellon (small macaroni cane), Spaghetti (standard spaghetti size), Spaghettini (fine spaghetti), Ziti (standard macaroni), Zitoni (large macaroni)

Short and shaped pasta
Farfalle (butterfly plain or fluted), Gramigna rigata (bent and fluted macaroni), Occhi di Lupo (large cut macaroni), Penne (large, slant cut macaroni), Pennini (pen nib macaroni cut at a slant), Ruote (wheel-shaped pasta), Tortiglioni (short, plump spirals)

Filled pasta (after boiling)
Cannelloni 1 (short fluted large tubes), Cannelloni 2 (flat squares or rectangles), Lumache (large snail shapes), Manicotti (large cannelloni)

Filled pasta (before boiling)
Agnolotti (dome-shaped, stuffed small cushions), Ravioli (square-shaped stuffed cushions), Tortellini (small, stuffed round shapes)

Typical fillings and garnishes

Meat-based fillings

The most widely used fillings are given on the following page. However, any finely chopped minced meat fillings can be used for filling or stuffing. Chicken, turkey, duck or game can also be used, as can blanched, skinned sweetbreads. Fillings and stuffings can also be made from vegetable proteins (e.g. TVP) for vegetarians.

Many chefs today are experimenting with unusual fillings, especially as customers become more discerning about the quality of food, looking for dishes that are fresh, tasty and value for money.

Ragu (Bolognese)

Ragu is the original name for Bolognese sauce made from lean beef, chicken livers, bacon (fat and lean), carrot, onion, celery, tomato purée, stock and wine (optional), butter, salt and pepper.

Bolognese is sometimes prepared without the livers or wine in the UK by commercial chefs (according to customer requirements). It is used as a sauce and filling for pasta (e.g. tagliatelle, cannelloni, spaghetti, lasagne) or as a stuffing for ravioli.

Pasta garnishes

Al forno	Ragu sauce, Mornay sauce, Parmesan cheese, layered and baked
Au beurre	Butter and Parmesan cheese
Au gratin	Mornay sauce, Parmesan cheese and nutmeg
Bolognese (ragu)	Tomato concassée, chopped shallots, minced beef, garlic, demi glace
Carbonara	Smoked bacon (pancetta) and eggs (sometimes ham and cream)
Florentine	Spinach and Mornay sauce, Parmesan cheese
Milanaise (Milan style)	Julienne of ox tongue, mushroom, truffle and tomato sauce
Napolitain (Naples style)	Tomato sauce, tomato concassée and Parmesan cheese
Niçoise	Tomato concassée, garlic, onion, olive oil, Parmesan cheese
Norma	Tomatoes, olive oil, onion, garlic, ground pepper, aubergine, basil
Romaine (Romana)	Melted butter and Parmesan cheese
Sicilienne	Butter, Parmesan cheese, purée of chicken livers, chicken velouté

Filling for cappelletti
Cappelletti are a form of small ravioli filled with meat. The filling is made from: lean pork, lean veal, ham, veal brains, carrot, celery, wine or marsala, egg and Parmesan cheese; seasoned with nutmeg and freshly ground salt and pepper. In the UK the veal brains are normally excluded.

An alternative cappelletti filling can be made from: chicken breast; butter; mortadella, Parmesan and ricotta cheeses; egg; and seasoning as above. The filling is placed onto small rounds of thin pasta and folded in the same way as for tortellini (see page 143). The cappelletti are then poached and served (either in a broth or on their own).

Vegetable-based fillings

Spinach and nutmeg
Cook some fresh or frozen spinach, drain and squeeze out the moisture. Season with salt, pepper and grated nutmeg.

Spinach and Ricotta cheese
This is used for filling tortelli (a ravioli made in the province of Parma). It consists of finely chopped minced spinach mixed with an equal quantity of Ricotta cheese, a sprinkling of fresh Parmesan cheese, 2 eggs, and seasoned with salt, pepper and nutmeg.

Cheese and herb
Used for filling Capri Ravioli (*Ravioli Capres*), this includes Caciotta cheese, a sheep's milk cheese from Tuscany and the southern regions of Italy. Gruyère can be used as a substitute. The mixture consists of an equal amount of Parmesan and Caciotta cheese, milk, eggs and season-

ing including nutmeg and basil and/or marjoram. For best results use fresh herbs (or freeze-dried if fresh herbs are unavailable).

Fish-based fillings

Fish and shellfish can be made into a filling for ravioli, cannelloni or tortellini. Seafood is now used for filling ravioli and makes a tasty meal or starter course. Smoked fish can also be used to make interesting and full-flavoured fish fillings for pasta.

Classical garnishes for pasta

The table given at the top of the previous page lists the most commonly used garnishes for pasta.

Preparing equipment

- Check that the cutting and shaping equipment is clean and in working order.
- Pasta machines should be kept clean, dry and well brushed. Stainless steel pasta machines should be cleaned after each production or between different types of pasta, e.g. spinach and white pasta. Some machines, especially the cheaper ones, are likely to rust if washed; clean these by brushing well after each use with a clean pastry brush.
- Keep pasta trays clean and and dusted with semolina flour to prevent the pasta sticking.
- Use cling film to wrap pasta doughs while they are resting prior to pinning.
- If cloth is used to stretch pasta it should be clean and never used for trays or wiping down.
- Small cutting and preparation equipment should be cleaned and dried after use and before storing away.
- Keep mixers, range tops and ovens clean. Make sure you clean them well each day after production.

The range of equipment used when making pasta is shown below.

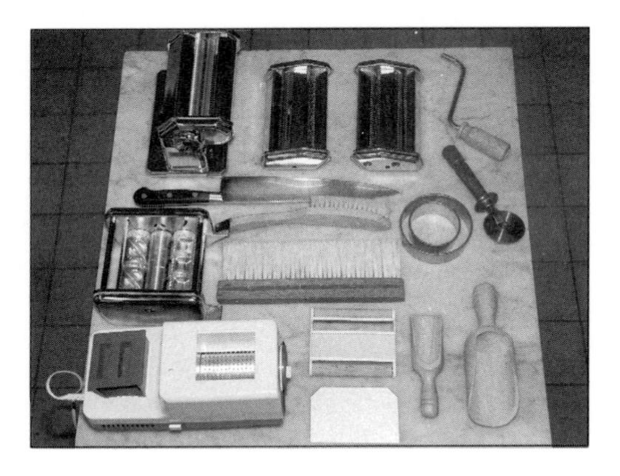

Equipment used for preparing pasta

Preparing the pasta ingredients

- Check that your scales are accurate and meet existing regulations.
- When ingredient items are low in stock inform the chef or pâtissier, who will re-order to prevent stocks from running out.
- Work cleanly and always replace lids on storage bins to prevent contamination.

- Check the cooking times of fresh and dried pasta.
- Check flour for weevil and contamination from foreign objects. Be vigilant. Always sieve flour in a clean and hygienic sieve: this can be a source of cross contamination.
- Remember that brown pasta will take take more time to cook.
- Always use strong flour in the production of fresh pasta.
- Ensure that portion sizes are correct. For a main course, allow 90–120 g (3–4 oz) per portion of fresh or dried pasta. Use smaller quantities for starter courses and even less if the pasta is to be used as a garnish for soups or other main dishes. When cooked, 90 g (3 oz) of dried pasta will produce three times the weight, i.e. 270 g (9 oz).

Mixing the prepared pasta ingredients

Mixed and prepared pastes

- Check the type of pasta being produced; weigh the ingredients carefully.
- Remember to sieve the flour and seasoning together.
- If using a mixing machine, observe the health and safety rules and regulations.
- Remember that pasta is best made the night before it is to be eaten.
- Pasta can be made by hand or by machine. If using a machine, mix on a slow speed using a dough hook.

Essential knowledge

The main contamination threats when preparing and producing pasta and pasta products are as follows:

- cross-contamination can occur between cooked and uncooked food during storage
- food poisoning bacteria can be transferred through preparation areas, equipment and utensils if the same ones are used for preparing or finishing cooked, shallow-fried or baked goods and raw, uncooked pasta
- food poisoning bacteria may be transferred from yourself to the food. Open cuts, sores, sneezing, colds, sore throats or dirty hands are all possible sources
- contamination will occur if flour or any foods are allowed to come into contact with rodents (such as mice or rats), or insects (house flies, cockroaches, silver fish, beetles). Fly screens should be fitted to all windows
- food poisoning bacteria may be transferred through dirty surfaces and equipment. Unhygienic equipment (utensils and tables, trays, mixers, cutting tools and pasta machines) and preparation areas (particularly egg and cheese based mixes) can lead to contamination
- contamination can occur through products being left opened or uncovered. Foreign bodies can fall into open containers and bins: metal objects, machine parts, string, wire, small plastic shavings, cigarette ends, etc.
- cross-contamination can occur if equipment is not cleaned correctly between operations
- contamination can occur if frozen pasta products are not cooked correctly according to the manufacturer's instructions
- incorrect waste disposal can lead to contamination.

Kneading, pinning and cutting pasta

- Dust each pasta piece with semolina or flour. Pin out evenly and cut into the required shape and thickness according to the dish requirements.
- Wrap and rest each pasta piece for the required amount of time to aid relaxation of the paste.
- Remove the mixed kneaded pasta cleanly onto a work bench and divide into the required weights.
- Cut the shape from the pinned pasta using the appropriate cutting tools. If required, form the pasta with the filling and fold to seal according to the pasta being produced.
- Finish the pasta with other ingredients according to customer and dish requirements.

Cutting pasta

Cutting and shaping by hand

It is very important that each pasta product is the same size, colour and shape. Dust ring cutters with semolina flour to prevent sticking and use any knives carefully. Ensure pasta shapes are of the same width and length. Cut and shape pasta in a cool area and store on trays dusted with semolina flour, covered and in the cool room or refrigerator until required for cooking. Pasta will stick together if not dusted and stored correctly.

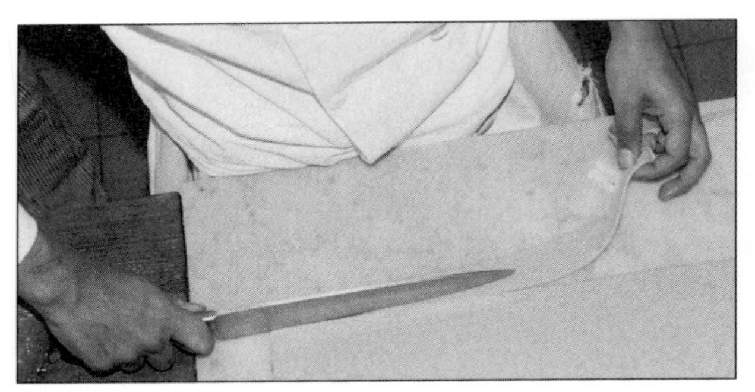

Left: cutting rolled pasta by hand. Right: cutting rolled pasta using a rotella

Cutting and shaping by machine

Pasta machines are often used in kitchens to speed up production. Use them carefully according to the manufacturer's instructions. When using small pasta machines, ensure that they are firmly secured to the workbench before starting.

Pin out pasta lengths by hand first. Keep your fingers away from the shaping rollers while working, and dust the rollers regularly during shaping to prevent the pasta from sticking to the machine parts.

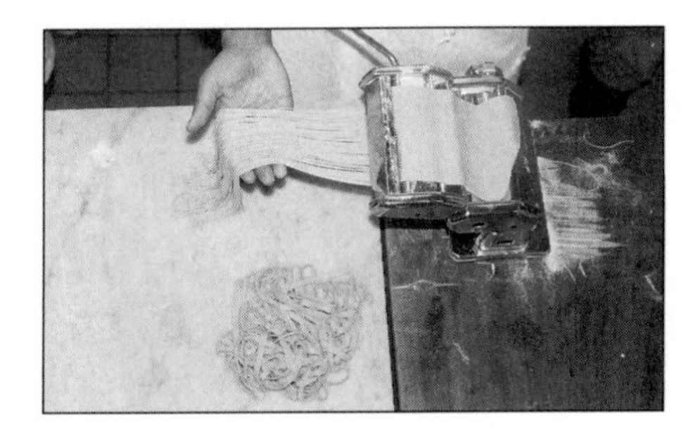

Fine noodles cut and shaped by machine

Cooking pasta

- If using dried pasta, allow a longer cooking time. Fresh pasta takes 3–5 minutes for fine pasta (e.g. noodles) and 8–12 minutes for thicker pasta (e.g. cannelloni and ravioli). Dried pasta needs to be cooked for approximately 12–15 minutes. Always check the manufacturer's cooking times for dried pasta products.
- Use a very large pan of fast boiling salted water.
- Always add the pasta when the water is boiling well and stir several times during cooking to prevent the pasta from sticking.
- Pasta is cooked *al dente:* meaning that it should be firm enough to bite into. (The literal translation is 'to the tooth'.) Do not over-cook pasta or the starch will become sticky and form a solid mass.
- When the pasta is cooked, drain well and use immediately. Finish according to customer and dish requirements.
- If the pasta is to be held back for service: cook, drain and refresh in cold iced water. Cover and refrigerate until required.
- Reheated pasta dishes must be heated thoroughly, to a temperature of 70–75 °C (158–167 °F). This temperature should be registered at the centre of the cooked pasta dish to ensure that any harmful bacteria are killed.
- Pasta to be used for cold salad should be cooled to below 5 °C (41 °F) within 90 minutes of being cooked or sooner.

Cooking pasta

Essential knowledge

Time and temperature are important in the preparation and cooking of pasta products in order to:

- ensure the pasta is mixed and rested for the required amount of time in order to prevent faults in the pasta products
- ensure that the pasta goods are correctly prepared, produced, cooked and finished
- prevent food poisoning from direct and indirect contamination
- eliminate the opportunity for shrinkage
- ensure that frozen convenience pasta dishes are thawed and cooked according to the manufacturer's instructions. Never re-freeze thawed convenience products
- ensure that pasta prepared for cook freeze is under-cooked before freezing
- promote and maintain customer satisfaction by maintaining consistent standards of quality.

To do

- Watch the chef or pâtissier producing pasta, particularly filled pasta.
- Check the resting and cooking times and make a list for each different pasta from the range.
- Find out how the chef or pâtissier knows when the pasta is *al dente*. What indicators do they use?
- Find out what the classical garnishes and finishing methods are for each pasta type.

Finishing and presenting cooked pasta

Whichever pasta dish you are finishing, certain key factors need to be considered.

- Always ensure that the pasta is fresh, cooked correctly and flavoured according to the dish requirements. Food needs to taste as good as it looks and look as good as it tastes; this balance needs to be maintained in order to satisfy customer expectations.
- Present pasta in the correct portion sizes.
- If finishing plain pasta, cook to order where possible. Drain it well and finish with butter or cheese (accoding to the recipe you are producing).
- Ensure the cooked pasta is cool before storing or assembling with other ingredients.
- Where pasta is to be used with other ingredients (e.g. sauces), ensure that a balance is maintained between ingredients.
- When using dried pasta, make up thinner sauces to allow for absorption by the pasta.
- Certain sauce ingredients require special attention. Always remove tomato seeds before using tomatoes in a sauce, as the seeds will develop a bitter taste if left to cook in the sauce. Cook any garlic carefully: if it begins to burn, the sauce will acquire a bitter taste.
- Once the pasta product is complete and ready for service, wipe the edges of the service dish with a clean cloth, removing any spillage burn.
- Decorate the finished dish with care, using a simple finish.

Storing cooked pasta dishes

Store finished cool products according to the *Food Hygiene (Amendments) Regulations 1993*. Store all products at a temperature below 5 °C (41 °F).

Health, safety and hygiene

Note all points in Units G1 and G2 of the Core Units book concerning general attention to health, safety and hygiene. When preparing pasta and pasta products the following points are particularly important:

- If using a pasta machine, make sure that the machine is set to the correct (slow) speed before switching it on.
- Check that all safety guards are in place and working correctly.
- Make sure that all of the machine's moving parts have stopped before removing the pasta from the mixing bowl. Do not attempt to remove any pasta while the machine is still mixing.
- Wipe up any spillages as they occur.
- Keep your fingers away from the machine rollers while forming pasta shapes.

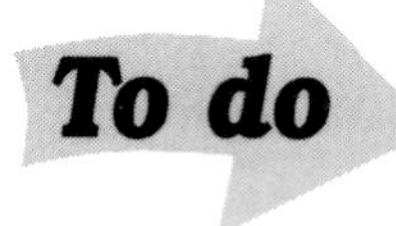

- Check to see if any cooked and uncooked pasta are stored together in your kitchen. If so, why?
- Find out where is the flour stored and how it is stored. Are flour bins used?
- How are the pasta machines cleaned in your kitchen? Look to check whether they need cleaning.
- Find out who is qualified in First Aid and where they work. Check what action you should take in case of emergency.

Planning your time

Refer to *Planning your time* (pages 1–4) concerning general points that you should always note when working in a kitchen or bakery. When preparing pasta and pasta products the following points are especially important to remember.

- Always check that you have used the correct flour and added the seasoning to the pasta. Errors here waste valuable time and can be expensive.
- Make sure that pasta is mixed according to production methods and for the required time.
- While pasta pastes are resting, use your time to make sure trays and tools are ready for cutting, shaping or filling the pasta items.
- While the pasta is cooking, clean and clear work surfaces ready for the next batch of work.
- Mix pasta with added ingredients such as spinach, tomato or saffron thoroughly.
- Check the length of time any pasta has been cooking. It should be cooked *al dente* (firm to the bite).
- When producing a large amount and range of pasta products, make sure you have enough pans of boiling salted water ready, adequate storage space and the correct storage equipment.
- Work in a clean and organised manner attending to any priorities and laid down procedures. Make sure that you know what to do in the event of unexpected situations, such as a rush order, customer request or emergency.
- Be aware of what your responsibilities are. If in doubt, ask your supervisor.
- As you work, clean up after yourself. Wipe down preparation and cooking areas between each task or job. Never use glass or oven cloths to wipe down work surfaces: this causes contamination.

Disposing of waste

- Dispose of waste in the correct manner, according to laid down procedures and ensuring high standards of hygiene.
- Waste materials should be disposed of cleanly and efficiently, to prevent contamination of uncooked and cooked pasta products and ingredients and to prevent health hazards.
- Remember that incorrect waste disposal can lead to food contamination.

What have you learned?

1 What are the main contamination threats when preparing and cooking fresh pasta?
2 Why is it important to keep preparation and storage areas and equipment hygienic?
3 Why are time and temperature important in the cooking of fresh pasta?
4 Why is pasta dough wrapped once it is prepared?
5 What are the dietary benefits of using wholewheat pasta?
6 Why is cooked pasta cooled quickly if it is not to be used immediately?
7 What are the advantages of using fresh pasta?
8 What are the advantages of using convenience pasta dishes?

Extend your knowledge

1 Many regions of Italy produce distinctive types of pasta which reflect the traditions of regional culture. Each area produces pasta dishes that identify with particular ingredients. Research these areas to find out what classical pasta dishes are produced in each one.
2 Find out what *pesto* is. How is it made and what does it contain?
3 Read up on *German Spätzle.* How does it differ from Italian pasta?
4 What range of pasta machines are available? How do electric machines form the pasta shapes?
5 Find out the history of pasta. Who is reputed to have discovered pasta, and why is this now doubted?
6 What classical dishes use pasta as part of their composition?
7 Why does veal have a prominent place in Italian culinary practice?
8 Discuss contemporary cookery using pasta with your chef or pâtissier.

UNIT 2D7

Preparing and cooking dough products

ELEMENTS 1 AND 2: Preparing and cooking dough products

What do you have to do?

- Prepare appropriate baking equipment and preparation and cooking areas ready for use.
- Prepare the dough ingredients correctly for each individual dough.
- Mix the prepared dough ingredients according to the product requirements.
- Divide, scale, knead, mould, shape and prove dough products according to product requirements.
- Bake or deep-fry dough products for the required time and at the correct temperature.
- Finish, present and store cooked dough products according to the product requirements, meeting food hygiene regulations.
- Plan and carry out your work in an organised and efficient manner taking account of priorities and laid down procedures.
- Dispose of waste correctly.

What do you need to know?

- What the main contamination threats are when preparing and cooking dough products.
- Why it is important to keep preparation and storage areas and equipment hygienic.
- Why time and temperature are important in the preparation of dough products.
- How to satisfy health, safety and hygiene regulations concerning preparation areas and equipment both before and after use.
- The type, quality and quantity of ingredients required for each dough and how to prepare it.
- How to prepare dough correctly for baking or deep-frying.
- What the correct procedure is for cooking and thawing frozen dough products to prevent contamination.
- How to plan your time to meet daily schedules.
- How to deal with unexpected situations.

Introduction

Freshly baked products like bread, buns, Danish pastries, croissants and brioche can be found in many catering outlets, hotels, restaurants and fast food operations. These goods entice customers with their sweet fresh smell and quality finish. The choice and variety of high fibre, soft-grain breads and speciality and new bread products has never been so great, particularly as we work towards a single European market.

We eat bread for breakfast as toast, for luncheon as sandwiches or filled rolls, for light snacks as hot filled rolls (burgers and hot dogs), cold filled rolls (French sticks) and as an accompaniment to our main meal. Filled croissants are also a popular snack, and filled bridge rolls are widely used for parties and weddings. Bread is used in many other dishes as an ingredient, such as bread and butter pudding, apple charlotte and bread sauce.

A wide variety of dough products can be produced from each type of dough. Bread, for example, can be white or brown, soft or crusty, risen or flat and baked in a variety of shapes. It is one of the oldest cookery products known, and is made all over the world in many shapes, sizes, flavours and textures. The basic bread dough can be enriched by the addition of fat, sugar, milk powder and flavouring to produce a bun dough which is also the basis for numerous products.

The basic ingredients for making dough

What is dough?

Dough consists of strong flour, water, salt and yeast kneaded (mixed) together to the required consistency at a suitable temperature. While proving, the yeast action occurs, which produces the gases necessary to aerate the dough piece. When baked, the yeast action works to produce dough products that are light and digestible with a good flavour and colour.

Basic doughs are made from strong flour, salt, yeast and water and are generally made into crusty bread products. By adding fat or milk powder you can obtain a softer range of products, as the fat content insulates the water molecules, keeping the moisture level higher during baking.

When fat, sugar and eggs are added in a variety of combinations the dough is said to be enriched. Examples of enriched doughs are buns, savarins and brioche. Doughnuts are also made from a basic bun dough. Enriched doughs which have the fat worked in by *layering* (as in puff pastry) are called *laminated doughs;* Danish pastries and croissants are made from laminated doughs.

Why is it important to us ?

Dough-based products add variety to the menu and have an important role in our daily diet. We need roughage to keep our digestive system healthy; this is provided by brown wholemeal, wholewheat, granary or stoneground breads. Flour-based products also play a key part in providing the energy, vitamins, and minerals that are required every day to keep our bodies fuelled and functioning. In the average daily diet, they account for 56 per cent of our carbohydrate intake, 25 per cent of proteins and 9 per cent of fats. Amongst the vitamins and minerals they contain, flour products are high in calcium (13 per cent), iron (19 per cent) and Vitamin B_1 (22 per cent).

Fermentation

Dough must undergo a fermentation process in order to rise. This is brought about by the action of the yeast and enzymes in the dough, which converts sugar into alcohol, producing the particular flavour of bread. The action of the yeast and enzymes also gives colour and bloom to the crust and produces carbon dioxide, which is necessary to cause the dough to rise.

In order for fermentation to work successfully, you will need to be aware of:

- *warmth.* A good working temperature for production of dough is 22–30 °C (72–86 °F)
- *moisture.* Water temperature can vary according to the kitchen temperature and weather, but it should be approximately 38 °C (100 °F)
- *time.* This is crucial for producing good quality doughs. Hand-made doughs need approximately 45 minutes to ferment before knocking back, in order to expel the gases and re-introduce oxygen, allowing the fermentation process to continue.

ELEMENT 1: Preparing dough products for cooking

In this chapter we will be dealing with the following types of dough: white basic dough (soft and crusty); brown basic dough; wholemeal and wholewheat dough; basic bun dough; fruited bun dough; enriched doughs (baba, savarin, brioche); laminated doughs (croissants, Danish); and speciality doughs (blinis, flat breads, pizza bases).

The basic ingredients for all doughs are given below, together with the preparation methods you will need to be familiar with. Details for making each particular dough are given on pages 163–8.

Basic ingredients

Flour

Strong flour contains approximately 14 to 16 per cent protein. When the flour is mixed with water, the protein is formed into gluten, an elastic substance that holds the gas within the dough, causing the dough to rise.

Yeast

This can be used in either fresh or dried form. Fresh compressed yeast is fawn in colour with a pleasant smell, and should be stored in a refrigerator. Dried yeast may be granular or powdered, and is also fawn in colour with a slight yeast smell.

Yeast is the agent responsible for the fermentation of dough and dough products. It is a living, single-cell micro-organism. When it is combined with food in a warm and moist atmosphere it ferments, producing carbon dioxide and ethanol (a type of alcohol), while at the same time reproducing itself.

The process of fermentation is a complex one, affected by many factors. Yeast dies if it is placed in direct contact with strong salt or sugar solutions. Its action is slowed if bun spice is used as an ingredient or if it used in a very enriched dough: the higher fat and egg content slow down the reaction.

When using dried yeast, you will need to use only half of the quantity specified for fresh yeast.

Water
This should be warm (see *Fermentation,* page 156). If mixing doughs by hand, dissolve the yeast in the water. If using a mixing machine, add the yeast dry, but do not allow it to come into contact with salt or sugar, as this can kill the yeast.

Salt
Salt provides flavour, acts as a preservative, improves the quality of the crumb and crust and strengthens the gluten. Check recipes for the correct quantity.

Preparing ingredients

- Check that your scales are accurate and meet existing regulations.
- When ingredient items are low in stock inform the chef or baker, so that they can re-order.
- Always work cleanly and replace lids on any storage bins to prevent contamination.
- Check the temperature of liquids used in preparation. In cold weather the water/milk may need to be warmed.
- Check fruit for stalks, stones and foreign bodies. Be vigilant: always wash and drain fruit well.
- Remember that brown doughs absorb more water than white doughs. The volume of water absorbed by flour also varies according to the strength of the flour: i.e. the protein and bran content.

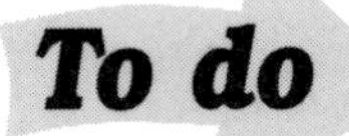

- Find out what fermentation is. How does it work? Why does it work? What are the ideal conditions for fermentation to occur?
- Find out what types of fat and other ingredients can be added to bread. Why? What effect do they have on fermentation?
- Check the storage temperature of fresh yeast.
- Find out what one block of yeast weighs.
- Ask your supervisor why the salt and sugar should be kept away from the yeast when placed in the mixing bowl.

Preparation methods

The following preparation methods are applicable to most dough products and you need to be familiar with all of them in order to produce dough goods successfully.

Mixing by hand

This is used for small dough mixes.

1 Weigh and check ingredients.
2 Place the dry ingredients into the mixing bowl.
3 Dissolve the yeast in warm water or milk (37 °C/98 °F).
4 Make a bay in the centre of the flour and add the warm liquid.
5 Mix with one hand and knead to a smooth dough.
6 Knead until the dough is soft and smooth in texture but not sticky.

Mixing by machine

This is used for large dough mixes and quick dough development.

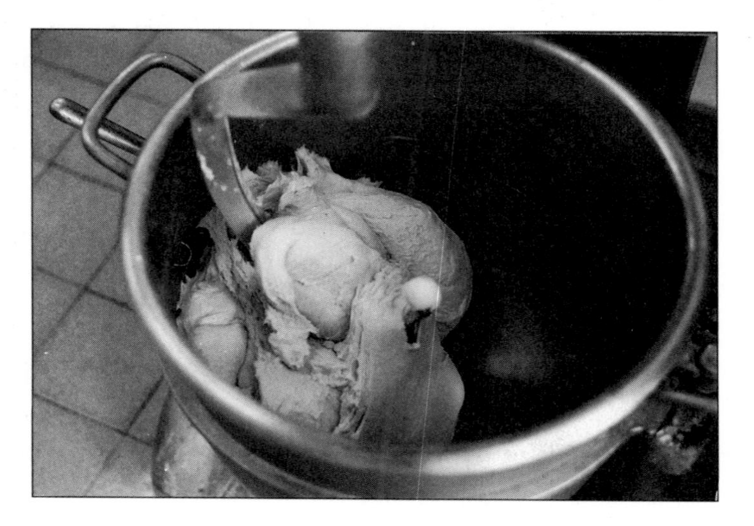

Dough being mixed by machine

1 Weigh and check ingredients, checking that all equipment is clean.
2 Place all ingredients into the mixing bowl, keeping salt and sugar away from the yeast.
3 Ensure the bowl and machine are set correctly.
4 Mix the dough for the required length of time.
5 If the dough is slack or tight, adjust as necessary.
6 Stop the mixer and remove the dough when safe to do so.
7 Handle the dough cleanly, placing it onto the bench for dividing and weighing.

Kneading

This is used to make the dough piece smooth.

Squeeze and rub (i.e. *knead*) the dough on the bench to give a smooth and silky finish to the dough piece. Note that machine-made doughs need only a light kneading to produce a clean smooth surface.

Dividing and scaling

This is used to portion dough into pieces of the correct weight.

Dividing and scaling dough

1 Once you have mixed and kneaded the dough, you need to cut (*divide*) it using a *dough divider;* i.e. a hand scraper or hydraulic cutting machine.
2 Take the divided dough pieces and scale them to the required weight; i.e. weigh each piece of dough on mechanical or electronic scales according to the legal requirements. Note that all dough pieces exceeding 300 g (12 oz) are subject to trading standards regulations. You will also need to remember that doughs lose up to 12.5 per cent of their moisture during baking and this needs to be taken into account when scaling. For instance, to obtain a finished, baked loaf weighing 800 g (1¾ lb), you will need to scale the unbaked dough to a weight of 920 g (2 lb). Crusty bread which requires a longer baking time will need to be scaled at a greater weight than soft bread, to compensate for the higher moisture loss through steam during baking.
3 Rest the scaled dough pieces for a short time.

Moulding

This is used to form the smooth shape of the dough.

1 When the dough has rested for a short time, mould the dough piece into a round ball shape and rest again for a short time prior to shaping or *moulding.*
2 Mould bread rolls by hand or using a roll moulder. Each bread product is moulded to obtain a very smooth, silky finish. The smoother the moulding, the better the proved and cooked product will look. If dough pieces are given a rough mould they will develop a rough finish when proven and baked.

Shaping

This is used to form the dough piece before the second/final proving stage.

1 Check that the moulded dough piece has been rested for sufficient time.
2 *Knock back* the moulded, proved dough piece to expel the gas and reintroduce oxygen for the final proving stage. To do this, knead the dough until it is 'knocked back' to its original size.
3 Check the shape required.
4 Shape according to the product requirement.

Shaping dough

Plaiting

1 Take a piece of dough and divide it into three equal pieces.
2 Mould into a smooth round ball and rest for 5 minutes.
3 Mould each piece into the shape of a rope using an even pressure with both hands.
4 Join the three rope lengths of dough at one end.
5 Starting with an outside length, fold it over the centre length, then fold the other outside length over the centre length.
6 Repeat Step 5 as many times as necessary, always folding the alternate outside length over the middle length.
7 Join the three ends by squeezing together and tuck the end under slightly.

Proving

This occurs when the moulded and shaped goods are put in a prover cabinet (or warm place) to double in size. Proving takes place over two stages. The initial or *primary* proof is the first time the dough is left to ferment and produce CO_2 (Carbon Dioxide) and Ethanol (Ethyl Alcohol). The secondary or *final* proof stage takes place prior to baking, and at this stage the shaped dough pieces double their original size.

The secondary proof is essential for giving dough products the necessary volume and a good flavour, and the method for this is given below.

1 When you have shaped the dough pieces, tray them onto silicone paper, spacing them evenly.
2 Place them in the prover. If you do not have a proving cabinet, place the tray of rolls in a warm place and spray with water to prevent a skin forming.
3 Prove until double in size. The dough should be light to the touch and smell fresh. An over-proved (*over-ripe*) dough smells of alcohol.
4 Remove carefully from the proving cabinet or warm shelf and place into the oven.

Essential knowledge

Time and temperature are important in the preparation of dough products in order to:

- ensure that the dough is mixed and proved for the required amount of time in order to prevent faults in the dough products
- ensure that the dough products are correctly cooked.

Folding

This is used when the flattened dough piece needs to be folded into a loaf shape, or when dough is folded to envelop (for example) a fat and fruit mixture, as in lardy cakes.

1 Weigh a piece of bread dough.
2 Mould the dough piece into a round ball.
3 Try to get a smooth surface as described on page 159, then rest the dough for 10 minutes.
4 Turn the dough piece upside down onto a lightly floured clean bench and flatten with a clenched fist. This will expel the gas and reintroduce oxygen for the secondary proving stage.
5 Fold the two outer (opposite) edges into centre, then roll and mould from a single edge towards yourself forming a cylinder shape, which will bake into a bread loaf.

The dough is now ready to be tinned.

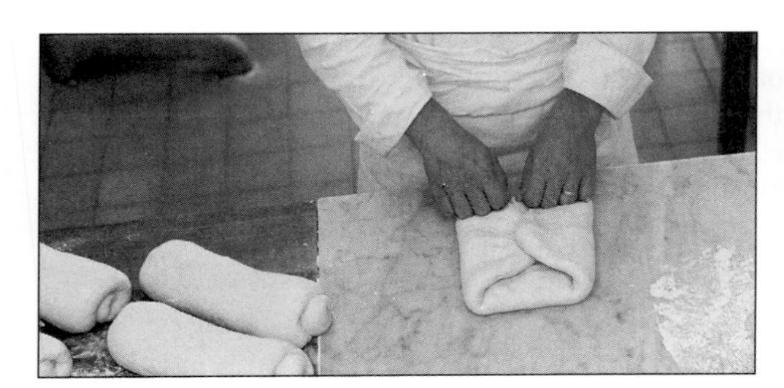

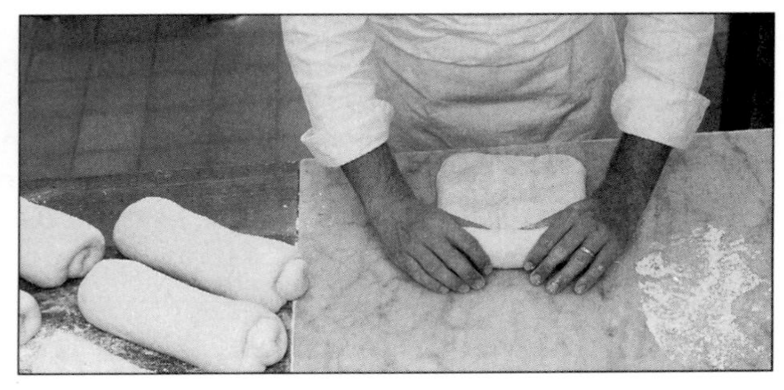

Top left: folding the two outer edges into the centre
Top right: rolling the dough into a loaf shape
Right: placing the dough into tins for baking

Rolling

Rolling is typically used for Chelsea and Belgian buns, baps, lardy cakes, pizza bases and unleavened bread. Dough is rolled (or *pinned*) to make it flat prior to shaping. The dough for Chelsea or Belgian buns is rolled into a flat rectangular shaped then rolled up by hand into a cylinder shape. Baps are rolled with a rolling pin after they have been moulded and shaped to give their characteristic shape.

1 Divide and scale a piece of dough to the required weight.
2 Using a rolling pin, dust the dough piece lightly and *pin out* (i.e. use a rolling pin to roll out) the dough piece to the required shape.

Glazing

Glazing with eggwash is carried out on raw unproved yeast goods to provide a brown colour. A sugar glaze (bun wash) is applied to baked goods to provide a shiny finish and seal the product (to prevent drying).

1 Check what type of glazing is required: eggwash, sugar or bun glazing.
2 If eggwash glaze is required, brush the formed dough product both before and after proving with eggwash containing a little salt. The salt will break down the protein and produce a better colour and glaze.
3 If sugar or bun glazing is required, brush the dough product with a boiled sugar syrup (bun wash), which can be flavoured, i.e. lemon. Apply the glaze after baking, while the products are still hot or warm.

Preparing equipment

- Check that ovens are preset and on, or that the deep fryer has clean oil, is filled to the correct level and preset.
- Keep bread tins clean but not washed.
- Use a bread emulsion to grease tins for bread.
- Keep baking trays clean, always removing any carbonised food before cleaning.
- Use silicone paper for trayed buns, Danish pastries or croissants and individual items. Silicone can be used a number of times to reduce costs, but remember that it will need replacing if it begins to break up.
- Cut a sheet of silicone into 6 pieces and brush with oil if you are making doughnuts and do not have a doughnut fryer.
- Keep mixers and ovens clean: make sure you give them a thorough clean each day after production.

Mixing and preparing the dough

1 Check the type of dough being produced. Make sure you have weighed the ingredients carefully.
2 Remember to keep the salt and sugar away from the yeast. If using dried yeast use only 50 per cent of the weight of any fresh yeast specified in the recipe; dried yeast is concentrated.
3 Mix the dough for the required length of time. If mixing by machine, check that it is set to the correct speed. *If the mix is tight:* stop the machine and add a little more liquid. *If the mix is too slack:* stop the machine and add more flour.
4 Remove the mixed dough cleanly onto a work bench and divide it into the required weights.
5 Scale each dough piece and check the weight carefully.
6 Knead each dough piece into a smooth round ball and allow it to rest.
7 Mould the rested dough piece into the required shape then tin or tray according to the product requirements. Prove the product for the required length of time: do not over or under-prove.

Essential knowledge

It is important to keep preparation and storage areas and equipment hygienic, in order to:
- comply with food hygiene regulations
- prevent the transfer of food poisoning bacteria to food
- prevent pest infestation in preparation and storage areas
- prevent contamination of food commodity items by foreign bodies
- ensure work is carried out efficiently and effectively.

Range of basic dough types

White basic dough: crusty

These products are made from refined wheat flour mixed with water, salt and yeast, sometimes with a little added bread fat and milk powder. They are prepared following the steps given in *Mixing and preparing the dough* (page 162).

White crusty breads and rolls are generally cooked on trays without any tin moulds. They are known as *oven bottom breads,* and the type includes breads such as crusty cottage loaf, crusty coburg and bloomer. Oven bottom doughs need to be tighter (firmer) than tin doughs as they have to support their own shape but not be too firm when cooked.

Crusty white rolls being trayed before baking

Softer basic doughs can be made for use in bread tins, where the tin supports the dough. The softer dough rises more quickly due to its lower resistance, resulting in a larger loaf.

White basic dough: soft

Soft white bread and rolls are made from refined wheat flour called *strong flour* (which has approximately 75 per cent of the wheat grain). The flour is mixed with water, salt, yeast, fat and milk powder, where the amount of fat and milk powder is greater than for basic crusty doughs. The addition of fat and milk powder prevents moisture loss and upon cooling results in a softer loaf or roll product. Note that when a dough contains six per cent or more of milk solids it is called *milk bread* and is subject to trading standards regulations.

These dough products are prepared following the steps given in *Mixing and preparing the dough* (page 162).

Brown basic dough

The following types of dough are mixed and prepared following the steps given in *Mixing and preparing the dough* (page 162).

Wholemeal or wholewheat bread or rolls

These types of bread follow the same basic ingredients and processes as for basic white doughs, using wholemeal or wholewheat flour. These flours contain 100 per cent of the bran and germ of the wheat with nothing added or taken away; we refer to this as *100 per cent extraction.* The following differences between wholemeal and white flours are worth remembering:

- wholemeal flour is particularly rich in Vitamin B
- doughs made from wholemeal flours need less proving than white flours, as the bran enzymes and germ react with the gluten to ripen the dough more quickly
- doughs made from wholemeal flours need more water, due to the absorbing properties of the bran
- less gluten is formed using wholewheat and wholemeal flours and therefore products have less volume
- because there is less gluten to develop, the doughs tend to be softer in texture and require a softer moulding. However, if the water content is too low and a tight dough is produced, the bread will lack volume and therefore develop a heavy texture

- if wholemeal doughs are over-proved the dough collapses, resulting in poor volume and a dense texture
- wholewheat and wholemeal flour will not keep as long as white flour and should be used as soon as possible to ensure good quality baked dough products.

Wheatmeal bread or rolls

Wheatmeal flour has at least 85 per cent of the wheat berry, all of the *endosperms* (starch content) and most of the bran from the wheat. A wheatmeal loaf must contain at least 0.6 per cent of fibre calculated on the dried weight of the flour. Many of the sliced brown breads and soft brown roll products found in modern bakeries are produced from fine wheatmeal flours where the flour has been ground finely, with the crushed bran giving breads a light brown colour.

There are several variations of wheatmeal flour:
- *stoneground* flour is simply wheat crushed between two stones in the mill
- *brown* flour generally refers to a finer ground wheatmeal flour
- *granary* flour is wheatmeal flour with added grains of malted rye and barley. It is used to produce popular products such as granary loaves, rolls or even pizza bases.

Basic bun dough

Basic bun doughs are made from strong flour, water, yeast, salt, eggs, sugar, fat, milk or milk powder and lemon essence or bun spice.

Bun spice provides the distinctive flavour noticeable in hot cross buns. It should only be added when the dough has been mixed and given the first proof, because it has a narcotic effect on the yeast and slows down the fermentation process. *Lemon oil* (or more often *lemon essence*) can be used for certain bun dough products instead of bun spice, and this does not effect the yeast.

The dough is made following the steps given in *Mixing and preparing the dough* (page 162), and is similar to the basic white dough, although it uses less salt and is enriched by the extra ingredients. Buns can be made in a variety of shapes and sizes, many of which are traditional and have local historical origins.

A bun dough should be clear and slightly sticky; it often contains more yeast in order to counteract the enriching effect of the butter and eggs. The range of goods made from this dough are sold mainly from bakeries, hot bread units and in-store bakeries, although some hotels and cafés offer them as *morning goods* with coffee.

Doughnuts

Doughnuts are made from a basic bun dough. For ring doughnuts, use a plain round cutter to remove the centre after the dough has been moulded and flattened.

1. Scale the dough and mould each piece to form a ball, then flatten each slightly.
2. Place the dough shapes onto a lightly oiled baking sheet, oiled silicone paper or a special doughnut tray.
3. Prove the dough to twice its original size.
4. Place the doughnuts carefully into a deep-fryer at 185–193 °C (360–380 °F).

5 The doughnuts will float, and need to be turned when the underside is golden brown.
6 Finish cooking, i.e. cook until both sides are coloured.
7 Remove the doughnuts from the fryer using a spider or perforated spoon. If using a special doughnut tray, lift the tray and drain the doughnuts well.
8 When cool, fill with jam and roll in caster sugar flavoured with cinnamon.

Special doughnut depositors can be bought for mass production along with special doughnut fryers (semi or fully automatic).

Fruited bun dough

When fruit is added to the basic bun dough (page 164), an extended range of bun dough products can be made.

Fruit should be added to the dough *after* mixing has taken place, to avoid crushing the fruit and therefore spoiling the colour and appearance. Currants, sultanas and cherries as well as mixed chopped peel can be used. These fruits add flavour, colour and texture to the basic bun dough.

Fruited bun dough is used to produce Chelsea, currant and Belgian buns, fruit loaves, plaits and bun rounds. Regional specialities such as Cornish saffron loaves, lardy and decorated fruit breads are delicious when eaten fresh with morning coffee or afternoon tea.

Ratios of ingredients for basic doughs

In the following list, the weight of flour is used as the benchmark for the ratios within a recipe. The flour is therefore always given at 100 per cent, with other ingredients given in percentages of the flour.

For example, if you are using 500 g of flour, 500 g becomes your 100 per cent marker. So if fat is given as 50 per cent, you would need to use half of the amount of flour used; i.e. 250 g of fat. In all instances listed below, use water at a temperature of approximately 38 °C (100 °F).

The ratios are as follows:
- *White tin dough:* strong flour 100%, water 57%, yeast 3%, salt 2%, fat 2%.
- *Wholemeal dough:* wholemeal flour 100%, water 57%, yeast 3%, salt 2%, fat 2%, sugar 1%.
- *Basic bun dough:* strong flour 100%, water 50%, yeast 7%, milk powder 3%, salt 1.5%, fat 10%, sugar 10%, eggs 5% (optional). Also add lemon flavouring and bun spice.
- *Granary dough:* granary flour 100%, water 64%, yeast 2.5%, salt 2%, fat 2%, sugar 1%.

Enriched doughs

Baba

This is a rich fermented yeast sponge made from flour, yeast, eggs, butter, milk, sugar and currants. The flour used can be either strong or medium (a mixture of strong and soft flours in equal proportions).

Method

1 Mix the flour, yeast, milk and sugar for ten minutes until a smooth sticky paste results.
2 Scrape the paste down well, pour in the melted butter and cover the mixing bowl.
3 Leave the paste to double in size (approximately 45 minutes).
4 Carefully mix the paste with a paddle or dough hook to incorporate the butter (which will have sunk to the bottom while the paste fermented).
5 Add the currants then pipe the mix into well-greased baba moulds, piping each mould one third full.
6 Prove until the dough has risen to two-thirds of the height of the moulds.

Babas are baked until each one has risen to the top of the mould. When cool, they are trimmed level and soaked in a rich fruit and spice soaking syrup.

Savarin

This follows the same recipe as for baba (above) but does not contain any fruit. When the paste is ready and mixed (Step 5), pipe it into well-greased moulds called *savarin rings,* which come in individual or multi-portion sizes. The paste can also be piped into a charlotte mould, then baked, sliced, dusted with sugar and toasted, decorated with fruit and glazed to produce *Croûte aux fruite.* Alternatively, the paste may be baked in a barquette mould, finished with a soaking syrup, apricot glaze, fresh fuit and cream and decorated with almonds of pistachio nuts. The finished products are known as *Marignans.*

Brioche

Brioche products are made extensively in France, where they are eaten for breakfast with coffee. Many British supermarkets stock a limited range of brioche products, but they are best eaten freshly baked. Made from a rich yeast dough paste with a high butter and egg content, the dough is soft and difficult to handle; it benefits from being placed in the refrigerator overnight. Once rested and moulded into small rounds, the paste can be placed into well-greased, fluted brioche tins and baked. The tins may be individual or multi-portion in size.

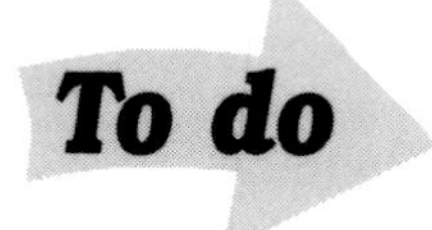

- Read the section on gluten in Unit 2D8 (page 181).
- Find out whether dough is mixed by hand or machine in your kitchen.
- Find out why enriched dough baked products have a longer shelf life than basic dough products.
- Ask your supervisor what effect milk has when used in dough making.
- Find out how the chef or baker knows when the goods are proved. What indicators do they use?

Laminated doughs

Croissant

Croissants, meaning *crescents,* are eaten mainly for breakfast, although they are also widely used today instead of bread to make filled snack items. They can be filled with fish, meat and salad ingredients.

Croissants are made from a dough in which the fat content has been *layered* (laminated) as in the production of puff pastry (see page 183). The dough is made from flour, yeast, milk, sugar and salt.

Method

1. Make up the initial dough using the ingredients listed above. Set aside to rest for 30 minutes.
2. Pin out the paste into a rectangular shape.
3. Place the butter or fat onto the dough, so that two thirds of the dough is covered. Fold as for puff pastry (page 183), giving 3 half or single turns. Allow to rest.
4. When completely rested, pin out the paste to a 2–3 mm thickness and cut (using a knife or special croissant roller) into triangular shapes, each measuring approximately 10–12 cm (4–5 in) wide at the base and 20–24 cm (8–9 in) long.
5. Roll each triangle, starting from the widest part of the base, then gently form into a crescent shape for proving. The proving stage should be carried out in dry heat, not in a moist prover as for bread and bun products. This is because of the high fat content of croissants, which reacts with moisture to give a poor finish.

Danish

Like croissant dough, Danish dough is a laminated dough. Danish pastries may be filled with fruit, frangipane, apple, mincemeat or other ingredients to produce a popular array of morning pastries. The basic paste is made using medium strong flour, yeast, chilled milk, egg and cardamon spice.

The dough is made and rested and the fat added as for croissant paste. When fully rested (after the three single or half turns: Step 3 above), the paste is cut into shapes, formed into Danish pastry bases and filled as required. Once formed, the pastries are dry proven and baked. When cool, an apricot glaze is brushed over the baked goods which can be finished with water icing and/or toasted nuts. High quality fondant piping improves visual appeal.

Speciality doughs

Blinis

Russian in origin, these are pancakes is served with caviar. The pancakes are made from flour, warm milk, a little yeast, eggs and salt. When fermented, the paste is mixed lightly and whipped egg whites are blended in (cream may also be added). The Blinis are then cooked in special small blini pans to resemble tiny pancakes.

Unleavened or flat breads

These breads are made without yeast. They include the following types:

- *nan bread* soft, Indian flat bread
- *pita bread* soft flat bread, Middle Eastern and Greek in origin. In Greece they are eaten with humous and taramasalata
- *chapati* soft, Indian unleavened bread made from Ata, a fine ground wholewheat flour

- *tortilla* crisp Mexican unleavened bread, made from Masaharina (a finely ground cornmeal flour, made from white maize soaked in lime water) or wholewheat flour.

Pizza

Pizza is one of the most popular baked dough-based range of products. A pizza base is made from basic bread dough (see page 163), made with cold water, less yeast than for bread and a little oil. White or brown flours may be used.

The dough is mixed well but does not need to be developed fully as the base should not be too thick, except in the case of deep pan pizzas. When rested, the paste is scaled off and moulded into round shapes (or large rectangles if catering for many people, e.g. for a buffet). The top of the paste should be brushed with oil to prevent the tomato topping from softening the dough. The topping (cheese, tomato, etc.) can then be placed on top and sprinkled with oil and herbs before baking.

What have you learned?

1. What are the main contamination threats when preparing dough products?
2. Why is it important to keep preparation and storage areas and equipment hygienic?
3. Why are time and temperature important during the preparation of dough products?
4. How does the process of fermentation work?
5. Describe *proving*. Why is it important?
6. What are the basic types of dough?

ELEMENT 2: Cooking dough products

Baking dough products

Oven bottom breads and rolls are generally cooked on trays without any tin moulds. The dough used for these breads needs to be firm enough to support the shape during baking (see *White basic doughs: crusty* on page 163). Softer basic doughs can be baked in bread tins, where the tin supports the dough. The softer dough rises more quickly due to its lower resistance during baking, resulting in a larger loaf. These loaves are often baked in *strapped tins,* i.e. a row of tins joined together for ease of handling.

Tinned loaves being placed into the oven

Crusty rolls are evenly spaced over baking trays with small holes, which allow the heat to pass upward and produce a brown, even, dry crust finish to the roll or loaf. The oven is *steamed* to aid the lift as the bread begins to bake and to assist the bloom finish.

Brown breads such as wholemeal are usually baked in bread tins or formed into rolls. Brown breads are baked as for white tin bread and rolls. If you are baking buns, tray them according to their type and bake them at a reduced heat to allow for the enriched dough. Always check the baking temperature for each individual bun product.

Key points: cooking dough products

- When the products are proved, handle them carefully and place into a preset oven for the required period of time.
- Remember that some baked goods need to be piped, cut or marked before baking.
- Try to fill the oven space when baking by planning your production carefully. This saves time, energy and money.
- Always remove baked products carefully and cleanly, making sure you have cleared a space to place the hot goods for hygienic storage while cooling.

Cooked bread placed on wires for cooling

- Many kitchens do not have special bakery ovens but use roasting ovens for baking bread. If using a roasting oven, you might need to brush dough products with eggwash or decorate with rice cones, flour or seeds to provide a decorative finish.
- Bread achieves a good colour when cooked by steaming, but this should only be done in ovens specifically designed for such use. Never steam in an electric oven (such as a convector) if it has not been designed for this: *it can be dangerous.*

Deep-frying dough products

Doughnuts are always deep-fried. The frying can be carried out in vegetable oil or special fats manufactured specially for deep-frying. Do not fry doughnuts in fat is used for frying fish as the flavour can transfer from the oil to the doughnuts.

Always use clean oil. Oil is usually strained each day after use; when the oil is very dark, make sure you strain out any carbonised debris. Keep a little of the old oil to add to new oil: this will help to colour the food.

The frying temperature needs to be monitored carefully. If the temperature is too hot the doughnuts will appear cooked (i.e. they will be the correct colour) but will be uncooked in the centre. If the temperature is too low the doughnuts will absorb the oil, making them unpleasant to eat.

Always drain the doughnuts well and allow them to cool before rolling them in sugar. This prevents the oil from dampening the sugar and affecting the visual appearance of the sugar coating. Drain the doughnuts on kitchen paper or a wire rack.

Always remove fried products carefully and cleanly, making sure you have cleared a space to place the hot goods for hygienic storage while cooling.

Thawing and cooking frozen dough

Unbaked products can be placed in the proving cabinet to thaw rapidly, providing the goods are thoroughly baked following thawing. However, frozen dough and dough products are normally thawed either by placing them in a refrigerator (allowing enough time for them to thaw at a temperature of 1 °C/33 °F) or by placing them in a hygienic area at room temperature (not exceeding 10 °C/50 °F). If the temperature is too high the dough can become damaged and the risk of contamination is increased.

Small, individual frozen dough items can be placed on a tray, thawed, finished and baked. Bulk dough items need to be thawed carefully; check that the centre of the dough has completely thawed.

All frozen doughs and dough products must be defrosted *completely* and baked well. If the goods are not thoroughly baked the centres may not be cooked and this can result in contamination and food poisoning. Follow the instructions provided by manufacturers for frozen products; some products do not need to be thawed and can be cooked from frozen.

Always follow the manufacturer's instructions exactly. Do not attempt to take shortcuts when under pressure: this can lead to contamination.

Essential knowledge

Time and temperature are important when cooking dough products in order to:

- ensure the dough goods are correctly cooked (either baked or deep-fried)
- prevent food poisoning from direct and indirect contamination
- prevent shrinkage
- promote and maintain customer satisfaction by the production of consistent standards of quality.

Finishing and presenting cooked dough products

Ensure the cooked products are cool before finishing, dusting or filling except for bun goods which are bun washed while still warm. Finish by filling, dusting or decorating the baked/fried cool products according to requirements. Present and store the finished cooked products according to the *Food Hygiene (Amendments) Regulations 1993*.

Essential knowledge

The main contamination threats when preparing and producing dough and dough products are as follows:
- cross-contamination can occur between cooked and uncooked food during storage
- food poisoning bacteria can be transferred through preparation areas, equipment and utensils if the same ones are used for preparing or finishing cooked, baked goods and raw, uncooked doughs and dough products
- food poisoning bacteria may be transferred from yourself to the food. Open cuts, sores, sneezing, colds, sore throats or dirty hands are all possible sources
- contamination will occur if flour or any foods are allowed to come into contact with rodents (such as mice or rats), or insects (house flies, cockroaches, silver fish, beetles). Fly screens should be fitted to all windows
- food poisoning bacteria may be transferred through dirty surfaces and equipment. Unhygienic equipment (utensils and tables, trays, mixers, dividers, provers) and preparation areas (particularly egg and cream based mixers) can lead to contamination
- contamination can occur through products being left opened or uncovered. Foreign bodies can fall into containers and sacks, and dough mixes, such as sack tape, metal objects, machine parts, string, etc.
- cross-contamination can occur if equipment is not cleaned correctly between operations
- contamination can occur if frozen doughs are de-frosted incorrectly (always check the manufacturer's instructions)
- incorrect waste disposal can lead to contamination.

Finishing methods

When finishing dough products you need to bear the following points in mind.
- Finishing is very important. Each dough product in a range should be the same size, colour and shape.
- Fondant icing, jam and clear glazes should be thin and glossy. Decorating media should *highlight* the product, not overwhelm it. Decoration should be attractive and simple, not too complicated.
- When filling with cream, fruit or custard, do it neatly and cleanly. Use the correct amount of filling: too much filling makes the product difficult to handle, too little might not maintain customer satisfaction.
- Ensure that cream is whipped correctly, in a hygienic manner, and that cream-filled products are stored according to the *Food Hygiene (Amendments) Regulations 1993*.
- Piping cream is a skill that develops with practice. Check that piping bags have been sterilised as they are a source of cross-contamination. Work carefully, making sure that the finished product looks clean and tidy.
- *Never* use cream to decorate a product that is still warm.
- When you are creaming cooled baked or fried dough products, work in a cool area of the kitchen.

- Check that you have used fresh cream in rotation.
- Ensure that any cream is whipped cleanly. All equipment used should be hygienically cleaned before and after use to avoid contamination.
- When whipping cream, watch the development of the cream carefully. It can easily be over-whipped and formed into butter. Cream is an expensive commodity: use it carefully and cleanly.
- Heat any fondant gently over a period of time to a temperature not exceeding 40 °C (105 °F), then spread thinly over products. Decorate the finished product with decorating media while the fondant is still sticky.
- If using water icing, do not make it up too thinly: it should hold firmly to the product, providing a glossy coating.
- When dusting with icing or caster sugar, dust the product evenly using a shaker or sieve.
- Handle and store all finished products carefully until required for presentation.

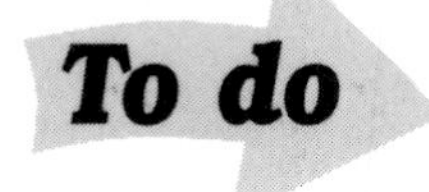

- Watch your chef or baker producing products made from bread dough and bun dough. Notice how they are prepared and cooked.
- Find out what convenience products (if any) are used in your kitchen.
- Check the cooking times and oven temperatures for each dough type.
- Watch your supervisor piping or filling with cream. Notice the techniques and equipment used.
- Check to see whether any cooked and uncooked foods are stored together in your kitchen. If so, why?
- Read Unit G2: *Maintaining a professional and hygienic appearance* in the Core Units book.

Convenience dough products

The range of convenience dough products is considerable, including fresh and frozen pre-proved dough products such as rolls, croissants, Danish pastries and French breads. These goods are available at competitive prices and can be used in the production of cost-effective quality products. They allow the contemporary craftsperson to concentrate on the flavour and decorative display of food items, without having to spend long hours in production or investing in a large variety of specialist equipment.

Bake-off products have invaded stores, shops and supermarkets, bringing a more diverse and interesting range of dough products to the public. These are products prepared ready for baking, in either frozen, fresh or Modified Atmosphere Packaged (MAP) forms. The MAP method replaces most of the oxygen around the product to slow down spoilage, and these products will need to be kept refrigerated.

We all readily use convenience dough-based products such as MAP garlic sticks and petit pain products, frozen danish squares and semi-finished pastries. It is essential for everyone training in our industry to become aware of the best selection of such goods, to seek out the best suppliers and to combine the best classical craft skills with the technology of the twenty-first century.

Storing cooked dough products

Damp, humid weather or bad storage can soften crusty rolls and loaves. Wrap and store dough products at room temperature, or freeze them for longer storage. Do not store dough products in a refrigerator: this would make them become stale more quickly. The staling process occurs fastest at just above freezing point, and slowest at temperatures below freezing point. Products with higher ratios of fat and milk tend to go stale more slowly than those made from a basic dough.

Health, safety and hygiene

Note all points given in the Core Units book concerning general attention to health, safety and hygiene. When preparing dough and dough products the following points are particularly important:

- make sure any machines are set to the correct (slow) speed before they are switched on
- make sure all safety guards on machines are in place and work correctly
- make sure all moving parts of machines have stopped before removing dough from the mixing bowl. Never attempt to remove any dough while the machine is still mixing
- wipe up any spillages as soon as they occur
- always lift and transport heavy loads such as dough mixes in a safe and hygienic manner.

Planning your time

Refer to *Planning your time* (pages 1–4) concerning general points to note when working in a kitchen or pâtisserie. When preparing dough products the following points are especially important to remember:

- always check that you have added the yeast to the dough mix. An error here wastes valuable time and costs money
- make sure that doughs are mixed according to the correct production methods and for the required time
- while doughs are mixing, use your time to make sure trays and tins are prepared and ready for the dough items
- while doughs are resting/proving, clean and clear work surfaces for the next batch of work
- doughs with added ingredients such as eggs, fat, milk powder or sugar take longer to prove. Remember that you will need to allow more time for the proving stages of these doughs
- be aware of how long doughs have been proving. If you allow proving to continue for too long, over-ripe doughs will develop and you may have to throw them away and begin again
- when producing a large amount and range of dough products, check

Removing baked dough products from the oven

that there is enough space in the prover and the oven
- check oven temperatures carefully. If the oven is not at the required baking temperature goods will be dry, and lack colour and volume
- always ask about the oven setting for each individual dough product. Doughs containing sugar need lower temperatures than bread goods. This is because the sugar caramelises and quickly browns the bun products. Breads have less sugar and need a higher temperature to colour and bake the roll or loaf
- work in a clean and organised manner
- clean up after yourself as you work.

Dealing with unexpected situations

An unexpected situation might be a rush order, customer request or emergency. Be clear about the priorities in any of these situations, and always follow laid down procedures.

Check that you know what to do in the case of accidents and emergencies. Note all the points given in Unit G1 of the Core Units book: *Maintaining a safe and secure working environment.* Be aware of your personal responsibilities within the kitchen. If in doubt, ask your supervisor.

Disposing of waste

- Waste materials should be disposed of cleanly and efficiently, to prevent contamination of uncooked and cooked dough products and ingredients and to prevent health hazards.
- Always replace lids on waste bins.
- Clean up after yourself as you work. Wipe down surfaces between each task or job. Do not use glass or oven cloths to wipe down work surfaces: this causes contamination.

What have you learned?

1 What are the main contamination threats when cooking dough products?
2 Why is it important to keep preparation and storage areas and equipment hygienic?
3 Why are time and temperature important in the cooking of dough products?
4 How would you thaw frozen dough before cooking it?
5 What are the ideal storage conditions for cooked dough products?
6 List four types of convenience dough product.

Extend your knowledge

1 Find out what effects fat, eggs and milk or milk powders have on the fermentation process.
2 Check the range of frozen pre-proved and ready-to-prove dough products available today.
3 Do some sample costings for a variety of dough goods then discuss this with your supervisor.
4 Ask your supervisor about flour improvers. What are they? How do they work? Why do we use them?
5 Keep a detailed records of all recipes, costings, cooking temperatures and production methods you use. These will all be useful at Level 3.

UNIT 2D8

Preparing and cooking pastry dishes

ELEMENTS 1 AND 2: Preparing fresh pastry and pastry dishes

What do you have to do?

- Prepare appropriate baking equipment ready for use.
- Prepare the paste ingredients correctly for each individual dish.
- Mix the prepared paste ingredients according to the product requirements.
- Mix, roll, fold, cut, shape, mould and rest paste according to product requirements.
- Bake or steam paste products for the required time and at the correct temperature.
- Fill, pipe, glaze, finish, present and store pastry dishes according to the product requirements, meeting food hygiene regulations.
- Dispose of waste in the correct manner according to laid down procedures to ensure standards of hygiene.
- Carry out your work in an organised and efficient manner taking account of priorities and laid down procedures.

What do you need to know?

- Which type of equipment you will need to use.
- The type, quality and quantity of ingredients required for each paste and how to prepare the different types of pastry.
- What the correct procedure is for thawing and cooking frozen pastry products to prevent contamination.
- How to satisfy health, safety and hygiene regulations concerning preparation areas and equipment both before and after use.
- What the main contamination threats are when preparing and cooking pastry products.
- Why it is important to keep preparation and storage areas and equipment hygienic.
- Why time and temperature are important in the preparation of pastry products.
- How to plan your time to meet daily schedules.
- How to deal with unexpected situations.

Introduction

Pastry is the basis for a useful range of baked and steamed products that fall under the category of *pâtisserie*. This includes both sweet and savoury pastry dishes, ranging from traditional products popular for morning coffee or afternoon tea (e.g. French pastries), to snack foods (e.g. sausage rolls, pasties, quiches) and desserts (e.g. fruit pies, tarts).

Well-made pastry is the basis of good pâtisserie. The basic pastes, made from a combination of flour, fat, salt/sugar, water, milk and/or eggs produce a range of flavours and textures. The types discussed here include short pastes, both sweet (sugar paste) and savoury (short paste); laminated paste (puff pastry); suet pastes; boiled pastes (e.g. English pie pastry and choux pastry); French pie pastes and German pastes. Making good pastry that is tasty and light with visual appeal takes time, practice and understanding of the essential principles and processes involved.

Pastry has a tradition of its own, from basic short pastry flans to the rich almond speciality biscuits from Italy, Denmark or Britain. Today, the convenience of purchasing ready-made pastry items or pastry pre-mixes from supermarkets has meant that we no longer rely on the traditional skills of home baking which were once a cornerstone of European culinary practice.

ELEMENT 1: Preparing fresh pastry

Pastry ingredients

Fats

Pastry is made from a range of fats used either singly or in a combination. These include butter, margarine, cake margarine, pastry margarine, shortening and lard.

- Butter provides a good flavour and rich colour when used in pastry, but it can be too costly for catering operations. Unsalted butter is ideal for making shortbread pastry.
- Margarine is produced for particular purposes and is a blend of oils and fats which have been hydrogenated (pumped with hydrogen gas and set with a catalyst). The blend can contain up to 10 per cent butterfat, but no more than 16 per cent water.
- Cake margarine is widely used for its emulsifying properties (i.e. it helps combine ingredients that would not otherwise easily mix together). It creams well and is quite soft at room temperature. As with margarine (above), it can include up to 10 per cent butterfat; this is the legal maximum. When preparing machine-made paste, you would cream together the fat and sugar, add cold water to the emulsion and finally blend in the sieved flour. This makes a good commercial paste for manual or semi-automatic pie machines, large fruit slices or individual tarts and sweet pies.
- Pastry margarine was developed for use as a substitute for butter in the production of puff pastry. It is less expensive and easier to handle. The margarine is a tough plastic or waxy fat which is easily formed into the fine layers essential for puff pastry. It melts at a much higher temperature than butter and is both tough and flexible (it is sometimes called flex). This fat is also used for fat sculpturing, now a reviving art in culinary circles.
- Shortening is made from a white edible oil manufactured for the production of 100 per cent fat shortening or compound. It has no protein or salt content and is hydrogenated. Developed originally as a substitute for lard, the modern forms of shortening have an excellent shelf life and produce good results when used for savoury paste.

- Lard is the rendered fat of the pig. It is used for savoury pastes such as hot water paste or hot savoury paste. It is also a main ingredient of lardy cakes, where the lard fat is mixed with sugar and fruit and placed into the centre of the rolled dough piece.

Flour

Short, sweet suet, flan, German, French and almond types of pastry use soft weak flour, i.e. one with a low protein content of approximately 7–8 per cent. This is milled from wheat grown in milder climatic conditions such as those found in the United Kingdom. Strong flour with a higher protein content is used for the production of puff and choux paste.

Toughness in pastry is caused by a substance called *gluten*. This is found in all flours, but to a higher degree in strong (hard) flours, which therefore yield a tougher paste. Gluten develops into long elastic strands when mixed with water, and these can be more or less tough depending on certain factors. To avoid strengthening the gluten (and toughening the paste):

- use as little water as possible. The more water used, the tougher the paste will be, as the strength of the gluten is increased by water
- make sure that you rub the fat well into the flour; this will prevent the water having direct access to the flour particles
- do not over-mix the paste or over-handle the pastry
- use a soft flour whenever indicated in recipes.

See also page 181: *The principle of shortening.*

Sugar

Caster sugar is normally used in pastry because of its fine grain and good creaming properties when mixed with fat. Some pastes can be made with icing sugar but it is unusual to use granulated sugar for the production of pastry of any type.

Sugar has two functions when used in pastry: it helps to prevent toughness by breaking down gluten and it caramelises to produce a rich golden colour. The colour of the cooked pastry is partly affected by the amount of sugar in the recipe: the higher the sugar content, the deeper the colour. A lower cooking temperature is required to prevent a high sugar product from burning. These kind of products, such as shortbread, have a much shorter texture.

Eggs

Some pastry types such as sugar pastry *(pâte sucrée)* include eggs as a basic ingredient. Fresh whole egg is mixed with sugar and/or water and added at the emulsion stage using the creaming method.

Water

This should always be used cold and added while mixing on a slow speed. The aim is to add as little water as possible while using enough to bind the paste. It is generally used for savoury paste, such as hot water pie pastry *(pâte à l'anglaise)*, suet pastry *(pâte à grasse de boeuf)*, puff pastry *(pâte feuilletée)* and moulding/lining pastry *(pâte à foncer)*.

Salt

Salt should be used carefully. It assists in the colour and finish of pastry, but is only used in small quantities. If you are using salted butter or margarine, you will not need to add as much salt as when using unsalted varieties of fat. Remember that shortening does not contain salt, so you will need to add some when making savoury pastes using this type of fat.

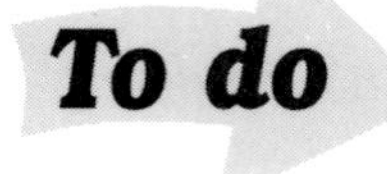

- Find out what type of fat high ratio shortening is. How does it differ from the other fats mentioned and what special properties does it have? What type of flour is it used with and why?
- What kinds of pastry fat are used in your kitchen? Which fats are used for which pastes?
- Ask your supervisor what function salt plays in pastry.
- Find out why fats become rancid. How can this be avoided, especially in butter?

Preparing and mixing the paste ingredients

- Check that your scales are accurate and meet existing regulations.
- Work in an organised and clean manner.
- Sieve the flour to remove any lumps or foreign bodies and to aerate the flour.
- Make sure that all eggs and fats are at room temperature.
- Always check that you have weighed the ingredients for the pastry mix correctly. Any errors here can be costly in both time and money.
- When ingredient items are low in stock inform the chef or baker, who will re-order to prevent stocks from running out.
- Work cleanly and replace lids on open storage bins to prevent contamination.
- Check the storage temperature of pastry fats. They should be stored in a cool place (preferably in the refrigerator) and kept covered and sealed. Bring the fat to room temperature before use; fat used straight from the refrigerator will be hard and therefore difficult to cream.
- When opening cans, remove the lids completely: sharp lid edges are very dangerous.
- Remember to rub in any fat finely, to form a light crumb.
- Cream the fat and sugar well before adding any liquid, and be careful not to over-work the paste once formed.

Mixing pastes

- Make sure that pastes are mixed according to production method/s and for the required time.
- Make sure that puff pastry is folded according to the type of production method used.
- Remember that pastes which are produced using the creaming or flour batter methods tend to be soft and so need careful handling or (preferably) to be chilled before use.
- Make sure that any mixing machines to be used for making pastry are set up correctly, that attatchments are fixed on and that mixing bowls are safely attatched according to the manufacturer's instructions. Never use wet hands near electrical machines.

Health, safety and hygiene

Note all points given in Units G1 and G2 of the Core Units book concerning general attention to health, safety and hygiene. When preparing pastry and pastry products the points listed at the top of the next page are particularly important.

- Ingredients used for making pastry and pastry dishes should be stored cleanly: always cover open boxes or packaging to prevent contamination or infestation. Flour should be stored in a flour bin with a tight fitting lid. Fats need to be kept cool until brought to room temperature for production. Dry ingredients should be stored in closed storage containers. Keep all pastry and pastry products chilled until required.
- Make sure that all mixing machines are set to the correct (slow) speed before they are switched on.
- Check that all safety guards on machines are in place and working correctly.
- Never remove pastry from the mixing bowl of a machine until all of the working parts have stopped moving. Make sure the correct machine attachment is used for each production method.
- Wipe up any spillages as they occur.
- Always lift and transport heavy pastry mixes in a safe and hygienic manner.

Short savoury pastry (*Pâte brisée*)

As the term implies, 'short' pastry should break easily when cooked and melt in the mouth; it should be tender, light and dry. A complete absence of toughness should be evident. This is achieved through the balance of fat and flour in the recipe together with the correct production method.

In order to avoid causing toughness in short pastry, ensure that the gluten content is low: make sure that the fat is rubbed into the flour well enough to prevent the water having direct access to the flour particles; add as little water as possible; use a soft flour; and do not over-mix the paste or over-handle the pastry.

Short pastry differs from short sweet pastry in that it contains no sugar and is used for the basis of savoury products. Flans, tartlettes and barquettes are prepared using thinly pinned short pastry, either baked blind or filled and then baked.

Method

The paste is made from soft flour, fat (50 per cent lard or shortening and 50 per cent margarine or butter), seasoning and as little water as possible (to prevent toughening the paste).

It is made by the rubbing in method (page 181) for quick use or the flour batter method (page 182) if prepared in advance (as the paste tends to be of a softer consistency and will need to be chilled before use).

When mixing by machine, take care when rubbing in the fat: if over-mixing occurs before the water has been added the paste will become crumbly and tough and of little use. Watch for the crumb occurring and the change of colour as the fat is broken down.

Add the water gradually and always stop the machine when testing the texture of the paste. If necessary, add a little more liquid rather than more flour to tighten the paste for handling. Use only enough liquid to bind and bring the paste together: the more water used, the more liable the mix is to form gluten, which will toughen the paste (see page 181).

Once the paste is prepared, you should wrap and cool it in the refrigerator until you need it.

Fats used

Fat shortening, a white commercial fat, produces good results when used in short pastry. A mixture of fats will give colour and texture, but remember that the use of butter will increase the cost of the paste and therefore the selling price of the product.

Make sure that any fats are at room temperature: this will make them easier to use and they will rub in well. Once the paste has formed, handle with care and bring the paste together; i.e. gently mould the paste until all the ingredients are mixed to form an overall colour and texture. Do not over-mix, as this will toughen the pastry. Remember that shortbread pastes can sometimes be difficult to work because of the high ratio of both fat and sugar in the recipe.

General rules for handling short pastry

- Make sure all equipment and work surfaces are clean.
- Ensure paste ingredients are at room temperature (21 °C/70 °F).
- Always sieve the flour and salt to aerate the flour and disperse the salt evenly.
- When rubbing in the fat, lift the mix to incorporate air.
- Add the liquid gradually, checking the consistency of the paste.
- When making short paste by creaming, occasionally scrape down the sides of the machine or mixing bowl using a plastic scraper.
- When a clear paste has formed wrap well, chill and rest before use.
- Scraps of short paste should be worked into fresh mixes to minimise waste.
- Break any eggs into a clean bowl before preparing the paste to prevent contamination. Wash your hands after cracking eggs and before starting the next task.
- Use the minimum amount of flour when dusting pastry to prevent the paste from sticking.

Short pastry can be made using a range of recipes and production methods; *The International Confectioner* (Virtue & Company, 1968) and *Complete Pastry Work Techniques* by Ildo Nicolello (Hodder & Stoughton, 1991) demonstrate many approaches.

Short sweet pastry (*Pâte sucrée*)

Short paste (as described above) is used for the production of savoury goods, but short paste can contain small ratios of sugar, and is then known as sweet or sugar pastry, depending on the sugar to flour ratio. Sweet pastry has approximately 50 g (2 oz) of sugar to 500 g (1 lb) of soft flour. Sugar pastry has 125 g (5 oz) of sugar to 500 g (1 lb) of flour.

Sweet pastry is used to make flans, pies, bandes, tarts and tartlettes which are either baked blind and filled or filled and then baked. The paste is made from soft flour, fat, sugar, eggs and/or water. The more fat that you use, the less egg will be required to bind the pastry together.

Key points

- When using the rubbing in method (page 181), mix the sugar with the egg/liquid and add it to the paste once the fat has been finely crumbed into the sieved flour.
- When using the creaming method (page 182), cream the fat and sugar until light and white, add the eggs one at a time, then finally blend in the sieved flour to form a clear, smooth paste. Do not over-mix the

paste. The flour batter method (page 182) can also be used and is often referred to as a *secondary creaming method.*
- Cake margarines, which have good emulsion properties, are used in bakeries for the production of sweet pastes to be tart stamped by machine, and for custard or fruit filled tartlettes.
- Sweet paste is best made the day before it is required and rested in the refrigerator until used. Wrap it in cling film for resting; do not wrap it in greaseproof paper or the paste will become dry and form a skin.
- It is possible to purchase a tartlette stamp and cooking machine which uses a high sugar paste recipe. The paste (usually made with icing sugar) is placed into the mould shape, the lid is closed, then the paste is shaped into the mould and cooked. This is based on the sandwich toaster principle and takes about 3 minutes to cook. The goods produced have a good shelf life, but note that they are quite sweet.

The principle of shortening

The purpose of shortening is to coat the sub-proteins within the flour, preventing the development of gluten (which can toughen the paste, making it unpleasant to eat and difficult to digest). Since the gluten in flour forms into tough elastic strands when mixed with water, this can be avoided by preventing the water coming into direct contact with the flour; in this case by mixing the fat and flour together first in a particular ratio.

In the dough unit we wished to develop the long elastic strands of gluten to hold in the expanding gases; with short pastry we wish to do the exact opposite. The gluten strands that are formed within a light short pastry should be short and easily broken when eaten.

The standard ratio of fat to flour is 50 per cent fat to 100 per cent flour, where the flour is soft (i.e. has a low protein content). As the ratio of fat is increased in the recipe, the texture of the pastry becomes shorter. Shortbread has one of the shortest textures of any biscuit and melts in the mouth (this is also due to being made with butter).

Short pastry can made with a range of fats:
- *butter* (best for flavour and colour but most expensive)
- *margarine* (inferior to butter but less expensive)
- *special fats,* such as shortening and cake margarine, which are used for commercial production (because of their excellent creaming qualities).

The shortening principle is used with sweet and savoury short pastry, using the rubbing in, creaming or flour batter method. Suet paste is generally produced by mixing or blending the fat into the flour before adding the water and mixing the paste (see page 185).

Methods of producing short pastry

Rubbing in method
Here the fat content is either rubbed into the flour by hand (or mixed into the flour using a paddle on a mixing machine) to produce a fine-textured crumb before the liquid is added. This helps to prevent the pastry from becoming tough (see *The principle of shortening* above). It is important that the pastry is handled lightly and worked as little as possible, as over-handling will also increase toughness. Pastry should be rested/relaxed well before being used, or the items prepared and formed then left to rest before cooking, to allow the gluten to relax.

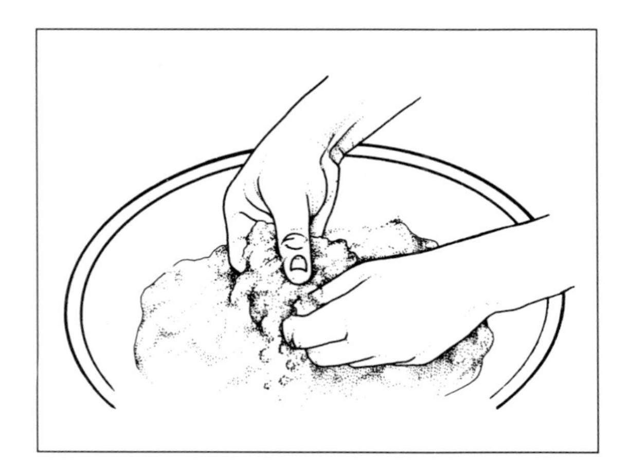

Rubbing in by hand

Rubbing in by hand

1 Sieve the flour to aerate it and remove any small lumps or particles.
2 Place the fat into the flour, then, using your fingers, rub the fat into the flour until you obtain a fine crumb texture.
3 Add the liquid (egg, milk or water), bringing the paste gently together with as little handling as possible until a clear paste is formed. If the paste is over-handled or developed the gluten will form into long elastic strands causing the pastry to be tough.
4 Rest the paste until required by wrapping it in cling film and placing it in the refrigerator.

Rubbing in using a mixing machine

Mixing machines do not have the facility to detect when the paste has been rubbed in sufficiently for the liquid to be added; the chef needs to determine when this point has been reached. This knowledge comes with practice, as you learn what signs to look for: such as the crumb texture, the colour of the mix and the way that the paste starts to come together as the fat content starts to re-join.

Very goods results can be achieved using the machine for mixing large quantities, but you need to pay close attention to the mixing, as over-mixing can waste food commodities and time.

Creaming method

Creaming is different from *rubbing in* because the fat and sugar are creamed together to form a light, white emulsion *before* the liquid is added and then the sieved flour folded in and cleared to a smooth paste.

Special fats such as cake margarine and shortening which have good creaming properties are used in bakeries and kitchens. If butter is used, it needs to be at room temperature to enable creaming to take place successfully.This method of production produces a softer and lighter paste but does require refrigeration to firm the paste before handling can begin.

Flour batter method

This is often defined as a *secondary creaming method.*

1 Cream the fat with an equal quantity of flour.
2 Mix together the sugar and liquid then slowly add them to the fat and flour mixture to form a smooth paste.
3 Finally add the rest of the flour and clear. Do not over-mix once all the flour has been added.

Puff pastry (*Pâte feuilletée*)

Puff pastry is a light, layered, baked pastry that is made from strong flour, butter and/or pastry margarine, cold water, salt and a little acid (vinegar or lemon juice). The lemon juice helps to extend the formed gluten, making the paste more malleable when folding.

Principle of aeration by lamination

Puff pastry lamination is the *layering of fat and dough*, so that when heated, moist air is trapped between these layers and converted to steam, causing the layers to rise and the gluten to blister. As the layers rise, the flour protein sets in the heat of the oven. Puff pastry can be full puff (i.e. 100 per cent fat to 100 per cent flour) or three-quarter puff (75 per cent fat to 100 per cent flour).

The layering is achieved by making a number of turns (or folds) of the pastry and fat together. The pinned out pastry and shaped fat are placed on top of one another in some form (or combined, as in Scotch pastry), then particular sections of the pastry (covered by fat) are folded over each other to create layers of fat and paste.

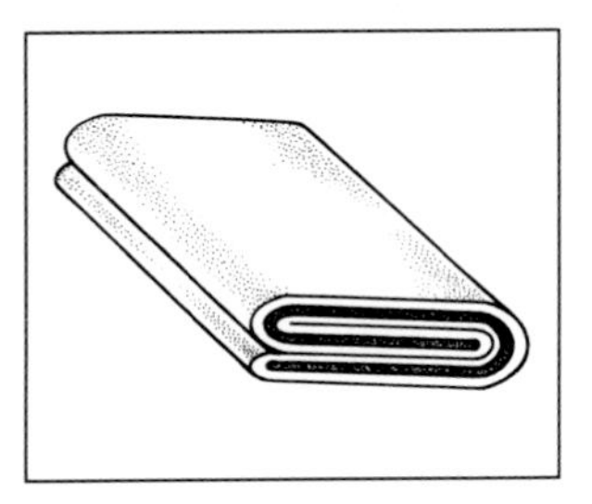

Single or *half* turn

Six single or *half* turns

The first turn occurs when the pastry is folded enclosing the fat (see the illustration in the margin). The chart below demonstrates how the layers occur.

Turns	**1st**	**2nd**	**3rd**	**4th**	**5th**	**6th**
Layers at start of turn	1	3	9	27	81	243
Layers at end of turn	**3**	**9**	**27**	**81**	**243**	**729**

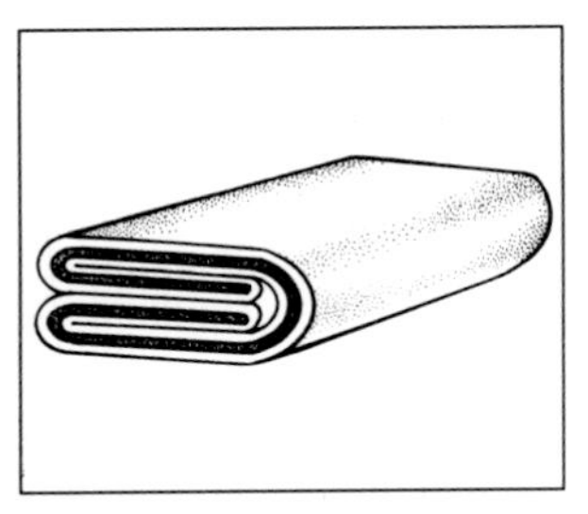

Double or *book* turn

Four double or *book* turns

These are made as shown in the diagram in the margin. The layers are formed as shown in the chart below.

Turns	**1st**	**2nd**	**3rd**	**4th**
Layers at start of turn	1	4	16	64
Layers at end of turn	**4**	**16**	**64**	**256**

The *six single turns* method produces pastry with a finer layered texture and is good for vol-au-vents and pastries requiring lift.

The *four double turns* method produces the same lift but the texture is more open, with larger layers and a coarser finish.

Three-quarter puff

This paste uses three-quarters the amount of fat to flour, i.e. 750 g ($1\frac{1}{2}$ lb) of fat to every kilo (2 lb) of flour and often only receives four of the six single turns. It is used for puff pastry goods that do not require too much lift. There are other types of puff pastry (such as continental) where the fat is folded on the outside of the dough; this is more difficult to handle but does produce a better baked finish.

Methods for preparing puff pastry

Puff pastry is made in a number of ways: by the French, English or Scotch (rough puff) methods.

French method

1 Sieve the flour and salt, rubbing some of the fat into the flour.
2 Make a bay and add the water and acid. Mix to a clear, soft smooth dough.
3 Rest for 20 minutes.
4 Cut a cross into the top of the rested dough, cutting half way through the dough.
5 Pull out the corners of the dough to form a star shape. Pin the points of the star so that the total thickness of the four star corners equals the thickness of the centre square of the star shape.
6 Knead the remaining fat until pliable and of a similar softness to the dough.
7 Shape the fat to a similar size and thickness as the centre square of the star, shaping it between silicone paper.
8 Brush off any excess flour from the dough. Place the square block of fat on top of the centre square of the dough, then fold the four star corners over the fat.
9 Pin out the dough using either 4 double turns or 6 single turns to produce the layered or laminated effect.

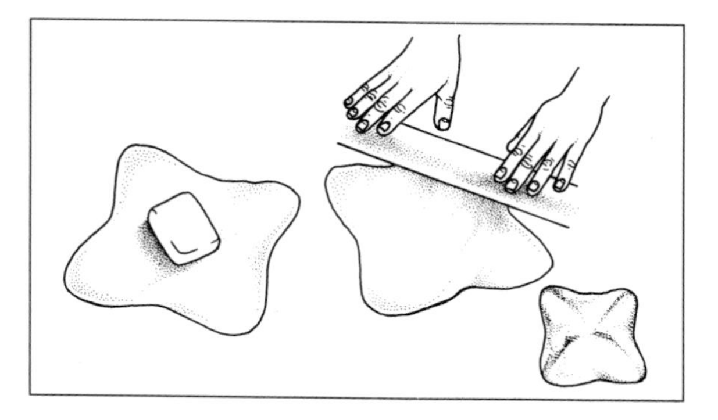

Rolling out the star shapes

English method

The English method only differs from the French in the way that the fat is combined with the rested dough.

1 Produce the dough in the same way as for French puff pastry (Steps 1 and 2 above) and rest for 20 minutes.
2 Pin the rested dough out into a rectangle shape: 45 × 30 cm (18 × 12 in).
3 Knead the fat until pliable. The dough and fat should be similar in consistency so that when pinned out they will move equally. If the fat is too firm it will break up, producing uneven fat and dough layers.
4 Pin the fat between silicone paper to a square shape 30 × 30 cm (12 × 12 in).
5 Place the fat onto the rectangle of dough to cover two-thirds of the paste. Fold the one third of paste without fat onto the middle third and then the final third (with fat) over the other two thirds. This forms three layers, which is the first turn. Rotate the folded paste 45° to the right and pin out the pastry; this time to 45 × 45 cm (18 × 18 in). Repeat this process until you have completed six turns. Rest the paste for 20 minutes between every two turns.

Scotch or rough puff method
This method of making puff pastry is very quick.

1 Cut the fat into small pieces and place it in a mixing bowl with the sieved flour and salt.
2 Add the water and mix the paste for approximately 60 seconds. The fat should still be intact; i.e. in small pieces.
3 Form the paste into a rectangle shape (45 × 30 cm/18 × 12 in) and fold to form a single or half turn. After two turns, wrap the paste in cling film and chill, resting it for 15 minutes. Repeat this three times to complete six single or half turns.

This paste is only suitable for recipes using scrap puff, for savoury and sweet pies, or for products not requiring too much lift.

Key points: puff pastry

- Always pin out using even pressure.
- Ensure the fat and dough are of similar consistency so that they will pin evenly.
- Always dust off excess flour between folding, otherwise the layers will not stick and a large pocket will form in the cooked puff pastry.
- Pin the folds in both directions to ensure an even layer effect occurs; i.e.fold the paste to form a turn, then rotate the paste by 45° before rolling again before making the next turn.
- Rest the pastry between folding, to relax the gluten and aid even lamination.
- When lifting the folded pastry, pick up the paste on a rolling pin to avoid stretching the fine folded layers; they will stretch unevenly if lifted by hand.
- Mark the paste with a finger for each fold to remind you how many folds the pastry has been given. Too many folds will result in too fine a lamination.
- Wrap the paste in cling film between folds to prevent a skin forming.

Convenience puff pastry

Puff pastry today is often purchased frozen in sheet or block form. Many kitchens use frozen puff or preformed puff products because this saves time and labour (increasing profitability) and provides consistency of standard. The quality of lift and cost effectiveness of sheet puff is far superior to many of the block frozen puff goods available, and it is widely used in the bakery industry for turnovers and cream slices, with scraps used for eccles, coventries, etc. Cooked puff horns with a long shelf life can also be purchased.

Suet pastry (*Pâte à grasse de boeuf*)

Suet is the protective fat that surrounds the kidneys of cattle and sheep. The fat is covered with a fine membrane (skin) which is removed along with any traces of blood or kidney to leave the suet only. Fresh suet is made by passing cold suet through a mincing machine on slow speed. It is also available in a pre-shredded form in commercial units.

Suet pastry is used for sweet and savoury dishes such as steamed suet roll, steamed steak and kidney or steak and mushroom pudding and as an ingredient of sweet mincemeat.

Method

1 Sieve the flour, salt and baking powder together into a mixing bowl. Always use fresh baking powder: mix 1 part bicarbonate of soda to 2 parts cream of tartar and sieve this at least 10 times. Store any excess in an airtight container.
2 Mix in the shredded suet.
3 Make a well in the mixture then slowly add cold water. Bring the ingredients together to produce a paste. Wrap and rest for 10 minutes and use as required. Do not make more suet paste than is required for each task, as this paste should not be stored.

Essential knowledge

The main contamination threats when preparing and producing pastry and pastry products are as follows:

- cross-contamination can occur between cooked and uncooked food during storage
- food poisoning bacteria can be transferred through preparation areas, equipment and utensils if the same ones are used for preparing or finishing cooked, baked goods and raw, uncooked pastry and foods
- food poisoning bacteria may be transferred from yourself to the food. Open cuts, sores, sneezing, colds, sore throats or dirty hands are all possible sources
- contamination will occur if flour or any foods are allowed to come into contact with rodents (such as mice or rats), or insects (house flies, cockroaches, silver fish, beetles). Fly screens should be fitted to all windows
- food poisoning bacteria may be transferred through dirty surfaces and equipment. Unhygienic equipment (utensils and tables, trays, mixers, moulds, pastry brakes) and preparation areas (particularly egg and cream based mixes) can lead to contamination
- contamination can occur through products being left opened or uncovered. Foreign bodies can fall into open containers and flour bins, such as sack tape, metal objects, machine parts, string, egg shell, cigarette ends, etc.
- cross-contamination can occur if equipment is not cleaned correctly between operations
- contamination can occur if frozen doughs are de-frosted incorrectly (always check the manufacturer's instructions)
- incorrect waste disposal can lead to contamination.

Boiled/hot pastry

Choux pastry *(Pâte à choux)*

Choux paste is made from boiling water, fat, salt and sugar, strong flour and eggs. The method given below shows how to produce fresh choux pastry. However, pre-baked choux eclairs and choux buns can be purchased that have a six month shelf life. These are used extensively in the bakery and catering industry.

Method

1 Boil the water and fat with the salt and sugar.
2 Stir in the sieved strong flour, mixing it into the boiling water and fat to produce a paste called a *panade* (a type of roux). The heat gelatinises the starch in the flour, which swells, absorbing the water to form the panade. Cook for approximately 1 minute on top of the stove. The panade should readily leave the sides of the pan when shaken.
3 Remove from the heat and allow to cool by spreading the panade onto a clean plastic tray.
4 When cool, crack the eggs into a bowl or jug then mix them into the panade to form a smooth paste of medium dropping consistency (it should be firm enough to hold its own shape when piped). The eggs should be at room temperature (21 °C/70 °F); the amount of egg used varies according to the strength of the flour and the size and age of the eggs. If too much egg is added the choux paste will be soft and not hold its shape when piped. If this happens, mix up a new panade and add the soft choux paste to the new panade when cool.
5 Pipe the choux paste to form items such as eclairs and choux buns. These can then be baked in a sharp oven until golden and crisp.

The *BBB* rule

The best way of remembering how to produce good choux products is to think of the *BBB rule:*

- **B**oil the water, fat and salt/sugar, without evaporating the water; this would alter the recipe.
- **B**eat the panade well and cook for 1 minute on top of the stove, stirring well.
- **B**ake the piped choux products thoroughly; if under-cooked the products will become soft and collapse. Choux pastry expands in the oven as the water and fat liquids are converted into steam, then the egg and flour proteins set to maintain the shape. As a general rule, do not open the oven for at least 20 minutes. The exceptions to this are very small items such as swan heads or petit profiteroles for soup garnish.

English pie pastry (Pâte à l'anglaise)

This uses the same basic method as choux paste (above) omitting the eggs. Boil the water, fat and salt, then stir in the flour to form a pastry. The paste is used warm and moulded into pie shapes.

Shape either by hand blocking around traditional wooden blocks, or by using a pie stamping machine (manual, semi or fully automatic). The machine operates by heating a metal dye, which is then pressed onto the paste to form the shape of the pie tin required, such as those used for individual pork pies.

Hot savoury pastry

This is another variation of boiled pastry.

Method

1 Rub the fat into the flour to form a fine crumb.
2 Add the boiling water to this flour and fat mixture, and clear to a smooth paste.
3 Allow the paste to cool, then shape according to dish requirements. It can be shaped, filled and topped for pies and savoury goods.

Brown/wholemeal pastry

Pastries can be made with wholemeal flour to increase the fibre content. Short paste works particularly well. Many pastry dishes today are made with wholemeal flour, such as quiches, individual savoury tartlettes, and meat and vegetable pies. Wholemeal flour and white flour (soft or strong) can be mixed 50:50.

- Watch your chef or baker producing puff and short pastry.
- Ask your chef to demonstrate making pastry using the flour batter method.
- Check the method of mixing and make a list for each different pastry from the range.
- Find out what types of convenience pastry products are purchased frozen for your kitchen. Which types of pastry freeze well and how long can they be kept?
- Find out what filo and strudel pastes are. How are they made and what are they used for ?
- Watch your supervisor mixing pastry on a machine. Ask what type of pastry is being made, and note the ingredients and methods used. Also notice what safety precautions are followed when using the machine.

Equipment

- Keep mixers and ovens clean: make sure you clean them thoroughly each day after production.
- Baking beans should be replaced when they begin to break up through age and use.
- All moulds should be clean and dry before storing away; this prevents rust and contamination.
- Rolling pins should be wiped clean after washing, never left to soak in water.
- If rolling pastry using a pastry brake, the machine should be stripped down and cleaned well after use.
- Check that ovens are preset and on.
- Tartlette moulds, flan rings and barquette moulds should be wiped clean while still warm with a clean cloth, not scoured by washing with a detergent or cleaning agent.
- Baking trays should be kept clean with any carbonised food removed before cleaning.
- Use silicone paper for trayed flans, quiches and baked puff goods. Palmiers and papillions are cooked on well buttered clean trays to half fry and half bake the pastry (see page 198).
- Silicone can be used a number of times to reduce costs. Replace it if it begins to break up.

Pastry types: ingredients and ratios

In the table given at the top of the next page, the weight of flour is used as the benchmark for the ratios within a recipe. The flour is therefore always given at 100 per cent, with other ingredients given in percentages of the flour; i.e. if using 500 g of flour, 500 g becomes your 100 per cent marker. So if fat is given as 50 per cent, you would need to use half of the amount of flour used, i.e. 250 g of fat.

Pastry types: uses and ingredient ratios

Pastry type	Ingredients (with per cent of recipe)	Uses
Moulding/lining pastry (*Pâte à foncer*)	Soft flour (100%), fat (butter/margarine) (50%), water (30%), pinch salt	Pasties, quiches, savoury tarts and tartlettes, tartlettes, small individual pies, baked jam roll, baked apple dumplings
French pie pastry (*Pâte à pâté*)	Medium flour (100%), fat (lard/butter) (50%, water (25%), egg (12%), salt (3%: optional)	Pâtes, croustades and timbales, game and meat pies
Sweet moulding/lining paste	Soft flour (100%), fat (butter/margarine) (50%), egg (12%), sugar (10%), water (5%), pinch salt	Petit four bases, small tartlettes, flans, barquettes, pastry bases
Sugar pastry (*Pâte sucrée*)	Soft flour (100%), fat (butter/margarine) (50%), egg (12%), sugar (25%), pinch salt	Fruit flans, sweet tarts and tartlettes, bakewell goods, lemon meringue, flans, barquettes, custard tarts, mince pies, tranche and bandes
Shortbread pastry (*Pâte sablée)*	Soft flour (100%), butter (75%), caster or icing sugar (33%)	Finger biscuits, small and large blocked shortbread biscuits, petit fours sec, gâteau and torte bases
Piped shortbread pastry (*Sablés à la poche*)	Soft flour (100%), butter (80%), icing sugar (30%)	Petit four sec, tea biscuits (plain and fruited)
Sweet almond or German pastry	Soft flour (100%), fat (butter/margarine) (60%), sugar (20%), ground almonds (15%), egg yolks (8–10%)	Linzer tart, torten bases, biscuits, petit fours, tartlettes
Choux pastry (*Pâte à choux*)	Strong flour (100%), fat (butter/margarine) (50%), water (120%), sugar (1%), salt (1%: optional)	Choux buns, profiteroles, eclairs, kidney buns, carolines, cream buns, gnocchi, religieuse
Hot water pastry (*Pâte à l'anglaise*)	Medium flour (100%), fat (lard) (50%), water (25%), salt (3%: optional)	Pork pies, veal, egg and ham pies, hand-raised pies
Suet paste (*Pâte à grasse de boeuf*)	Soft flour (100%), fat (suet) (50%), water (60%), baking powder, salt (pinch)	Steamed rolls, steamed sweet and savoury puddings, golden syrup roll, fruit rolls
Full virgin puff pastry	Strong flour (100%), fat (butter/margarine) (100%), water (50%), lemon juice, vinegar or cream of tartar (2–3 drops), salt (pinch)	Vol-au-vents, bouchées, palmiers, papillons, turnovers, cream puffs, gâteau pithiviers, jalousie
Scotch/rough puff pastry	Strong flour (100%), fat (butter/margarine) (75%), water (50%), lemon juice, vinegar or cream of tartar (2–3 drops), salt (pinch)	Sausage rolls, cheese straws, fleurons, mille feuilles, eccles, banbury, dartois, mince pies, bande or tranche aux fruits, meat/vegetable pies, pasties, meat/fish en croûte, chausson, allumettes glacés, cream horns

Preparation methods

Mixing

- *Rubbing in:* short and sweet pastry (see page 181).
- *Creaming:* short and sweet pastry (see page 182).
- *Flour batter:* short and sweet pastry (see page 182).
- *Blending/mixing:* suet pastry (see page 185).
- *Folding* by layering dough and fat layers: puff pastry (see page 183).
- *Boiling:* choux and English pie pastry (see page 186).
- *Hot mixing:* savoury pastry (see page 187).

Folding
This is for puff only (see page 183).

Kneading
This term is generally used when producing doughs. However, the initial dough made for puff pastry is kneaded with the base of the palm of the hand to produce a clear soft elastic dough. Light kneading of paste such as short or sweet paste can be given but this must be done carefully so as not to toughen the paste.

Relaxing
Most pastry benefits from relaxing (resting) before being cooked. Short type pastes can be shaped before being rested, although generally once the paste is formed it is wrapped (to prevent a skin forming) then relaxed in the refrigerator. Relaxing will not help to soften a tough or tight paste.

Pastry is best made the day before it is required for use. If pastry is made correctly, short paste items will relax within the hour, while puff paste items benefit from resting for approximately one hour prior to baking. Convenience puff pastry is widely used because of the shorter relaxing time required. All pastry items should be rested in a cool environment, preferably in the refrigerator.

Cutting
- When cutting pastry, cut it cleanly and evenly.
- Raw paste should be cut using a clean, damp knife with a sharp edge such as a cook's knife.
- If cutting/trimming paste from pies, tarts or flans, use a small office knife.
- When cutting short paste discs for tartlettes, use a lightly floured ring cutter, ensuring the cutter is floured every two or three cuts to keep the cut sharp and neat. Remember that fluted cutters have a tendency to trap paste in the gap between the ring frame and the flute; this is important to remember because contamination can occur if the gaps are left uncleaned.
- When cutting puff pastry, use a warm, damp knife or preferably, a hot oiled cutter to melt through the layers rather than pressing through them.
- Only use a lattice cutter on pastry that is cool and firm. If the paste is too soft you will have problems when trying to lift the cut latticed paste into position.
- When cutting cooked pastry (especially mille-feuille slices), use a sharp pointed cook's knife, wiping the blade between each cut.
- When preparing barquettes, cut the pinned paste into rectangles slightly larger than the boat. Alternatively, pin out the paste thinly, pick it up on the rolling pin and place it over the rows of barquette moulds. Use a small piece of paste to depress the paste into the mould, then hold two rolling pins together and roll them over the top to cut the paste. Each boat should be *thumbed up* using your thumb and index finger to finish shaping the boat. This 'thumbing up' also applies to tartlettes lined with short paste (either sweet or savoury).
- As a general rule, plain edges are used for savoury goods while fluted edges are used for sweet goods. Dishpapers and doilies are used as underliner decoration for dishes.

Rolling

- When rolling short types of pastry, the technique is to dust, pin and turn the pastry to prevent it sticking to the work bench. Only handle the amount of pastry needed for each flan or pie. Form the paste into a circle for flans, dusting lightly then pressing with your hands to flatten prior to pinning. Pin lightly, turning and dusting the pastry regularly, always keeping the rolling pin free from paste.
- Puff pastry will require a firmer pinning pressure because of the toughness of the gluten in the dough and the plastic pastry margarines used. If the pastry is too short or soft problems with pinning can occur.
- Paste of any type should be rolled with care. Puff pastry needs careful handling when rolling so as not to destroy the fine lamination built up during production. Convenience sheets of puff pastry do not require rolling as they are manufactured to be used directly. Try to avoid being heavy-handed when rolling pastry, and keep dusting and turning the paste as it is rolled, or the fat will be pressed against the bench surface and cause the paste to stick. Always roll soft paste lightly and heavier paste more firmly. Do not roll paste that is too soft or still frozen.
- When cutting a large number of shapes for tarts and tartlettes, divide the bulk paste into a number of portions, then pin, dust and turn the paste to prevent sticking.
- Always roll paste by applying even pressure to the rolling pin with both hands spread slightly. Run your hand lightly over the paste to determine the areas of thickness.
- When rolling puff pastry there is a tendency to pin a 'saddle shape', where the paste is rolled unevenly. Always pin the edges as well as the middle or the saddle effect will occur.
  Remember: pin – dust – lift – turn; keep the paste on the move and dust lightly.
- Defrost block paste or frozen pastry in the refrigerator to defrost naturally; do not force thawing in a warm environment.

To do

- Draw up a chart of faults that can occur with pastry from any of the standard text books available.
- Find out what kinds of pastry cutters are used in your kitchen. What products are they used for?
- Watch your supervisor rolling out puff and short pastry. Are the techniques used different?
- Check that you know where to store pastry for resting.

Shaping

- Shaping of flans, barquettes, tartlettes, horns, banburies, etc. requires dexterity to form the required shape. In commercial kitchens this needs to be done neatly, quickly and efficiently to ensure the batch of products being shaped are identical. The techniques will develop with practice.
- Keep work benches clean and fingers lightly dusted with flour when shaping short paste goods.
- Ensure that cream horn tins are greased and set before you attempt to shape puff strips onto the mould.
- When hand raising pies, shape the paste first on your hand before transferring the paste to the floured mould. Shape the paste so as to keep the base of the pie thicker to support the weight of the filling and

top. When shaping by crimping, use the back of a small knife and the thumb technique; i.e. use the finger and thumb of one hand and the first finger of your other hand to flute a crimp shape on pie, flan or tart edges.

- When cutting either plain, round shapes or more difficult shapes handle the cut paste with care; the shape is easily lost by heavy handling.

Glazing

The glazing of pastry dishes takes a number of forms: glazing with hot and cold process gels; melting a fine dusting of icing sugar on products in a hot oven; using an eggwash glaze; glazing with boiled apricot jam to seal fruit flans; or using a fondant glaze.

- Hot process gel is a clear gel which is boiled with a little water then brushed over dishes to both seal out oxygen and provide an attractive finish with a high sheen. Hot process gels set while they are still hot. Cold process gels are used for the same reasons as hot process gels but set when they are cold. Any air bubbles should be removed with a fine point of a knife or a cocktail stick.
- A fine dusting of icing sugar will produce a glaze. This should be carried out using a fine meshed sieve or muslin duster. The oven should be hot enough to melt the thin dust of sugar and produce a glorious shine and golden finish.
- Eggwash will glaze most pastry products. Here the pastry goods are glazed by brushing with an egg yolk and water glaze. If a little cream, salt and water are mixed with the egg yolk a high gloss brown finish will result; this is often used on puff pastry goods. Do not leave eggwash glaze out of the refrigerator when not in use. Keep it covered with cling film and only mix in small quantities.
- Apricot jam (containing fruit) can be boiled with a little water and strained to provide a good amber glaze. This will seal the dish from any oxidation, so improving the shelf life and appearance of the product.
- Fondant is used extensively to glaze and decorate cakes, pastries and gateaux. The fondant is warmed to blood temperature (37 °C/98 °F) and softened using stock syrup. To warm the fondant: cover it with boiling water, allow it to stand for 10 minutes, drain off the water then alter the consistency using the stock syrup. Do not over-heat, or sugar crystals will form and the fondant will lack sheen. Keep the fondant covered at all times to prevent a skin forming on the surface.

Preparing pastry for baking

- Remove the mixed paste cleanly onto a work bench and divide into the required weights.
- Lightly knead each piece of paste into a smooth round ball, wrap and rest.
- Roll each paste piece to the required thickness and form into the appropriate shape carefully.
- Mould the rested pastry into the required shape and tin or tray according to the product requirements.
- While pastes are resting use your time to make sure trays, flan rings and moulds or pie dishes are prepared and ready for the next stage of production.

What have you learned?

1 What are the three main methods for producing short pastry?
2 Which fat is most suitable for making puff pastry? Why?
3 How is the principle of *lamination* used in puff pastry?
4 What is the ratio of fat to flour when making short pastry?
5 Why might pastry become tough and unpleasant to eat?
6 What does BBB indicate when you are making choux pastry?
7 Why should flour be brushed off puff pastry during folding?
8 What benefits are gained by *relaxing* or *resting* pastry?

ELEMENT 2: Preparing pastry dishes

Short pastry dishes

Fruit pies

Pies can be either sweet or savoury, and may or may not have a base. Apple pie, the classic English pie, can be made with sweet or sugar pastry, preferably using the creaming method of production.

Fruit pie

Method

1 Place the fruit into a china, earthenware or metal pie dish, filling to just above the rim of the dish. Sprinkle with caster sugar.
2 Wet the rim of the pie dish, place a 2 cm (4/5 in) band of sugar pastry around the edge and seal well to the rim of the pie dish. Eggwash this rim of pastry.
3 Evenly pin out the sugar pastry until just slightly larger than the dish size. Pick up the paste onto a lightly dusted rolling pin, then place the rolling pin at one end of the dish and gently unroll the paste, laying the pastry on top of the washed pastry rim.
4 Seal by pressing the two pastes together, mark using the finger and thumb technique, brush with milk or eggwash and finally sprinkle with sugar.
5 Make a 5 cm (2 in) hole in the centre of the pie to allow the steam to escape during baking.
6 When formed, allow the pies to rest for 20–30 minutes.
7 Bake in an oven preset to 200–205 °C (390–400 °F), until light and golden brown. Insert a small sharp knife into the hole in the centre of the pie to check that the fruit is cooked.
8 Serve with a pie frill and an underliner on a doily, with sauce anglaise.

Notes

- When using apples early in the season (September–December), remember that they will be sharp and require more sugar. As apples age the natural sugars develop producing a less sharp taste.
- Other fruit pies include: apple and blackberry, cherry, blackberry, plum, rhubarb, blueberry, pumpkin, gooseberry, blackcurrant, greengage, apple and raspberry, damson and quince.

Pork pie

Pork pies

These can be hand raised using hot boiled pie pastry, moulded while still warm around a wooden pie block (either individual blocks or larger 500 g/1 lb and 1 kg/2 lb blocks). Most pork pies, however, are moulded by machine, where the paste is stamped into a pie mould by a warm dye stamp on a rotating pie stamp machine.

Gala, veal, and egg and ham pies are made by lining a hinged tin frame mould with hot water paste, then filling it with the appropriate filling to just below the top edge of the lining. The top is sealed with a strip of paste and a second finish layer of pie pastry is added and decorated. Holes are cut using a hot, oiled plain cutter and pie funnels are used with kitchen paper to absorb any excess juices during the cooking process.

Meat or vegetable pies

Unlike the pork pie, meat pies classically do not have a base. The meat (e.g. steak and kidney or chicken) is placed with some seasoning, chopped onion and stock into the pie dish. The edge of the pie dish is lined with a strip of paste some 2 cm (1 in) wide, then the whole covered with cold savoury paste, sealed well with a hole in the centre and baked.

Alternatively, you can make a slightly thickened filling first, by cooking the meat slowly, correcting the seasoning and colour, and then filling the dish. The pie is covered in the usual way and baked in a moderate oven until a light golden crust results. This method is more popularly used in Britain than the thinner version given above.

Variations

Meat and potato pies and lamb and mutton pies are also produced by this method. Vegetable pies are very popular for those on vegetarian diets, where the filling consists of a crisp seasoned mix of vegetables, bound with a light cream sauce, vegetable velouté, portugaise or other flavoured sauce.

Meat and vegetable pasties

Meat pasties must legally contain at least 12.5 per cent meat for the total weight of the pasty (Meat Pie and Sausage Roll Regulations 1967). Regulations for particular meats are as follows:

- Pork, mutton, veal and ham: total meat content 25 per cent minimum.
- Steak and kidney: total meat content 25 per cent minimum. Kidney must account for 15–20 per cent of the total steak weight.
- Meat and potato pie and pasties: the total meat content of 25 per cent refers to the raw meat before processing.

Pasties can be made using either a short savoury pastry or more generally today, puff pastry: scotch rough puff or three-quarter puff. The pasty has a special place in culinary history, as tin and copper miners would traditionally take pasties down the mines as their main meal. Opinions on the original filling vary: some maintain that the pasties were filled with fish, while others think that one end was filled with a meat and potato filling, and the other with a sweet fruit filling (generally apple).

The pasty needed to be strong enough to take down the mine and it is said that if it were dropped down the mine shaft it should not break. Today, minced beef, potato and onion is the standard filling, though pasties can be filled with other meats, fish or vegetables. Pasties are generally eaten as a snack food item.

Method

1 Place the raw minced meat, seasoning and potato in the centre of an 13–16 cm ($5^{1}/_{4}$–$6^{1}/_{2}$ in) circle of cold savoury or rough puff paste.
2 Eggwash one half of the circle of paste; i.e. on the inside of the pastie.
3 Fold the pasty in half, then seal by folding the outer edge back on itself. This is a traditional way of sealing pasties.

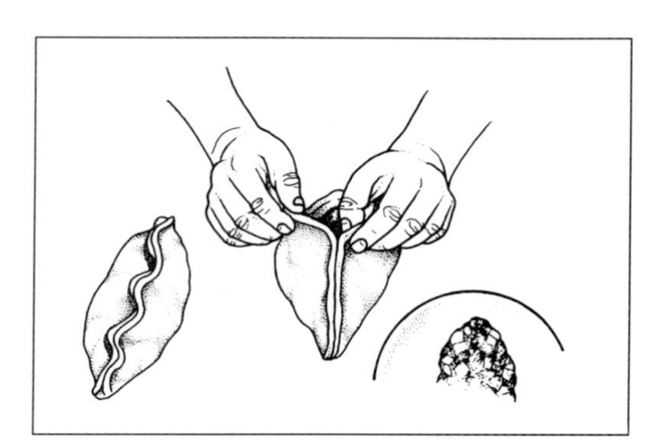

Folding the outer edges to seal the pasty

Tarts

Tarts can be open or closed, and may be cooked blind and then filled or filled and then baked depending on the dish and recipe. Fruit tarts can be made to an individual size (e.g. mandarin, strawberry, pineapple fillings), or be multi-portioned.

Open tarts such as jam, treacle and lemon curd, apricot and fruit are made by lining the flan ring or tartlette moulds with sweet short paste, then spooning in the filling. They may also use a pastry cream base (see page 180), and are usually finished with a clear glaze or boiled apricot purée.

Closed tarts (such as apple tart) are flan rings lined with a sweet or sugar paste and closed with a pastry top or lattice decoration. When cooked the flan ring or mould is removed and served whole on a round salva, with semi-whipped fresh cream or hot sauce anglaise.

A third form of tarts may be made on plates, where the cooked pastry cases are filled with fresh, tinned or frozen fruits. In this case, it is important to keep any juice to a minimum.

One of the most famous of tarts is the Bakewell tart.

Method

1 Make a flan of uncooked sweet paste.
2 Spread jam over the inside base and then pipe or spread frangipane over the jam.
3 A lattice may be used but this is not essential.
4 Bake until cooked, then brush the top with boiled apricot purée and spread on some warm fondant.
5 Decorate with piped chocolate or coloured fondant, drawing the fondant with a fine point to form the typical spiders web decoration.

Variations

Types of tarts made with sweet short paste (some may be made using puff paste trimmings): Linzer tart, Brittany tart, treacle tart, jam tart, Belgian tart, Dutch apple tart, curd tart, maid of honour tart, congress tart, conversation tart, strawberry cream tart and assorted fruit tarts.

Sweet flans

A flan is a mould of pastry usually formed inside a flan ring.

Method

1 Pin the paste (sugar or sweet short pastry) then lift it onto the rolling pin and gently place it over the flan ring. The paste should not be too thin; otherwise the flan case will collapse once filled and spoil the finish.

2 Lift the edges inside the ring and mould to shape, ensuring the paste is formed into the corners of the flan ring shape.

3 When pressed into place, trim away the excess pastry, then use floured fingers to *thumb up* the paste to form a pointed ridge. The top edge of the flan ring should be visible, so that when the paste is cooked it shrinks slightly away from the flan ring edge, aiding the removal of the ring when cool. Crimp the pointed ridge using your fingers or a crimping tweezer to produce a decorative top edge.

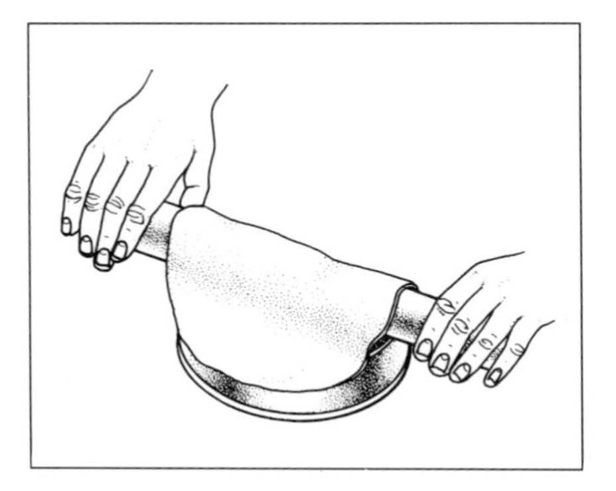

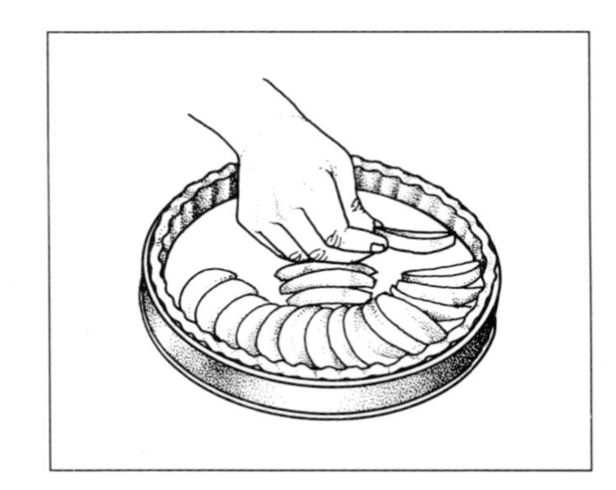

Left: using a rolling pin to lift and place the pastry over the flan ring
Right: Adding fruit to the pastry base

Preparing a flan for *baking blind*

4 Certain flans are baked *blind*. This is necessary when the filling does not require cooking or where the flan is to be filled with pastry cream and finished with soft fruits.To *bake blind:* line the flan ring and edge, lift the flan onto the baking tray, dock the base of the paste with a fork, then place a paper cartouche (3 cm/1 $^{1}/_{4}$ in larger than the flan ring size) into the flan and fill with baking beans. Cook the flan until it is set but has not yet developed colour. Once baked, remove the cartouche and beans,wash the semi-cooked flan with egg and return the flan to the oven to colour and finish cooking.

Variations

Types of flans made with sweet short paste: apple flan, Alsacienne flan, Manchester flan, Normandy flan, cherry flan, lemon meringue flan, apricot flan, Jeanette flan, pear flan bordalou.

Baked blind: banana, peach, cherry, pineapple or mandarin flans. The flan cases for these are baked blind, filled with pastry cream and topped with drained tinned fruit, decorated with boiled apricot purée, and finished with hot or cold process gel.

Savoury flans

The most famous savoury flan is Quiche Lorraine, a cheese, bacon and onion flan which can be served hot or cold as a snack, luncheon or buffet savoury. It can be made in individual sized portions or as a large multi-portioned flan.

Method

1. Line a flan ring with a good short paste that has been thinly pinned out.
2. Spread some lightly cooked (sweated) onion over the base, then place small ham or bacon pieces placed over the onion and top with grated cheese.
3. Pour over a savoury strained egg custard (see page 87).
4. Bake the flan in a medium oven, approximately 180–185 °C (340–350 °F) until the custard is set and light brown in colour. If the oven is too hot the custard will boil and the displaced water content will soak the base of the flan, producing a wet and unpalatable waxy pastry.

Essential knowledge

Time and temperature are important in the preparation of paste products in order to:

- ensure the paste is mixed and rested/relaxed for the required amount of time, in order to prevent faults in the paste products
- ensure that the pastry dishes are correctly baked or steamed
- prevent food poisoning from direct and indirect contamination
- eliminate the possibility of shrinkage
- promote and maintain customer satisfaction by the production of consistent standards of quality.

Puff pastry products

Sausage or vegetarian rolls

Sausage rolls use rough or Scotch puff pastry, as this is quicker to produce and the rolls do not need the quality of lift essential for products such as vol-au-vents and bouchées. The method given below is for producing sausage rolls in quantity.

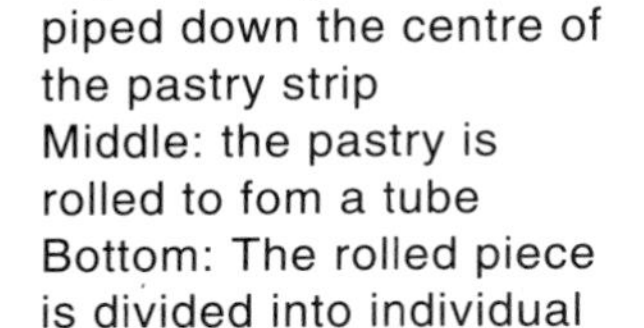

Top: sausage meat is piped down the centre of the pastry strip
Middle: the pastry is rolled to fom a tube
Bottom: The rolled piece is divided into individual sausage rolls

Method

1. Pin the puff paste into lengths of approximately 100 × 10 cm (40 × 4 in).
2. Eggwash a 3 cm (1¼ in) strip to one side of the length of pastry and pipe the appropriate filling (sausage meat or vegetable) down the centre of the pastry strip.
3. Roll from the side that is not eggwashed, forming the filled pastry into a tube.
4. Flatten the top of the tube slightly using a rolling pin and then brush it with eggwash.
5. Mark out the cutting places for the individual sausage rolls, then cut into the required length.
6. Bake in the oven at 220–225 °C (420–450 °F) for 20 minutes.

Storing uncooked sausage rolls

If not eggwashed, the uncooked sausage rolls can be frozen and stored provided the meat or vegetable filling was not previously frozen.

Vol-au-vents and bouchées

Vol-au-vent translates from the French as *puff of wind*. These are made from full virgin puff pastry given four full book/double turns or six single turns (see page 183). The vol-au-vents are cut from thin (5 mm/$^1/_4$ in thick) sheets of puff pastry and when baked form a light case for sweet or savoury food items. They are served as hot entrées with fish, meat or vegetable/salad fillings. Oval shaped vol-au-vents are used for the service of fish; usually sole or salmon.

Top: docking the pastry circles
Middle: the double layered circles ready for baking
Bottom: baked vol-au-vent cases

Method: vol-au-vents

1 Cut the prepared puff pastry into circles using a plain or fluted ring cutter which has been dipped in hot oil. This is to allow the cutter to melt through the layers rather than press through them, which would happen if using a floured cutter. Dock the pastry circles.
2 Eggwash one of the circles, then place a second circle on top and seal gently.
3 Using a smaller cutter (approximately 5 mm/$^1/_4$ in smaller) which has been dipped in hot oil, cut half way through the top disc of puff pasty.
4 Repeat Steps 2 and 3 for all of your vol-au-vents, then eggwash them. Apply the eggwash carefully to make sure that the wash does not flow down the sides of the pastry; egg protein coagulates at a lower temperature and glues the pastry, preventing an even rise.
5 For best effect and even lift, place the vol-au-vents close together on the baking tray so that they will support each other. Cover them with a sheet of lightly oiled paper, (preferably silicone) to control the lift of the paste and prevent any from *flying* (lifting unevenly while baking).
6 Place the vol-au-vents in a refrigerator to rest for about 30 minutes prior to baking.
7 Bake in a hot oven: 220 °C (430 °F) until light and golden. Remove the lid and inner part of the baked vol-au-vent while the pastry is still hot.

Bouchées

Bouchées and petits bouchées are smaller in size than vol-au-vents. Bouchées are about 5 cm (2 in) in diameter and petits bouchées are approximately 3–4 cm (1$^1/_4$–1$^1/_2$ in) in diameter. These can be made and frozen raw until required. They can also be purchased ready formed for baking or pre-baked in assorted sizes from catering suppliers.

Sweet pastries

Sweet pastries can be produced using sweet short paste, sugar paste, almond short paste or biscuit paste (see page 207). Puff pastry is used for a range of French pastries. The range of sweet pastries is vast and we have covered many of the popular categories for each paste mentioned. However, you may like to look at the sweet pastries given in the table below to extend your knowledge:

Sweet pastries made with puff pastry and sweet short pastry

Puff pastry
Dartois, Mirliton, Lampions, Allumettes, Ramequins, Mince pies, Puits d'amour (wells of love), Turnovers (*chaussons*), Palmiers (pigs' ears), Batons glacés (Iced stick), Papillions (butterflies), Cream horns, *Mille-feuilles* (thousand leaves)

Sweet short pastry
Congress tarts, Sablés à la poche, Fruit barquettes, Fruit tartlettes, Egg custard tarts, Honey and almond barquettes, Tartlette Lorraine, Tartlettes Evesham, Tartlettes forestaire, Jam tarts, Curd tarts, Treacle tarts, Mince pies

Suet pastry dishes

Suet pudding

Suet pudding is made with suet paste (see page 185). The following method is for a sweet pudding, although savoury versions can be made (e.g. steak and kidney pudding).

Method

1. Mix the paste lightly until firm.
2. Roll out the paste to form a rectangle 30 × 20 cm (12 × 8 in) to a thickness of 5 mm (1/4 in). Spread the jam over the surface, leaving a 5 cm (2 in) strip down the length of one side. Do not over-handle the paste, or it will become tough.
3. Roll up the pastry and place it in a buttered and sugared steaming sleeve.
4. Steam for approximately 60–90 minutes. Note that if the steaming temperature is too low the paste will be heavy and soggy.
5. When cooked, remove from the sleeve and serve with a jam and/or custard sauce.

Variations

Fruit can also be used in the suet roll, such as apple, sultanas or currants. Suet paste is also used to make dumplings, where small pastry pieces are poached in the sauce of the stew or plain boiled in salted water.

Convenience pastry dishes

The range of convenience pastry products is considerable and these assist in the production of cost effective quality products. The range of frozen pre-formed pastry products such as tarts, pies, flans and cooked puff pastry shells available both ready to bake and pre-cooked make the task of producing pastry products easier at times of severe skill shortage.

The provision of such goods enables the contemporary craftsperson to concentrate on the flavour and decorative display of food items, without having to spend long hours in production or investing in a larger variety of specialist equipment. Puff pastry made with butter has no equal when freshly made and baked, but the place of convenience pastry products is well established and must be used alongside the skill and knowledge of traditional craft and contemporary innovations.

Key points: cooking pastry dishes

- When the products are rested/relaxed and docked where necessary, handle them carefully then place into the preset oven for the required period of time.
- While pastry dishes are baking or steaming, clean and clear work surfaces ready for the next batch of work.
- Check the length of time pastes and products have been relaxing or cooking: overcooked products might mean you will have to start again.
- Transport boiling liquids such as hot jam glazes with care and attention.
- Check oven temperatures carefully: if the oven is not at the required baking temperature goods will have a poor colour and the pastry product will lack volume and be dry (especially puff pastry items).

- Always ask about the oven setting for each individual pastry product: pastes containing more sugar need lower temperatures than savoury goods. This is because the sugar caramelises and colours the products, particularly those with external sugar coating such as puffs, cornets, palmiers and papillions.
- Check products that are being baked blind. Do not over-cook them before applying the eggwash.
- Try to fill the oven space when baking by planning your production carefully; this saves time, energy and money.
- Remove baked/steamed products carefully and cleanly, making sure you have cleared a space to place the hot goods for hygienic storage while cooling.

Essential knowledge

It is important to keep preparation and storage areas and equipment hygienic, in order to:
- comply with food hygiene regulations
- prevent the transfer of food poisoning bacteria to food
- prevent pest infestation in preparation and storage areas
- prevent contamination of food commodity items by foreign bodies.
- ensure that standards of cleanliness are maintained.

Finishing methods

The finish to any pastry product is important for a number of reasons. All products should be the same size, finish and shape for consistency of standard. The finish should appeal to the eye but still taste good. Aim for decoration that is delicate and balanced, where the colours are not too bright and any piping is neat and tidy. Strive for an artistic balance, being careful not to overdo the decoration, which would detract from its saleability.

Dusting

Dusting is the sprinkling of icing sugar onto a pastry product. The dusting should be light and delicate but enough to decorate the top of the particular dish. The sugar may be flavoured; for example, with sugar.

A fine sieve or shaker is normally used but you can achieve a very good end result by using some muslin formed into a puff ball for fine finishing. This is especially good for the famous Gâteau Pithiviers, where a fine glass finish is only achieved by the fine dusting of sugar before the gâteau is placed into a very hot oven.

Piping

Learning to pipe well takes practice. Piping any product requires skill both in holding the bag and controlling the flow by hand pressure to create the particular design required for each dish. The following points are important to remember:
- fresh cream should not be too stiff for piping or the buttermilk will run, especially in a warm environment such as a kitchen, pastry or bakery
- chocolate should be thickened with either a drop of stock syrup, a spirit such as brandy or (best of all) a few drops of glycerine. Melt the chocolate to approximately 45 °C (110 °F) and stir in the glycerine. It will take a few minutes for the chocolate to thicken. Half fill a paper cornet bag and seal well, always folding into the centre to seal the liquid chocolate in the bag

- fondant also provides a useful medium for piping. Chocolate motifs can be made when you are not busy and stored in a cool place for decorating trifles, torten etc.

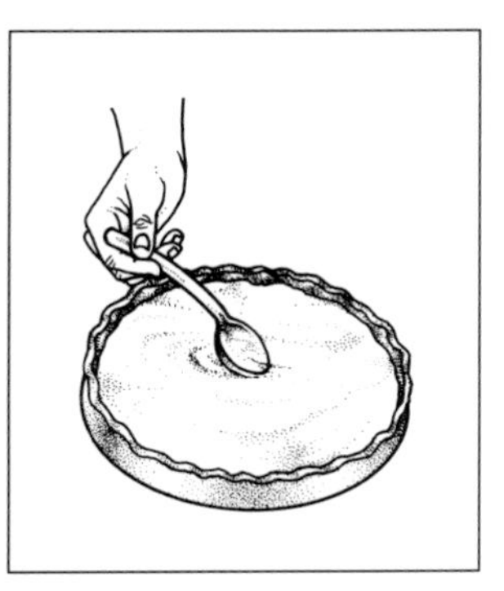

Filling a baked flan

Filling

Note the following points:

- make certain that all products have the same percentage of filling. This should not be so much as to cause the pastry to burst or the filling to spill over, so spoiling the finish and undermining professional standards. Leave room for expansion, especially with products like turnovers, pasties, pithiviers and puffs
- when filling with fruit, make sure the design produced is even in shape and colour, and balanced to make the dish attractive. Use complimentary colours and sizes. These factors are very important for customer satisfaction and therefore sales
- when filling with cream, try to pipe an even design using star pointed tubes. Remember that star tubes have different numbers of teeth for a variety of piped designs, as do royal icing tubes.

Finishing, presenting and storing pastry products

- Ensure the cooked products are cool before finishing, dusting or filling. Finish by filling, dusting piping and/or glazing products according to requirements.
- Present and store the finished cooked products according to the *Food Hygiene (Amendments) Regulations 1993.*
- Finishing is very important; each pastry product should be the same size, colour and shape.
- Fondant icing, jam and clear glazes should be thin and glossy. Decorating media should highlight the product, be attractive and simple.
- Cream, fruit or custard fillings should be added in a neat and clean way. Remember that too much filling makes the product difficult to handle, while too little might not maintain customer satisfaction.
- Piping of cream, chocolate and fondant is a skill that develops with practice; aim for a product that looks clean and tidy.
- All equipment should be hygienically cleaned before and after use, to avoid contamination.
- Carry out creaming of cooled baked pastry products in a cool area of the kitchen. Check that you have used fresh cream in rotation and that it is whipped cleanly.
- *Never* use cream to fill a product that is still warm.
- Ensure cream is always whipped correctly, in a hygienic manner, and that cream-filled products are stored according to the *Food Hygiene (Amendments) Regulations 1993.*
- Keep alert when whipping cream: it can easily be over-whipped and formed into butter. Cream is an expensive commodity, and should always be used carefully and cleanly.
- Check that piping bags have been sterilised; they are a possible source of cross-contamination.
- Glazing with hot jam and hot gels is dangerous; handle pans carefully.
- Make sure fondant is heated gently over a period of time to a temperature not exceeding 48 °C (120 °F). Spread it thinly over products and decorate with decorating media while the fondant is still sticky.
- Water icing should not be made up too thinly; it should hold onto the product to provide a glossy coating.

- Dusting with icing or caster sugar should be done evenly using a shaker, sieve or muslin bag.
- Handle and store products carefully once finished until required for presentation.
- Make certain storage trays are clean and stacked safely.

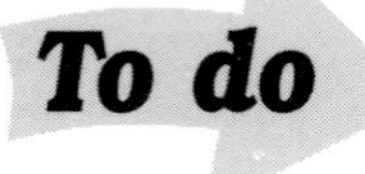

- Check the storage temperature of pastry products that are fresh and frozen.
- Watch your chef glazing pastry products. Notice the different techniques used for the different types of glaze.
- Practice dusting using a fine sieve, a shaker or muslin.
- Look at some filled pastry products that are ready for service. Notice how the filling is carefully added to complement the product, in colour, texture and amount.
- Watch your chef whipping cream. How do they know when it is whipped to exactly the right consistency?

Planning your time

Refer to *Planning your time* (pages 1–4) concerning general points that you should always refer to when working in a kitchen or bakery. When preparing paste and pastry dishes the following points are especially important to remember:

- Work in a clean and organised manner attending to any priorities and laid down procedures.
- Know what your responsibilities are. If in doubt, ask your supervisor.
- As you work clean up after yourself. Wipe down between each task or job. Do not use glass or oven cloths to wipe down work surfaces, as this causes contamination.
- Use spare time while pastes are relaxing or products are cooking to prepare for the next task. Check that all production requirements have been or will be met. Notice whether you need to clean down work areas or equipment.

Disposing of waste

Waste materials should be disposed of cleanly and efficiently, to prevent contamination of uncooked and cooked pastry products and ingredients and to prevent health hazards.

Dealing with unexpected situations

An unexpected situation may be anything from a rush customer order to an accident in the kitchen. Make sure that you are familiar with all the points in Unit G1 of the Core Units book: *Maintaining a safe and secure working environment*. Be aware of what your personal responsibilities are, and any laid down procedures for coping with unexpected situations.

What have you learned?

1. What are the main contamination threats when preparing, cooking and storing pastes and cooked pastry products?
2. Why is it important to keep preparation and storage areas and equipment hygienic?
3. Why are time and temperature important when preparing and cooking pastes and pastry products?
4. Why might you *bake blind* when preparing a flan?
5. Why is care needed when preparing fondant for glazing?
6. What considerations need to be taken when finishing pastry products?

Extend your knowledge

1. Read up on other types of pastry such as strudel and filo pastes.
2. Try making paste with different types of flour such as wholemeal or wholewheat. What differences are there compared to using soft bleached flour ?
3. Examine the range of tea fancies, gâteaux and tarts that use short and puff paste as a base or decoration.
4. Find out how blocked shortbread is produced, annd what shortbread is usually served with.
5. Discuss with your supervisor some of the faults that can occur with pastry and how they can be prevented.
6. Look at other text and recipe books and compare recipes. Try different recipes to develop your competence and understanding in the preparation, production, finishing and presentation of fresh pastry and pastry dishes.
7. Read up on other products and dishes using choux pastry, such as kidney buns, Carolines and polka.

UNIT 2D9

Preparing, cooking and decorating cakes and biscuits

ELEMENTS 1, 2 AND 3: Preparing, cooking and decorating cakes and biscuits

What do you have to do?

- Prepare appropriate baking equipment ready for use.
- Prepare the sponge, biscuit or cake ingredients correctly for each individual product.
- Prepare the fillings, icings or toppings correctly for each individual product.
- Mix the prepared ingredients according to the appropriate product method.
- Bake cake, biscuit or sponge products for the required time and at the correct temperature.
- Identify faults during preparation, baking and cooling of cake and sponge mixtures.
- Finish, present and store baked and finished products according to the product requirements, meeting food hygiene regulations.
- Dispose of waste in the correct manner according to laid down procedures to ensure high standards of hygiene.
- Carry out your work in an organised and efficient manner taking account of priorities and laid down procedures.

What do you need to know?

- The type, quality and quantity of ingredients required for each mixture, filling or topping and how to prepare it.
- How to prepare cakes, sponges and biscuits correctly for baking.
- The correct procedures for thawing and cooking frozen cakes and biscuits; in order to prevent contamination.
- What the main contamination threats are when preparing and cooking cake and biscuit products.
- Why it is important to keep preparation and storage areas and equipment hygienic.
- Why time and temperature are important in the preparation of cake and biscuit products.
- How to satisfy health, safety and hygiene regulations concerning preparation areas and equipment both before and after use.
- How to plan your time to meet daily schedules.
- How to deal with unexpected situations.

Introduction

Cakes, biscuits and sponges and the products made from each range make up the sweet dessert, confectionery and snack elements of our daily diet. In many ways, these items are not conducive to a balanced nutritional diet but do, however, meet the requirements of a modern day fast food market and life style. We can enjoy the modern skills and vast production capabilities of supermarkets, bakeries and patisserie shops, and tea and coffee houses through an enormous range of products.

The term cakes covers a range from light fruit cakes to rich wedding and celebration cakes; and individual cup cakes to large layered cakes such as high ratio angel or Battenburg. Sponges form a light alternative to cakes and also come in a variety of forms. They can be produced from egg foams (such as the range of genoese sponges), creamed mixtures (such as the Victoria sponge), or blended batters (such as high ratio sheet sponge).

Biscuits come in many shapes, sizes, flavours and finishes and vary from country to country. Regional specialities are common, with recipes indicative of an area's particular tradition and practice: such as short-bread from Scotland, tuilles from France, cookies from the USA and rout biscuits from Holland.

Supermarkets, specialist wholesalers and retailers to the bakery, catering and hospitality industries are able to offer a vast range of cake, sponge and biscuit products in various forms: fresh, frozen and ready to mix. These convenience products are high quality goods with an extended shelf-life due to advanced technical preservation methods.

Cakes, sponges and biscuits can be very complicated or extremely simple. All of them require a thorough knowledge of preparation and production techniques, and skill in formation and artistry, in order to produce a decorative, visually appealing pastry.

ELEMENT 1: Preparing cake, biscuit and sponge mixtures

Preparing cake mixtures

Cake ingredients

Flour
Use a soft, low protein flour when making cakes. A proportion of the soft flour can be replaced by cornflour. Always sieve any flour before use and mix in any spices, powders or dry ingredients at this stage.

Sugar
To ensure a good batter, use caster sugar, which has a fine grain and good dissolving properties. Rich fruit and celebration cakes can be made with soft light or dark brown sugar to enhance colour and flavour.

Fat
Butter provides a good flavour and imparts a natural yellow colour to cakes, but has poor creaming properties. Use unsalted butter if possible. Cake margarine and shortening agents trap more air in the cake batter

and are manufactured to have very good creaming qualities. They produce quality cakes with good volume, an even structure and a fine and balanced crumb and texture.

Butter and cake margarine can be used together to balance the cost and aeration qualities: in this case, use 50 per cent butter and 50 per cent cake margarine and/or shortening.

Dried fruit

Many cake mixtures contain fruit, mainly dried. Such fruit needs careful preparation to ensure that the resulting cake is moist with the fruit evenly set throughout the baked cake. Wash and drain any dried fruit well; hard dried fruits such as currants, raisins or sultanas can be soaked in hot water for 10 minutes but must be left to drain for at least 12 hours.

Once drained, lay the fruit out on a clean working surface and check for stalks, stones and foreign bodies. Dried fruit is expensive to buy and many cheaper brands are available, but buying these is false economy as they often contain more foreign bodies. Good quality dried fruit is more expensive but reliable, and will take less time to check.

Methods of producing cake mixtures

There are three methods for making cakes: the sugar batter method, the flour batter method and the blending method.

When producing cake batter mixtures the temperature of the batter is crucial; aim for a working temperature of 21 °C (70 °F) by ensuring the ingredients are at room temperature.

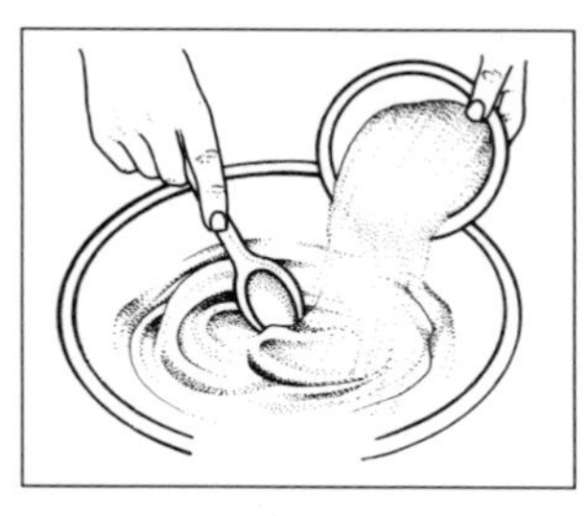

Creaming the butter and sugar

Sugar batter method

1 Use a machine paddle to beat the cake margarine, butter or shortening to a light emulsion with the caster sugar. Scrape down the sides of the mixing bowl at frequent intervals to incorporate any fat and sugar that has not been mixed.
2 Add any colour, essences or flavourings to the light batter.
3 Add the liquid egg in a steady flow over approximately 2 minutes. If the eggs are added too quickly the mix can curdle, in which case a little flour can be added to bring the mixture back.
4 Carefully add the sieved flour to the batter mixture, mixing it in slowly to produce a smooth clear cake mixture, free from lumps and not over-mixed.
5 Add any other ingredients (e.g. milk, fruit and nuts), blending them in slowly to distribute them evenly throughout the batter.

Cakes: production methods and examples

- *Sugar batter method:* Dundee, Slab cake, Angel cake, Fruit cake, Cup cake.
- *Flour batter method:* Madeira, Slab cake, Angel cake, Fruit cake.
- *Blending method:* High ratio genoese, Slab cake, Angel cake, Fruit cake, Madeira.

Flour batter method

1 Blend in any dry commodities (such as cocoa, coffee powder, ground almonds or baking powder) with the flour.
2 Whisk the eggs and sugar to a half sponge: this is to foam the two ingredients until half the potential volume is achieved.
3 Add any ingredients providing colour and/or flavour.
4 Cream the cake margarine with an equal quantity of flour.
5 Fold the half foam mixture into the fat and flour batter in three or four equal portions, folding and blending each portion carefully.
6 Lastly, fold in the residue of the flour to produce a smooth cake mixture. Make sure all other ingredients (such as fruit, milk or nuts) are blended to distribute them evenly throughout the cake mixture.

Glycerine is usually added at Step 2 to assist in moisture retention during baking as it has high hydroscopic properties (i.e. it attracts moisture).

Blending method
This applies to the making of high ratio cake mixtures. Use a special flour (high ratio cake flour) which has been fractionated to very small particle size to increase its surface area and its ability to absorb liquid. The fat for this method must be a high ratio fat or shortening, i.e. a very soft white fat that has been fully hydrogenated and superglycerinated. This will hold and balance the high level of liquids in the form of milk and dissolved sugar, giving this type of cake a fine stable crumb, good eating quality, extended shelf life and excellent freezing qualities.

Prepare a high ratio cake as follows:

1. Sieve together the flour and dry ingredients (e.g. baking powder).
2. Use a mixing machine to mix the fat and flour to a crumbly texture.
3. Dissolve the sugar in the milk and add it to the mixture slowly (taking approximately 1 minute) on slow speed.
4. Scrape down the bowl, then mix for a further 2 minutes on medium speed.
5. Lastly, add the liquid eggs in the same way, i.e. add them slowly over 1 minute, scrape down well, and mix on medium speed for 2 minutes. It is essential that each stage of batter is blended into the next to produce a smooth batter free from any lumps.

Using a machine for mixing
When beating a cake mixture, use the paddle attachment on a medium speed. When blending, use the paddle attachment on a slow speed. When blending high ratio mixes, always clear the mix from the bottom of the machine bowl to ensure any first and second stage batter does not remain in the bowl.

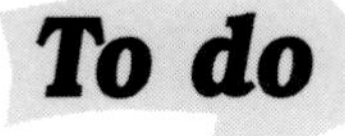

- Find out what sultanas, currants and raisins are made from and how they are made.
- Ask your supervisor about high ratio recipes and products made from them.
- Find out where the mixing machines are kept in your kitchen. Make sure you know how to use and clean them correctly.
- Find out why fruit cakes are so costly. Investigate the cost of celebration cakes from bakery and cake artistry outlets in your area.

Preparing biscuit mixtures

Biscuits can be sweet, savoury or unsweetened. The term *biscuit* means *twice baked*, although commercial biscuit manufacture is subject to a legal definition of the term. In catering, a range of biscuits are produced as garnish or snack foods, petit four bases or main confectionary items. We will deal here with those biscuits produced for catering and bakery operations but not those which are traditionally the practice of the biscuit industry.

Biscuit ingredients

Flour

Use a flour that is soft with a low protein content. Special biscuit flour is available but plain or soft type flour is adequate for many biscuit recipes. Where a recipe suggests the use of a medium flour, use a mixture of soft and strong flour in equal quantities. Always sieve flour with any spices or dry ingredients (such as ground almonds) before production begins.

Sugar

Biscuits are best made with fine caster sugar because of the low moisture content, although brown or icing sugar is used for certain recipes. Syrup and treacle are commonly used in ginger biscuit recipes, where they should be warmed slightly to allow them to flow. When weighing glucose (as sugar or syrup), use a wet scale pan with a cold wet scraper to prevent sticking.

Butter

Butter provides the flavour base especially noticeable in shortbread biscuits, and gives a texture also accentuated in shortbreads: a melt-in-the-mouth, short texture. Biscuits can be made with cheaper margarine and shortening agents but these will not have the richness that butter provides. Some biscuits do not use fat at all: such as macaroons, champagne and cuiller biscuits.

Weighing biscuit ingredients

Where recipes use light powders or spices these should be measured and weighed carefully. If possible, use an electronic balance, as most cheaper mechanical scales tend not to be balanced for very fine measurement.

Methods of producing biscuit mixtures

There are five methods for producing biscuit mixtures: the sugar batter method, the flour batter method, the blending method, the foaming method and the rubbing-in method.

Biscuits: production methods and examples

- *Sugar batter:* Palettes des dames, Langues des chats, Langues des boeuf.
- *Flour batter:* Cookies.
- Blending: Decor, Brandy Snaps, Ratafia, Russian Cigarettes, Tuilles.
- *Foaming:* Biscuit à la cuillère, Biscuit perles, Biscuit champagne, Japonaise.
- *Rubbing in:* Shortbread, Shrewsbury, Wine biscuits, Ginger biscuits.

Sugar batter method

1. Beat the butter and sugar to produce a light cream.
2. Mix in the eggs and beat well.
3. Add the sieved dry ingredients, then blend carefully to produce a clear biscuit mixture.

Flour batter method

1. Cream half the flour with the butter.
2. Mix the sugar and eggs together, then beat into the flour and fat mixture.
3. Fold in the sieved dry ingredients and blend to a smooth biscuit paste.

Blending method

In certain biscuit recipes the ingredients are simply blended together to form a paste, e.g. decorating biscuit paste (made from flour, egg white and icing sugar). These pastes do need to be rested once mixed. Brandy snap biscuits can be made either in this way or by using the creaming method.

Foaming method

This method can used to produce either a single whites foam (where a meringue is made, as for japonaise) or two foams: one of egg yolk and sugar and another of egg white and sugar (as for biscuit à la cuillère). The two-foam method is given under *Sponges* on page 210. To produce a single egg white foam, follow the method given below.

1 Whisk the egg whites with a portion of the sugar from the recipe to produce a light meringue.
2 Fold in the remainder of the sugar, the cornflour and any ground almonds or hazelnuts. Nuts contain oil which will break down the foam structure, so when folding in the nut content of japonaise, do it very carefully, capturing large pockets of air by cutting and folding gently. Do not stir the mixture.

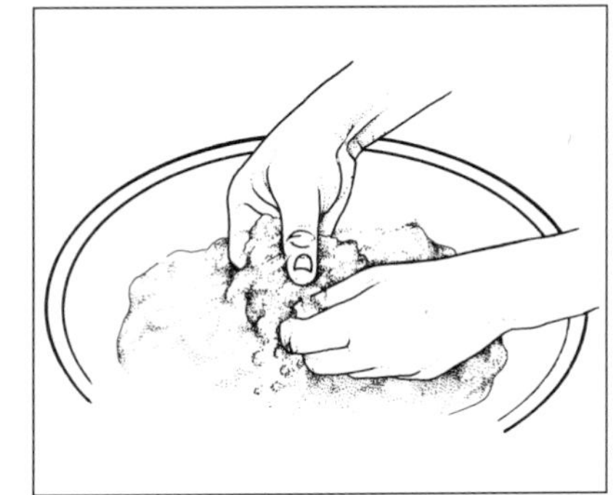

Left: whisking by machine
Right: Rubbing in

Rubbing in method

The rubbing in method for biscuits is the same as for producing basic short and sweet pastry.

1 Rub the fat into the flour, using either your hands or a mixing machine to form a fine crumb.
2 Add the liquid and sugar ingredients.
3 Clear the mix to produce a smooth biscuit paste. Do not over-work the paste, as this would toughen the mixture.

To do

- Find out why most biscuits are made by specialist producers.
- How are biscuit and biscuit products best stored? Where are they stored in your kitchen?
- Draw up a chart of petit four sec products made from biscuits or biscuit bases. How many can you find?

Preparing sponge mixtures

Sponge mixtures can be light medium or heavy, and are usually produced from a foam of eggs, sugar and flour. The eggs may be used whole or separated. Sponges form the basis of gateaux, torten, sponge fingers, sponge drops, Swiss rolls, roulades and other sheet and layer sponge products.

Sponges tend to be light aerated mixtures where the egg (white) content of the recipe traps air bubbles by forming a semi-rigid membrane structure. As the eggs and sugar are whisked, the liquid mixture thickens, until it reaches peak volume, when the sieved flour is *cut in* using a stainless steel spoon. When folding or cutting in the flour it is essential to lift the flour (preventing it from reaching the bottom of the mixing bowl), without stirring the mix. If the flour falls to the base of the mixture it will become lumpy and difficult to clear.

Sponge ingredients

Flour
Flour should be soft and well sieved. When producing coffee or chocolate sponges, sieve the powder two or three times to blend the two powders evenly.

Sugar
Caster sugar is ideal for making foam sponges because of its aeration properties and fine grain. Remember that sugar can become damp, developing a crust or lumps; if so, warm it in the oven and then sieve it to produce a free-flowing fine grain.

Eggs
It is important to use fresh hen eggs for making sponges. As eggs age the white can become watery (the egg shell is porous), and this can affect the volume of the foam. Once eggs have been taken from the shell, they should be used immediately, or the yolk will develop a skin.

Glycerine
This is a clear, syrupy, colourless, odourless liquid which is added to sponges to prevent them from drying out. Glycerine is very hydroscopic (i.e. it attracts moisture).

Stabilisers
These are added to sponges to help prevent them from collapsing. Types such as EMS (Ethyl Methyl Cellulose or Edifas) or GMS (Glycerol Monostearate) are commonly used. These are available as flakes, powder or a ready prepared gel and they are added to the eggs and sugar prior to whisking. Egg yolk (lecithin) is a natural stabiliser, and its main use in sponges is to stabilise the foam.

Methods of producing sponge mixture

Foaming method
Prepare any equipment such as trays, tins and moulds for sponge making before mixing begins. Clean all equipment such as bowls and whisks thoroughly; sterilise if possible. This is to remove any traces of fat or grease which would prevent the egg from trapping air.

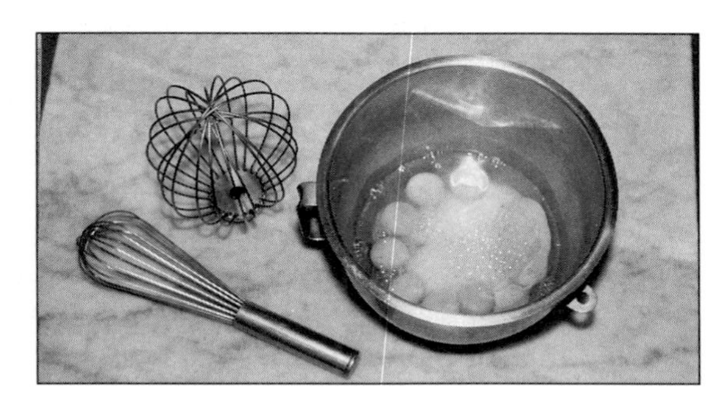

Eggs and sugar prepared for whisking

1 Weigh the ingredients carefully then mix the egg and sugar together in the machine bowl using a hand whisk.
2 Warm the mixture to 32 °C (90 °F). This can be done by placing the bowl over a pan of boiling water or by heating the sugar beforehand on a tray in the oven. Do not overheat the sugar and egg mixture, as this might cause the eggs to begin to coagulate and cook.
3 Whisk the warm eggs and sugar on Speed 3 until they are cool and the foam is light with a thick creamy consistency. Be careful not to over-whisk the foam: when dropped on itself the foam should leave its own mark. This stage is known as the *ribbon stage*. Any colour or flavour is added at this stage.
4 Sieve in the flour so that it is spread evenly over the top of the foam, then carefully fold in the sieved flour and dry ingredients by hand using a stainless steel spoon. It is important to keep cutting to prevent the flour from reaching the base of the bowl (this would cause lumps to occur). Cut and fold by turning the mixing bowl with one hand and cutting from the centre base of the foam in a circular motion upwards.
5 When the flour is clear, always tap the spoon, as there is usually a pocket of flour trapped in the spoon head.

Sponges: production methods and examples

- *Foaming:* Swiss roll, genoese, plain sponge.
- *Melting:* French genoese.
- *Boiling:* Genoese.
- *Blending:* High ratio sheet, cup and layer sponge.
- *Creaming:* Victoria.
- *Separate yolk and white:* Sponge fingers, drops and sacher.

Melting method
This follows the same procedures as the foaming method, adding melted clarified butter after the flour has been folded and cut (Step 5). The mixture is then carefully blended to incorporate the melted butter. The addition of the melted butter enriches the sponge, improving the flavour, texture and crumb structure and extending the shelf life of the sponge. French genoese sponges are made using this method.

Boiling method
Boiled genoese sponges have a stable crumb texture that is stronger than foam type sponges, crumbling less when cut. They will also keep longer than plain foam sponge mixtures. This method is particularly suitable for petit fours glacé.
1 Mix the eggs and sugar together, then heat to 37 °C (98 °F) and whisk to a thick sponge.
2 Place the butter into a mixing bowl and heat until boiling.
3 Stir the flour into the glycerine and beat to form a smooth paste.
4 Add the egg and sugar sponge to the butter and flour paste in 3 or 4 portions, beating each portion well to ensure a clear, smooth batter.
5 Pour the mixture into a lined baking tin or sheet according to the recipe and product requirements. Bake this type of mixture at 190 °C (375 °F) for approximately 35–40 minutes.

Blending method
High ratio mixtures can be used to prepare sponges following the same procedures as for cakes (see page 208). The method of blending the batter is the same, and, as before, it produces a sponge with a fine stable crumb, even texture, good shelf life and freeze-thaw stability. Many catering operations do not use such mixtures but they are very easy to produce if the mixing rules are followed. They will also keep well for a long time if stored correctly because of their high liquid sugar content.

Many of the confectionery cakes made with sponge are of this type. High ratio flour and high ratio fat must be used according to recipe specifications.

Creaming method
Victoria sponge, a traditional mixture where two sponges are sandwiched together with jam or jam and cream or butter cream, is produced by this method.
1 Sieve the flour and baking powder are sieved together.
2 Cream the butter and sugar together, adding any glycerine if required.
3 Add the eggs to the well-creamed butter and sugar.

Separate yolk and white method
Continental sponges are made by using the egg yolk and egg white separately, such as biscuit à la cuillière, biscuit perlés, sponge drops, and sponge brickettes for trifle. This sponge needs careful preparation.

All equipment should be to hand and trays or tins prepared, either with silicone paper or by being buttered and floured before production begins.
1 Beat the egg yolks well with half the sugar to form a light mixture.
2 Mix the egg whites with the rest of the sugar to make a meringue.
3 Carefully and lightly fold half of the meringue with the egg yolk and sugar mixture.
4 Sieve all of the flour onto this mix and carefully cut in.
5 Finally, fold in the rest of the meringue to lighten the mixture. As much volume as possible should be preserved by careful folding and cutting using a stainless steel spoon. If this mixture is handled by a stirring or vigorous action when folding, the mixture will collapse and become soft.
6 When mixed, the mix can be piped to form sponge fingers, sponge drops, or sponge shapes depending recipe requirements.

This type of sponge is very useful for many different pastry dishes. Charlotte russe uses sponge finger biscuits; Othello, Desdemonas, Jagos and Rosalinds are made from sponge drops sandwiched with fresh cream or pastry cream, coated with boiled apricot jam and covered with warm fondant. Special sponges such as the *Sacher torte* rely on a separate yolk and white sponge mixing method to produce a light delicate sponge.

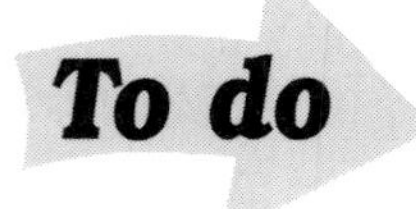

- Find out what range of premix sponges are available.
- Investigate the types of stabilisers used in your kitchen; how do they work and why?
- Watch your supervisor mixing a chocolate-based sponge. Why is particular care needed?
- Watch a genoese sponge being prepared in your kitchen. How far in advance of service is it made? Why?

Equipment and work areas

- Check that ovens are preset and on.
- Ensure that your work bench is clean, clear and all specialist tools are ready to hand.
- Check that cutters, cake tins, hoops and baking trays are clean and prepared according to the product requirements.
- Check that your scales are accurate and meet existing regulations.
- Sterilise all machine bowls and attachments before beginning production.
- Line cake tins, frame or hoops with a double or triple layer of paper.

- Place a water tray in the oven for plain or fruited cakes to moisten the baking atmosphere.
- Check that cooling wires and racks are set to receive baked products for cooling.
- Ensure mixers used for the production of mixes are set up correctly and that attachments and mixing bowls are safely attached according to the manufacturer's instructions. Never use wet hands near electrical machines.

Preparing equipment and ingredients for sponges, cakes and biscuits

Preparation: points to remember

- Check that all ingredients are to hand, with all the small tools ready for use: such as spoons, whisks, scrapers and bain-maries.
- Always sieve flour to remove any lumps, and blend any powders or spices at this stage.
- Make sure eggs and fats are at room temperature.
- Check that you have weighed the ingredients according to the recipe and dish requirements. Any errors here are costly in both time and money: you may waste valuable ingredients, produce a flawed product and miss schedule deadlines.
- Wash, drain and dry fruit for fruited cakes. Check for stones, stalks and foreign bodies by laying the fruit on a clean bench.
- Check that piping bags and tubes are clean and that bags have been sterilised.
- Make sure that the mixtures are mixed according to production method/s and for the required time, and that high ratio blending times are adhered to.
- Follow the rules for each production method carefully.
- Make sure that all the products produced for baking will fit into the available oven space. Check the baking times for each product.

Mixing: points to remember

- Never leave pastry brushes in the fat while it is melting (to be used for greasing tins or trays).
- Always scrape down the sides of the mixing bowl when creaming mixtures.
- Beat cake batters and mixtures on a medium speed.
- Test for ribbon stage when whisking foams. Do not over-whisk egg whites.
- Tap the spoon at the end of folding in to release the trapped flour pocket.
- Do not over-rub the fat when producing biscuit pastes.

What have you learned?

1 What are the main contamination threats when preparing cake and biscuit mixtures?
2 Why is it important to keep preparation and storage areas and equipment hygienic?
3 What are the three methods used to produce cakes? Describe each of them in detail.
4 What are the primary ingredients for sponges and what are the main methods used to prepare them?

ELEMENT 2: Cooking cakes, sponges and biscuits

Equipment: cake tins and hoops

Cake tins come in all shapes and sizes, including heart, oval, hexagonal and horseshoe shapes, but the round or square varieties are commonly used. Traditionally, wooden frames were used for square and slab fruit cake, but cake tins today are made from a light steel or aluminium.

Cake tins, hoops and baking trays should always be lined with paper to protect fruit cakes during baking. If you are using a hoop (which has no base), cover the baking tray with two or three sheets of paper and place silicone or greaseproof paper on top. Line the cake hoop in the same way, using fat to glue the layers of trimmed paper into place. Another way to prepare tins, trays and moulds is to grease the tin with fat and dip it immediately into flour; this provides a protective lining for the sponge or cake.

If sponge or cake tins are washed or scoured, the film which prevents the mixture from sticking is destroyed and the cake or sponge will stick. After you have turned out the cakes or sponges, simply wipe the tin or hoop clean and store it hygienically.

Points to remember when baking cakes, biscuits and sponges

- Check biscuit bases to test for cooking.
- Test sponges with your finger (the sponge should feel springy) or a needle (it should be clean when withdrawn from the sponge) to see if the centre is baked.
- Cover fruit cakes with paper once the surface is sealed and coloured.
- Once biscuits are rested/relaxed and have been docked or decorated where necessary, handle them carefully then place them into the preset oven for the required period of time.
- Spread or tin sponge mixtures carefully and then place them in the oven immediately. Be careful not to knock light sponge mixes once they are in the oven. Do not slam oven doors.
- Try to fill the oven space when baking by planning your production carefully; this saves time, energy and money.
- Check the length of time products have been baking or cooling; over-cooked products may mean that you have to start again.

- Check oven temperatures carefully: if the oven is not at the required baking temperature products will have a poor colour, lack volume and be dry; or, in the case of cakes and sponges, develop *M* or *X* faults.
- Always ask about the oven setting for each individual cake, biscuit or sponge product; recipes containing more sugar need lower temperatures than savoury goods. This is because the sugar caramelises and colours the products, particularly those with external sugar or nut coatings.
- Remove baked goods from the oven carefully and cleanly.
- Make sure you have cleared a space for turning out the hot goods and for storing them hygienically while they are cooling.

Essential knowledge

Time and temperature are important in the preparation and baking of cakes, sponges and biscuits in order to:

- ensure that the sponge, cake or biscuit mixture is mixed for the required amount of time in order to prevent faults during baking
- ensure that the sponge, cake or biscuit mixtures are correctly baked to impart the correct degree of colour at the same time as cooking the mixture, in the minimum period of time
- eliminate the opportunity for shrinkage during baking
- promote and maintain customer satisfaction by producing goods of consistent standards of quality.

Baking cakes

Time and temperature

Try to bake cakes as quickly as possible, bearing in mind that you want to obtain a good colouring of the cake crust but also need to ensure that the cake is cooked in the centre. Many factors will influence the baking of a cake and need to be understood if good baked cakes are to result.

Shape and size

The shape and size affect the degree of cooking required to ensure the centre of the cake is cooked.

- Smaller cakes take less time to cook than larger or deeper cakes.
- Larger cakes need a lower baking temperature and a longer baking time to make certain the middle of the cakes are not under-cooked or still raw.
- The shape of a cake will also determine the cooking time and temperature: the wider the cake, the longer and slower it will need to cook.
- The temperature range for baking cakes is wide: from 175 °C (350 °F) for rich fruit or wedding cakes to 230 °C (450 °F) for small sponge cakes such as fairy cakes or Swiss roll.

Richness

Where cake mixtures contain a high proportion of sugar in the recipe, the oven temperature will need to be cooler, and the cooking time longer. This is due to the caramelisation of the sugar in the mixture; sugar forms a caramel at approximately 155 °C (320 °F). When cooking a rich cake, cover it with sheets of silicone or greaseproof paper once it has coloured sufficiently.

Steam/moisture
Many cakes become too dry whilst baking because the oven temperature is too high and the atmosphere is dry. Ideally there should be some steam in the oven to delay the formation of the crust until the cake batter has become fully aerated and the proteins have set. A full oven of cakes will generate sufficient moisture to steam the oven, but if you are baking small batches of cakes, place a tray of hot water into the oven to generate steam. If the oven is too hot the cake crust will form early and the cake batter will rise into a peak.

Additions to the basic mixture
When the tops of cakes are covered with almonds or sugar, the mixture is enriched, so the baking temperature will need to be lowered to prevent over-colouring of the cake crust. Generally, a 5–10 °C (40–50 °F) reduction in temperature will compensate for the enriched mixture.

When adding ingredients such as glycerine, glucose and invert sugar, honey or treacle to cake mixtures, remember that these all colour at much lower temperatures than sugar, and the oven temperature will need to be lowered accordingly. Always check that the oven temperature is correct for each type of cake mixture.

Spreading and rolling

Once the sponge or cake has been mixed and is ready, pour the sponge mixture onto a tray which has been greased and lined with silicone paper. Using a stepped palette knife, *spread* the mixture with a light action over the area of the tray. Note any resistance against the knife which indicates areas where the sponge is thick. Do not use all the mix on one tray if a thin sponge layer can be spread over two trays.

When making Swiss rolls, you will need to roll the baked cake. This can be done simply by rolling the baked mixture up carefully after spreading with jam. Wait until the roll is cold before serving. (See the method illustrated in the diagram shown below.)

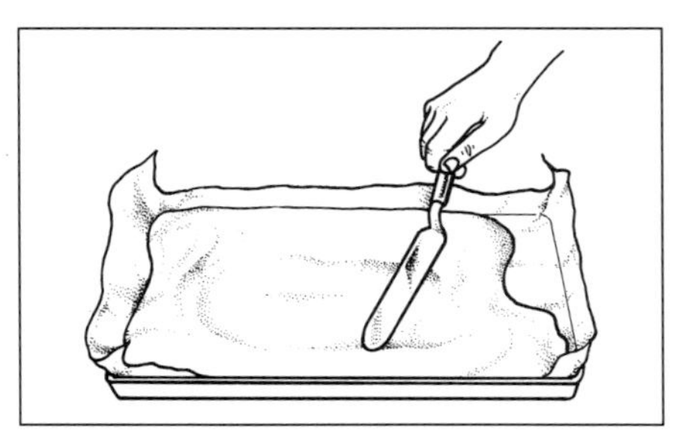

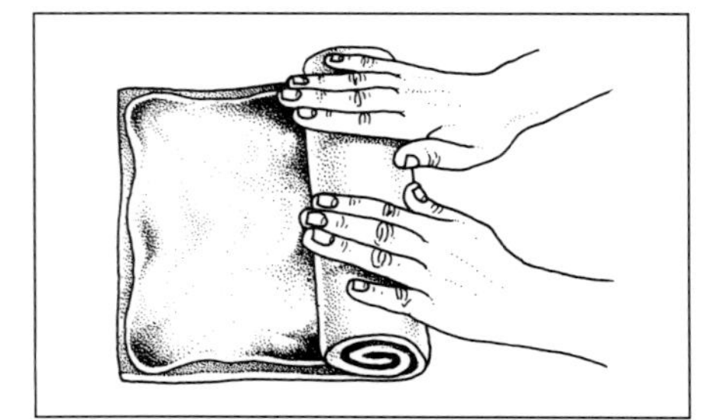

Left: spreading the mixture onto a baking tray
Right: rolling Swiss roll

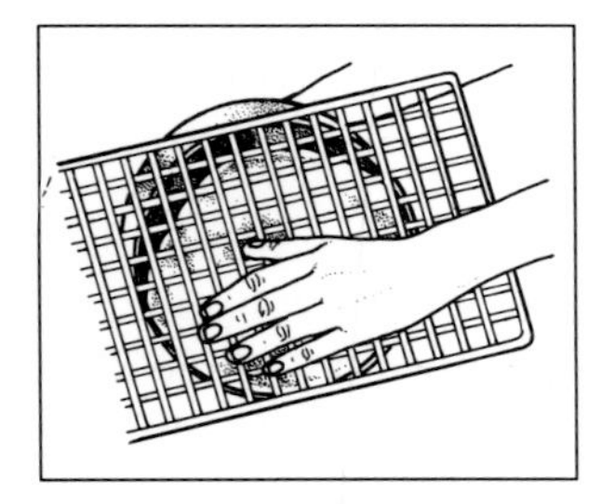

Turning out the cake

Turning out

Always turn out sponges onto a wire rack to allow the product to cool quickly. Place the wire rack on top of the baking tin and turn the tin upside down to prevent the sponge or cake breaking. Leave any protective paper on until the cake or sponge is cool unless you need the product urgently. Heavier fruit cakes should cool in the tin; if turned out while hot they may break or crack. Swiss roll should be turned out onto sugared paper and left with the tray on top until cool.

Cooling

If sponges are too fresh or still warm when you attempt to cut them, the sponge or cake crumb will tear and damage the slices or layers. If sponges need to be cooled quickly, place them by an open window, which should be fitted with fly screens.

Essential knowledge

It is important to keep preparation and storage areas and equipment hygienic in order to:

- comply with food hygiene regulations
- prevent the transfer of food poisoning bacteria to food
- prevent pest infestation in preparation and storage areas
- prevent contamination of food commodity items by foreign bodies.

Baking plain cakes

Plain cake mixes benefit from being produced by the flour batter (see page 206) or high ratio blended (see page 207) methods of production.

Madeira cake

Using unsalted butter for cake recipes is best.

Madeira is a lemon flavoured plain cake produced either as a round cake or in slabs to be cut into 500 g (1 lb) or 1 kg (2 lb) pieces. It should be baked at approximately 180–185 °C (350–360 °F) for 40–50 minutes according to the size and shape. Test for cooking either by pressing gently on the top of the cake (it should feel resistant but slightly springy) or by piercing through the centre with a needle (the needle should be clean when withdrawn, without any mix attached to it).

A slice of lemon can be used to decorate the cake prior to baking.

Slab cake

Slab cake is the term given to plain or fruited cakes cooked in frames which are usually rectanglar in shape. They are are baked at 177 °C (350 °F), and the tops should develop a light golden crust. The cooked cake is divided into portions by cutting the baked cake slab across to form bars of cake 10 cm (4 in) wide and 20 cm (8 in) long. Slab cake can be sliced and sandwiched with jam or buttercream when cool or decorated on top with almonds or sugar prior to baking.

You can also make fruited or flavoured slab cake by adding sultanas, currants, mixed fruit, cherries, coffee, walnuts, chocolate, coconut, orange or lemon to the mixture.

Angel cake

Angel cake refers to a light plain cake of various colours: white, pink, yellow, orange or chocolate brown. The differently coloured mixtures are layered: usually into white, yellow and pink layers sandwiched with buttercream. Angel cake is baked at 171 °C (340 °F).

Cup cakes

These are made using the sugar batter method (see page 206). The mixture is piped into small paper cases using a savoy bag and a plain 12 cm (5 in) tube. The piped, filled cases are baked at 204 °C (400 °F) for 15 minutes.

Cup cakes can be plain or fruited and the tops of plain cup cakes are usually finished with an icing or fondant: chocolate, lemon, orange, coffee or vanilla flavoured. Commercial cup cakes today are produced using a high ratio blending method and mixture.

Queen, butterfly or fairy cakes
Produced from a sugar batter method (see page 206), these can be plain or fruited. If you are making fairy cakes, remove the centre from the cooked cakes, pipe in some buttercream, then replace the top of the centre section. For butterfly cakes, cut the top into two pieces and place on top of the cake to form the wings of the butterfly. Queen cakes are made by placing a little prepared dried fruit in the base of the paper cup before piping in the cake mixture. These cakes are baked at 204 °C (400 °F) for 20 minutes.

Baking fruited cakes

This term covers various types of cake: mixed fruit, sultana, currant or Genoa. The type of fruit cake, i.e. light, medium or heavy, is determined by the ratio of total fruit weight to that of the total batter weight. Fruit cakes are produced by the sugar batter method (see page 206), with the fruit added last of all. The fruit cake mixture is then deposited into a prepared cake tin and dropped (carefully) onto the work bench to dispel any air pockets. The top may be levelled with a clean, damp hand or plastic scraper. The cake is then baked according to size and type (see below). Remember to place a tray of hot water in the oven when cooking fruit cakes.

Fruit ratio for fruit cakes

Cake type	Weight of fruit per 500g (1 lb) of batter
Lightly fruited	125g (4 oz)
Medium fruited	250g (8 oz)
Heavily fruited	500g (1 lb)

Cake tin/hoop sizes

Weight of batter/fruit	Size of cake tin/hoop
500g (1 lb)	14 cm (5 1/2 in)
700g (1 1/2 lb)	15 cm (6 in)
1 kilo (2 lb)	17.5 cm (7 in)
1.5 kilo (3 lb)	20 cm (8 in)
2 kilo (4 lb)	22.5 cm (9 in)
2.5 kilo (5 lb)	25 cm (10 in)

Light fruit cake
A light fruit cake contains 125 g (4 oz) of fruit per 500 g (1 lb) of batter. Bake at 182 °C (360 °F) for approximately 75 minutes (depending on the size and depth of the cake).

Medium fruit cake
A medium fruit cake contains approximately 250 g (8 oz) of fruit per 500 g (1 lb) of batter. As more fruit is added than when making a light fruit cake, the baking temperature should drop slightly to compensate for the denser mixture during baking. Bake at 182 °C (360 °F).

Heavy fruit cake

A heavy fruit cake contains approximately 500 g (1 lb) of fruit per 500 g (1 lb) of batter. Use a medium flour when making heavy fruit cake as this will make the batter stronger, helping it to support the fruit and prevent it from sinking during baking. Always protect heavy or rich fruit cakes well by lining the inside and outside of the cake tin with layers of paper. The baking temperature is reduced to compensate for the denser mixture and to prevent the outside of the cake from burning. Heavily fruited cakes are baked at 170–175 °C (340–350 °F).

Faults that can occur in cakes

Fault	Possible cause	Errors in method
Cake sinking in the centre	Too much aeration	Excessive sugar in the recipe (it will also have an over-coloured crust), *or* over-beating of the batter before adding the flour or beaing on fast speed, *or* excessive baking powder.
	Too much liquid	The sides and top of the cake cave inwards after it is removed from the oven. There will be a seam above the base crust, which shows that some of the moisture within the cake has recondensed to water and sunk to the base of the cake. As a result the texture collapses.
	Under-cooking	There will be a sticky, wet seam below the top surface of the crust. Either the oven was on too low a setting or the cake was not cooked for long enough.
	Cake tin being knocked during baking	The result of the tin being disturbed before the cake is set is that the mixture will cave inwards, particularly at the the top of the cake.
	Oven door being opened during baking	This causes cool air to rush into the oven, causing the cake to cave inwards.
Peaked top	Toughened cake mixture	Usually due to over-mixing the batter once the flour and dry ingredients have been added. This causes gluten to develop, toughening the mixture and trapping gas or oxygen. The crumb structure and eating quality will also be affected.
	Strong flour used for mixture	This has the same effect as over-mixing, as the strong flour contains too high a level of gluten. Weak, low protein flour is required for cake mixtures.
	Oven too hot and insufficient steam during baking	The surface proteins coagulate and set before the cake has completely risen. The uncooked internal mixture cooks and expands breaking through the baked surface, resulting in a peaked top.
Bound (stiff) appearance, little volume	Lack of aeration	not beating the cake batter sufficiently before adding the flour, *or* insufficient sugar in the recipe, *or* insufficient baking powder.
Sinking fruit	Batter too soft to hold fruit	Too much baking powder, *or* too much sugar, *or* batter not toughened sufficiently, *or* flour was too soft/weak, *or* wet fruit was soaked too long and poorly dried, *or* cooked at too low a temperature: this melts the fats and sugar, allowing the fruit to slip or slide in the mixture to the base of the cake.

Dundee cake
Traditionally a Scottish cake, this is medium to heavily fruited with a delicate moist crumb texture. Produce it using the sugar batter method (see page 206), and decorate the top with split almonds (round side facing up) prior to baking. Deepen the colour of the cake with blackjack (a caramelised sugar syrup used for colouring the original gravy browning).

Cherry slab or cake
The most common problem when baking cherry cake is preventing the cherries from sinking to the base of the mixture during baking. The following tips will help to avoid this:

- wash the whole or chopped glacé cherries in warm water to remove the preserving syrup, then dry them thoroughly. Coat them in flour or ground rice to keep them up in the cake mixture during baking
- use a medium strength flour to increase the strength of the batter
- it is possible to over-mix the cake batter slightly before adding the cherries to toughen the mixture, but this is not primarily recommended. The same effect may be obtained by adding an acid to the batter, e.g. cream of tartar
- start the cake in a hot oven to set it, then reduce the heat. This causes the egg and flour proteins to coagulate slightly and prevents the cherries sinking through the melted liquid batter.

Genoa
This is a mixed fruit cake or slab using mixed peel, currants, sultanas and cherries. The top of the cake is finished with flaked almonds and baked in a humid oven set at 171 °C (340 °F).

Wedding, birthday or celebration cake
Made from a rich fruit cake mixture, these three types of cake are often produced following one recipe (making no distinction between the types of cake), but you can also find individual recipes for each type. Note the following points:

- the cost of producing rich fruit cakes is high due to the amount of fruit used
- the ingredients for these cakes need to be prepared well in advance
- prepare the mixture using the traditional sugar batter method (page 206)
- when placing the mixture in the baking tin, flatten it out and made a slight depression in the centre of each tier. This will counteract any rising of the batter in the middle of the cake and produce a flat top
- these cakes require a long slow bake. Bake medium sized cakes at 177 °C (350 °F), and larger ones at 165 °C (330 °F)
- during baking, cover the cakes with silicone paper to prevent the tops from burning
- when cooked and cool, store the cakes for a minimum of 6–8 weeks to mature
- if you want to enrich the flavour by adding brandy or rum to the cakes, brush it on after baking, as alcohol will evaporate during baking. Brush the cakes with a warm solution of stock syrup and brandy or rum every 3–5 days and keep the cakes wrapped in greaseproof paper. When the paper becomes wet the cake has reached a point where it will not absorb any more liquid
- never store cakes in a damp atmosphere. Wrap them in greaseproof paper and store in a cool dry place to mature until required for decoration.

Christmas cake
A traditional measurement of quality for Christmas cake is the *pound all round ratio.* This is where a pound of each of the main ingredients is used in the recipe. Many reliable recipes for Christmas cakes can be found in pastry, confectionery and bakery textbooks. Christmas cakes can be soaked with rum and stock syrup while maturing. Baking should be slow as for celebration cakes.

Essential knowledge

The main contamination threats when preparing and producing cakes, sponges and biscuits are as follows:

- cross-contamination can occur between cooked and uncooked food during storage
- food poisoning bacteria can be transferred through preparation areas, equipment and utensils if the same ones are used for preparing or finishing cooked, baked goods and raw, uncooked mixtures, icings, creams and toppings
- food poisoning bacteria may be transferred from yourself to the food. Open cuts, sores, sneezing, colds, sore throats or dirty hands are all possible sources
- contamination will occur if flour or any foods are allowed to come into contact with rodents (such as mice or rats), or insects (house flies, cockroaches, silver fish, beetles). Fly screens should be fitted to all windows
- food poisoning bacteria may be transferred through dirty surfaces and equipment. Unhygienic equipment (utensils and tables, trays, mixing bowls and tins) and preparation areas (particularly egg and cream based mixes) can lead to contamination
- contamination can occur through products being left opened or uncovered. Foreign bodies (e.g. sack tape, metal objects, machine parts, string, cigarette ends, etc.) can fall into open containers and flour bins
- cross-contamination can occur if equipment is not cleaned correctly between operations
- contamination can occur if frozen cakes and sponges are not de-frosted correctly (always check the manufacturer's instructions)
- incorrect waste disposal can lead to contamination.

Cooling cakes

Fruit and slab cakes
Cakes such as these are fragile while still hot and are liable to break or crack if turned out straight from the oven. Leaving the cakes in the tin or cake frame will also help retain moisture.

- Fruit and slab cakes benefit from being left to cool in the tin once they have been removed from the oven.
- Plain sponges should be left to cool for 10 minutes in the tin and turned out when risk of the sponge breaking has passed; as the sponge cools it is less liable to be fragile. If sponges need to be cooled quickly, place a wire rack over the top of the cake tin and invert, then remove the lining paper and cool on the wire rack.

- When celebration cakes are cool, trim the top if slightly risen and turn the cake upside down. Check for level and use the base as the top surface for decoration purposes.

Plain or lightly fruited cakes
Light, plain or lightly fruited cakes can be turned out from their tins and placed on cooling wires after they have been removed from the oven. Stored them on cake racks with any lining paper left on until the cake is required: this helps retain moisture and protects the cake during storage. Cakes should be allowed to cool naturally at room temperature; if stored in an enclosed environment while still warm they will become sticky and create mould growth over time. When cool, cakes should be film wrapped if possible or wrapped in greaseproof paper.

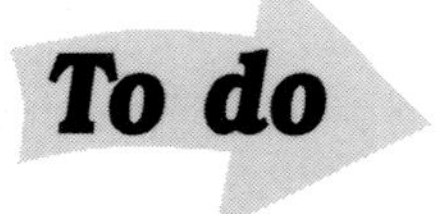

- Investigate the range of cake products available from local shops.
- How are these commercial cake products packaged?
- Ask your supervisor where rich cakes should be stored. For how long, and why?
- Find out what an '*M*' fault is. Why does it occur?
- Find out what an '*X*' fault is. How can it be prevented?
- Ask your supervisor why roasting ovens are not suitable for baking cakes, biscuits and sponges.

Baking biscuits

Biscuits vary considerably in shape, size, texture, ingredients, colour and flavour. Such diversity makes it difficult to apply general baking principles.

Biscuit baking needs care and attention, as many biscuit varieties take a short time to bake, are thin and prone to burning if left for even one or two minutes too long in the oven. Place them in the bottom of the oven to cook, where it should be slightly cooler. Use of a double tray or a few sheets of silicone paper will protect the base of the biscuits.

Moulded biscuits such as cigarettes, tuilles, cornets, etc., need careful attention to bake them to the required colour and to mould them without allowing them to become brittle.

A selection of biscuits

The following types of biscuits are those most usually produced in commercial, pastry and bakery kitchens for sale or as part of other dishes and products. Shortbread is often served with poached compôte of fruit.

Types of shortbread biscuits

Finger and blocked shortbread biscuits (*Pâte sablée*)
Uses: finger biscuits (plain and fruited), small and large blocked shortbread biscuits, petit fours sec (small, dry, delicate baked biscuits: plain or decorated), gateaux and tortes (e.g. gateau MacMahon bases).

Piped shortbread biscuits (*Sablés à la poche*)
Uses: petit fours sec and tea biscuits (plain or fruited).

Shortbreads: finger, blocked and piped

Made from unsalted butter, flour and sugar, the mixed shortbread paste is a similar texture to that of wet sand. The French for shortbread is *sablés,* which translated means *sand.* The paste is very short (giving the biscuits a melt-in-the-mouth quality) because of the high fat ratio.

Key points

- When dusting the paste before pinning out, use ground rice to prevent the paste from sticking.
- Shortbread can be shaped to form fingers, blocks or circles. Crimp and dock the paste with fine holes to allow air to escape; this keeps the biscuit flat.
- Shape moulded blocked shortbread using a block. Dust the design (which is cut into the wooden mould) with ground rice and pin the paste into the mould. Tap the mould onto a baking sheet covered with silicone paper and *dock* the shortbread; i.e. pierce it with holes using a docking roller, fork or cocktail stick. This allows trapped air to escape, preventing the product from rising. The shape can be marked with a knife or scraper to portion.
- Dredge all shortbread with fine caster sugar after removing it from the oven while still hot.
- Bake shortbread at 199 °C (390 °F).

Piped shortbread *(Sablés à la poche)*

These are piped using a savoy bag and tube and produced using a creaming method (see page 212). They can be decorated prior to baking with glacé cherries, nuts or angelica. Bake at 180–190 °C (360–380 °F) until light golden brown. These biscuits can be whipped or dipped in chocolate or sandwiched with ganache or buttercream. Individual shaped cutters can be used to produce animal or geometrical designs. Used as a petit four sec or tea pastry.

Viennese biscuits

Viennese biscuits are served as an afternoon fancy or, if small, as a petit four sec. They are light and delicate to eat. The mixture is made using icing sugar and unsalted butter.

Method

1. Cream the sugar and butter until very light.
2. Add the eggs and half of the flour, blending into a cream.
3. Fold in the remaining flour.
4. Pipe the mixture onto a cooking tray using a savoy piping bag and star tube. Either pipe each biscuit to form a rosette and decorate with half a glacé cherry; pipe into fingers, shells or scrolls; or simply make a small depression in the top of the biscuit before baking (and fill with raspberry jam when baked).
5. Bake at 204 °C (400 °F) until golden brown.

Flapjacks

Here melted butter is mixed with rolled oats, treacle, brown sugar, nuts and any other optional ingredients. The mixture can then be spread on to a lined tray or into greased rings and baked at 180 °C (360 °F) for 20–25 minutes until golden brown. The flapjack is then cooled and cut to shape: i.e. into fingers, or squares if cooked as a sheet (at least 6 mm/1/4 in thick).

Tuilles

Tuilles is the French for *tile*. These delicate biscuits are made from nibbed almonds, icing sugar, flour and egg whites. The mixture is blended and left to rest for a few hours. Small heaps are placed onto silicone paper with enough room to spread slightly; the mix is best flattened with a damp fork prior to baking. Cook in an oven set at 230 °C (450 °F) until golden brown on the outside and lighter in the centre. Leave to cool for 2–3 minutes after baking then lift carefully to mould over a savarin ring or rolling pin, forming the biscuit into the curved tile shape. These biscuits are very fragile and need careful handling.

Cat's tongues (*Langues de chat*) and derivatives

Cat's tongues (*Langues de chat*)

Cat's tongues are thin biscuits are used as a decoration for desserts and gateaux or for petit four sec. Produced by the creaming method (see page 212), they are piped to form 3 cm (1 1/4 in) finger-shaped biscuits. Bake at 210 °C (410 °F) until a light golden edge forms on the biscuits. Remove from the tray and place on a fine wire rack to cool.

Bull's tongue (*Langues de boeuf*)

Bull's tongues are a larger form of cat's tongue. They are used as petit four sec or garnish for desserts. Bull's tongues can be sandwiched with buttercream or dipped with chocolate for a tea pastry.

Cigarettes and cornets

These are made from the cat's tongue mixture (above), piped into round coin shapes approximately 3 cm (1 1/4 in) wide. Bake and colour as for cat's tongues but only pipe a few biscuits at once. When cooked, form each hot biscuit into a cigarette shape around a spoon handle or roll by hand. For cornets, pipe the mixture to a slightly wider diameter (5 cm/ 2 in) wide and mould while warm into a cream horn tin, ensuring the top baked surface is showing on the outside of the cornet biscuit. These are used for gateaux and pastry desserts such as Creme beau-rivage.

Marquis

These are two cat's tongues biscuits sandwiched with chocolate ganache and fine crushed praline at a ratio of 3:1. Join the two biscuits and use a piping bag filled with warm chocolate to pipe the word marquis on top of each biscuit.

Brandy snaps

Brandy snaps are simple to make and provide a useful biscuit suitable as a petit four, dessert decoration, basket for a pastry dessert or individual hotel sweet. The mixture is made from soft flour, butter, caster sugar, golden syrup and ground ginger.

Mix the ingredients into a smooth paste, then place in walnut-sized pieces on silicone paper and flatten out using a damp hand. Make sure you leave enough room for the snaps to spread. Bake in a moderate oven (170 °C /340 °F) until golden and well honeycombed. Leave to cool and begin to set, then mould into the required shape. If the snaps set before being moulded they can be returned to the oven and softened. These biscuits are often formed into tubes; in which case dip the ends in chocolate and fill with fresh cream, or fill with a good quality ganache or buttercream.

Florentines

This nut and fruit biscuit of flaked almonds, cherries, sultanas and peel, fresh cream and flour tastes wonderful.

Method

1 Melt some butter, mix in the sugar and bring to the boil.
2 Remove from the heat and stir in the cream.
3 Blend in the nuts and flour and lastly, the fruit.
4 Deposit the mix into plain metal rings (8 cm/3 in) on silicone paper and bake at 182 °C (360 °F) to a chestnut brown colour, taking care not to burn the biscuit.
5 Remove from the oven and allow to cool.
6 When cool, turn each florentine over and return to the oven to finish cooking.
7 Remove from the oven and allow to cool. When cold, coat the underside with melted chocolate, allow to set, then apply a second coat, marking it with a comb scraper.

Palmiers and derivatives

Palmiers

Palmiers (literally: *Pig's ears*) are small biscuits, used as a petit four (dipped or piped with chocolate and filled with fresh cream chantilly), or as a morning or afternoon tea pastry. They are made from puff pastry.

Method

1 Pin out the pastry to form a rectangular shape 45 cm (18 in) long and dust it with caster sugar. The aim is to use a substantial volume of sugar when pinning to impregnate the pastry, so that when these biscuits are cooked on a very well buttered baking tray they are half baked and half fried. This produces a crisp sugar glass finish.
2 Fold the pinned out paste evenly from each outside edge towards the centre. Each fold should be approximately 6 cm (2½ in) wide.
3 When evenly folded to meet at the centre, fold one half onto the other and lay the paste on its side.
4 Use a cook's knife to cut the paste into 1 cm (¼–½ in) thick biscuits.
5 Butter up the trays thickly, using clarified butter.
6 Place the biscuits onto the trays, spacing out the pig's ears to allow for expansion. Rest for at least 30 minutes.
7 Bake in a sharp oven 220 °C (420 °F) until light brown in colour, then turn the biscuits and finish colouring. The finish should be a glaze of caramel, achieved by the sugar content added at Step 1.

Pâte à decor

This biscuit paste is used for decorating. It is a smooth paste mixture made of flour, icing sugar and egg white which needs to be rested for one hour once made up. The paste is piped or spread using a card stencil to form decorative shapes and patterns, e.g. tulip biscuits, coil spring biscuits and many other designs of animals or flowers. Bake at 220 °C (425 °F) for 3–4 minutes until lightly brown. Mould and shape while hot, adding a little water if necessary to soften the paste for piping.

- Find out the best method of storing baked biscuit products.
- Cut some leaf-shaped templates from card or plastic for pâte à decor.
- Visit your local cake artistry shop and look at the range of cutters.
- Find out what *biscuit rulers* are.

Baking sponges

Spreading, rolling, turning out and cooling

Note all the points listed on page 216.

Preparation methods for sponge types

- *Foaming:* Swiss roll, English and French genoese
- *Boiling:* genoese
- *Blending:* high ratio sheet, cup and layer sponges
- *Creaming:* Victoria
- *Separate yolk and white:* sponge fingers, drops and sacher torte

Swiss roll (Biscuit roulade)

Fine, small Swiss roll is used for Charlotte royale, while chocolate Swiss roll is used for yule logs and Swiss kirsch roulade tea fancies. Fruit and cream filled Swiss roll is used as a dessert or pastry. Swiss rolls can also be made from high ratio sheet mixtures. The ingredients are eggs, sugar, soft flour, glycerine.

Method

1. Whisk the eggs and sugar over hot water to produce a light thick foam.
2. Carefully fold (cut) in the sieved flour.
3. Spread the mixture thinly onto a Swiss roll tray or baking sheet prepared with silicone paper. Always make thin sponges; do not make one thick one then attempt to split it.
4. Bake at 238 °C (460 °F) for 3–4 minutes.
5. When baked, remove from the oven and sprinkle with caster sugar.
6. Turn the sponge upside down onto a sheet of silicone or greaseproof paper, leaving the tray on top until cool. Because the sponge has been baked quickly in a sharp oven, it will not be dried out and should be soft when cool.
7. Trim the edges and spread with jam, then use the sheet of paper the sponge was turned out onto to roll it up tightly. Do not roll sponge when it is warm, as the sponge can sweat and become sticky, causing it to stick to itself.

Genoese sponge

Genoese sponge is used as a base for gateaux, torten and petit four glacé bases. A range of sponges can be made using a variety of production methods. Sheet genoese can be used for layer cakes, fancies and Battenburg.

Method: Light English genoese

1 Whisk eggs and caster sugar over a bain-marie to blood heat, then whisk on Speed 3 until a light foam occurs at 'ribbon stage'.
2 Fold in the sieved soft flour.
3 Place the mixture into a lined gateau tin or tray which has been either buttered and floured or lined with silicone paper.
4 Bake at 180 °C (356 °F) for 30–40 minutes. Test for cooking by pressing your finger on the top of the sponge: it should feel springy and no finger mark should be left on the sponge.
5 When cooked, cool in the tin for 10 minutes then turn it out onto a cooling wire, leaving the lining paper on until cool.

French genoese

This sponge has an enriched texture and flavour due to the addition of butter, and it also has an extended shelf life. It is produced as for English genoese (above), but when the flour has been carefully cut in (Step 2), fold in some melted clarified butter. This should be done with care to avoid loss of volume. Tin, bake and cool as for English genoese.

Boiled genoese

This type of sponge is stronger than English or French genoese sponge, and is suitable as sheet sponges for petit four glacé, and layer and Battenburg fancies. It can be flavoured and coloured (e.g. chocolate, walnut, almond, coffee, lemon, orange, etc.). This sponge freezes quite well for up to one month, or will keep fresh for one week wrapped in greaseproof paper. Dry sponge trimmings can be used as a substitute for ground almonds or used in Queen of puddings.

Foam the eggs and sugar as for Swiss roll (above). Melt some butter in a separate pan and bring it to the boil, then stir in the flour and glycerine and beat until a smooth paste forms. Cut the sponge foam into the flour and fat paste in 3 portions, mixing well as each portion is folded in. Pour into the prepared trays or tins and bake at 191 °C (375 °F) for 35–40 minutes.

High ratio genoese

This blended batter makes an excellent sponge which has a long shelf life, is stable and can be used for many commercial sponge products (e.g. mini rolls, Swiss rolls, fondant fancies, gateau and layer cake goods). High levels of liquid and sugar are absorbed by the special fat and flour; note that high ratio genoese sponges cannot be made using standard fat or flours.

Accurate mixing speeds and times are essential. The batter absorbs colours and flavours well, and can be used for angel or layer sponges, Battenburg or as a sheet for petit fours and tea fancies. High ratio mixes are also used for cup and layer sponges. There is no general recipe for this type of sponge; each recipe is different.

To do

- Find out the difference between a gateau and torten and draw up a list and index of gateau and torten products
- Ask your supervisor which type of sponge you should use for frozen sponge products. Why?
- Find out how dried or excess sponge can be used up.
- How are striped sponges produced?
- Watch your supervisor flavouring sponge with liqueur. Notice the method, equipment and ingredients used.

Separate yolk and white biscuits

Biscuits à la cuiller

Biscuits à la cuiller (spoon biscuits) are sponge fingers made by separating the yolks and whites as described on page 212. Used for Charlotte russe, these can be served with mousses, creams and fools. They can be finished by dipping the ends in liquid chocolate and then sandwiching in pairs with cream or ganache to form dumb bell biscuits. Small biscuits of vanilla or chocolate are used to decorate the sides of gateau and torten.

Method

1. Using a savoy bag and a plain piping tube (1 cm/½ in diameter), pipe finger shaped biscuits 5–6 cm (2–2½ in) long.
2. Dust with caster sugar then bake at 190 °C (375 °F) until lightly golden.
3. Allow the biscuits to cool before removing them from the paper. These biscuits store well if left attached to their baking paper until required.

Sponge drops and brickettes

Sponge drops are also made from a separate yolk and white sponge mixture. They are used as the base for Othellos, Jagos, Rosalinds and Desdemonas. Sponge brickettes can be baked in special tins for trifle sponge.

Biscuits perlés

These are made in a similar way to cuiller biscuits (above) but are dredged with icing sugar instead of caster sugar. Leave them to stand until the sugar dissolves and give them a light dusting of icing sugar prior to baking. When baked they have a fine pearl finish, and can be used as sponge fingers, drops and as petit fours sec.

Convenience cake, biscuit and sponge products

The range of convenience products is vast. The modern chef uses frozen gateau and cake products when under pressure to meet a high demand and also maintain the scope of products for customers. There are also other advantages. Modern technical premixes enable many of the costs to be calculated more accurately than cakes made entirely in the kitchen, help to reduce labour costs and limit the range of stock items that need to be held. Fresh cake and biscuit items of high quality and consistent standard enable smaller businesses (who cannot afford a variety of skilled chefs) to offer a wider range of products and remain competitive in the demanding markets of hotel, catering and hospitality.

The diversity of sponge, cake and biscuit ingredients, and the way that they have been modified to be more stable or have a longer shelf life, is a valuable aid to the contemporary chef, pâtissier or baker in the development of new products and lines.

Convenience gels, stabilisers, fudges, fillings, decorating creams and icings speed up production times and ensure consistent standards and quality. Coupled with modern purchasing consortium arrangements, convenience products and ingredients deliver greater choice and cost effectiveness to even the smallest of businesses.

What have you learned?

1 What are the main contamination threats when cooking cakes and biscuits?
2 Why is it important to keep preparation and storage areas and equipment hygienic?
3 Why are time and temperature important in the cooking of cakes and biscuits?
4 List the possible reasons why a cake might sink in the centre during baking.
5 What is the difference between French and boiled genoese sponge?

ELEMENT 3: Decorating cakes, biscuits and sponges

Decoration methods

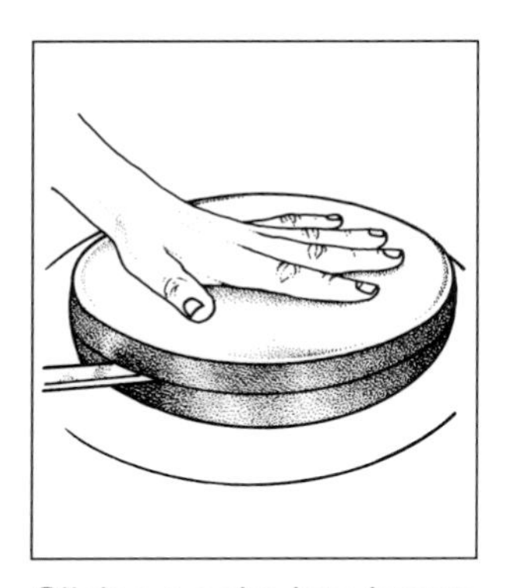

Slicing a cake into layers

Trimming and cutting

The trimming of sponges and cakes or biscuits should be carried out in a safe manner, taking care to hold the knife correctly and trim effectively according to the product requirements. Always ensure the bench where trimming is done is clear, and on no account cut products near someone else who is working. Always hold the knife with a firm grip, keeping the hand clasped around the knife handle with all your fingers tucked around the knife handle.

Pastes

You will need to trim both cooked and uncooked pastes and products. To do this, use a cook's knife: keep the point of the knife on the bench and use the knife in a guillotine action.

Cooked cakes and sponges

Use a serrated carving knife to trim cooked sponges and cake. When trimming the edges, keep a firm grip on the cake or sponge, and cut with a sawing action using the length of the knife. Keep the hand holding the cake behind the knife edge. Cutting small portions is safer than trying to cut a large section in one action.

If trimming a sponge by cutting into slices, as you would for a gateau, begin by making small cuts into the edge of the sponge to mark the required slice thickness. Use a slow, sawing action to score around the outside of the sponge to a depth of 2.5 cm (1 in) until you return to the starting cut. Once the entire sponge has been marked, use a gentle sawing action to cut the sponge into slices, following the initial cuts as a guide. Keep your other hand flat on top of the sponge, using it to hold the sponge in place and to turn the turntable.

Points to remember

- Hold the knife with all fingers wrapped around the knife handle.
- Cut with care and consideration for other people.
- Trim slowly, watching the leading edge of the knife at all times.
- Do not try and talk to colleagues while trimming.
- When in doubt ask for help. Fingers can be severed in seconds. *Be sure, be safe, be certain.*

Filling

Cakes, sponges and biscuits can be filled with an array of mixtures, which can be cream based, fruit based or paste based (e.g. chocolate ganache). The filling should be sufficient to flavour and complement the cake, sponge or biscuit and be pleasing to the eye. Never use filling in such a way that the product becomes difficult to handle or portion, the filling overpowers the flavour, or so that the final appearance is badly affected. Any of these might affect sales or service to customers.

Cream based fillings
Fresh semi or fully whipped cream (double or whipping); clotted cream; pastry cream; butter cream; set creams and mousses.

Fruit based fillings
Apple and assorted fresh fruit purée; tinned fruit pie fillings; fruit pastes and conserves such as raspberry or apricot confit; jams and preserves; whole fresh fruit or portioned sliced fruits; fruit creams, mousses or set custards.

Flavoured fillings
Chocolate ganache; fondant and pastry cream pastes; nut pastes, powders and products such as praline; coconut or almond pastes; fudges; caramels; butterscotch sauces and pastes; mallow; curds; gels of assorted flavours and textures.

Spreading and smoothing

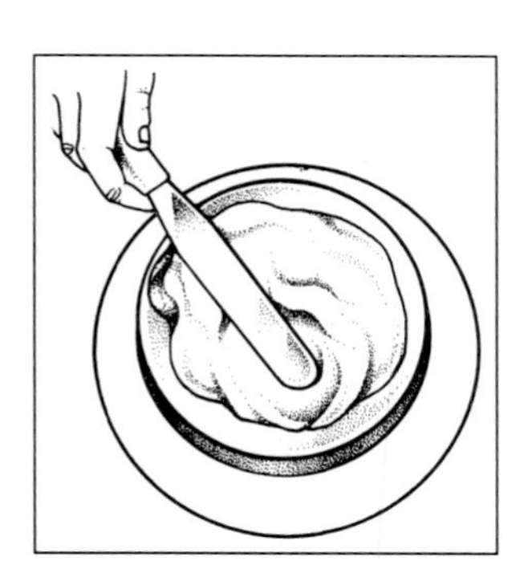

Spreading

The skill of spreading and smoothing mixtures to fill sponges or gateaux requires practice and patience. The technique of spreading is best learnt by watching an expert. Notice how they hold the palette knife. How do they place the mixture to be flattened onto the cake surface? What technique is used to distribute the icing or cream or topping onto the cake evenly?

Some simple procedures can be followed when spreading or smoothing soft surfaces.

- Make sure that any tools you will need to use are clean and ready to hand.
- Work cleanly; always wash your tools when they become sticky.
- Blend and work coatings and fillings before trying to achieve a smooth surface.
- Spread mixtures with an even pressure of the hand using a small palette knife.
- Never use too much or too little filling or coating to spread and smooth.
- If fresh cream is the medium being used, never over-whip the cream; a softer semi-whipped cream will smooth easily when worked. If cream is over-whipped, the action of smoothing will cause further coagulation of the butter fats, producing a rough and unsightly finish.
- If you are trying to obtain a clean buttercream or icing finish, with no air bubbles or scratches, first prepare the mixture by stirring and working with a small cake palette to remove the air.
- When spreading fondant or chocolate mixes, the action must be swift and skilled to spread and smooth such mixtures before they set. If some areas begin to set before you have finished smoothing, they will drag and spoil the finish.

Smoothing

- Royal icing should be a day old before being used to coat cakes.
- The marzipan on celebration cakes should be applied a few days before the flat icing, to prevent oil from the paste rising through the wet icing and spoiling the finish.
- Warm any fondant slowly over time and do not allow it to become too hot; this would produce large sugar crystals which set quickly with a dull finish.
- Never try to smooth a very large surface on cakes unless you are confident. Always ask for help and advice.
- Use plastic scrapers for smoothing sides of gateaux and torten or other cake products. Keep these wrapped in cling film to protect them from being scratched; a scratched scraper will leave a mark or line in the smooth surface when drawing royal icing.
- Straight steel edges are available for flat coating buttercream and icing. Protect them from being scratched by keeping them wrapped in cling film. These are best used slightly warm: rinse them in warm water and dry just before use.
- Sugar paste can be effective for achieving a smooth sugar surface on celebration cakes. Pin it out, then polish it with a clean soft cloth to buff the paste once in place and trim, crimp or flute to add decoration to the edges.

Piping

Piping with icing or chocolate
Piping icing or chocolate from a nylon savoy or paper bag requires control, balance and coordination. The skills required to pipe a shape with flair and speed can only be learned through practice.

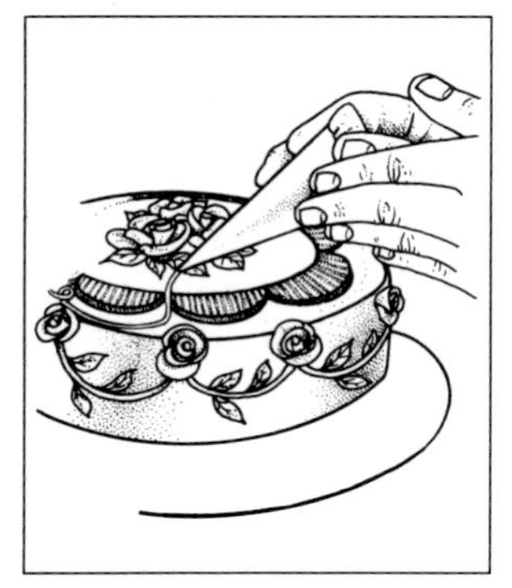

Piping with icing

There are a few basic rules:

- never fill the bag more than half full. If the bag is over-filled the chocolate will leak and become a problem, as you will tend to concentrate on the leaking liquid rather than the design being piped
- once the bag is filled, fold the top tightly inwards and down towards the point of the bag. Make certain the paper of a paper bag is not creased, as this can cause problems in both making the bag and keeping a fine sharp point
- ensure basic preparations are carried out before piping starts
- hand coordination is essential. Always use your spare hand to guide the bag: it is not possible to produce high quality piping without using both hands
- always try the design first on paper. When writing a name or message on cakes, check the balance of the words by sight, by writing the name on paper and comparing the spacing with your cake surface
- you might like to try piping icing onto waxed paper, allowing it to dry and then transferring it to the cake. This is a good way of decorating and will give you practice
- make sure that royal icing is not too stiff, and that the nozzle used is not too small. Otherwise your hand can become cramped if the pressure is constant even for a few minutes
- chocolate can be thickened with a little glycerine, alcohol or stock syrup and piped onto silicone. These decorations set quickly and can be stored until required. When you have some spare time, prepare some of these mixtures and practise creating neat rows of designs, both large and small: try some little birds, butterflies or palm trees
- see also *Piping royal icing* (page 234).

Piping with cream

Fresh or non-dairy cream can be used to finish cold desserts, cakes, puddings and pastries. Never over-whip fresh cream, as this will cause it to lose its buttermilk, resulting in liquid dripping from the piping bag and spoiling the product. It will also become difficult to pipe. The cream will look pale yellow and have a coarse texture.

Piping can be carried out using a plain or star tube made of plastic or metal, and a nylon piping bag (this is more hygienic than cotton). Do not place too much whipped cream in the bag: half fill the bag and then twist, at the same time working any trapped air from the bottom of the bag to the top.

When the bag feels comfortable to hold, pipe the shells by squeezing the bag with an even and steady pressure to shape the forms. Use your spare hand to guide the savoy bag. You can usefully practise with meringue and choux pastry on a work bench to develop control and accurate design.

Dusting, dredging or sprinkling

Pastries and other dessert items are dusted with sugar either to add to the final design or to form a glaze during cooking. The sugar can be caster, icing or granulated for pies, biscuits or gateaux. For dusting, sieve the sugar onto the product, using a fine mesh sieve held approximately 20 cm (8 in) above the product. This will deposit an even dusting of sugar which should not be unsightly. Dredging is used where a heavier dusting is required, such as when you use icing sugar for a white coated effect. If a recipe asks for a sprinkling of sugar, simply dust the product very lightly for effect, allowing the product surface to show through.

Coating

Coating products with liquid mixtures such as fondant, chocolate or hot jam requires basic preparation and techniques to be effective. Having spent time producing a good quality cake, sponge or biscuit, the coating it receives should be clean, provide a sheen, be of an even thickness, and cover the food item without showing the base product.

Coating with fondant

When prepared correctly fondant will give a high gloss finish, an even smooth surface and provide a good foundation for the final decoration. The fondant icing should never be heated above 38 °C (100 °F). Break the fondant into small pieces and place into a copper pan or saucepan, cover with hot water and leave to stand for 10 minutes. Finally, pour off the water and stir the fondant well. This method is effective for warming the fondant sufficiently without overheating it. If overheated, the fine small crystals in the fondant enlarge and reflect less light while setting quickly and producing a thick unpleasant coating.

Sponge items to be coated should be dipped or brushed with boiled apricot purée or covered in marzipan before coating. Evenly space out the items to be coated on a fine wire rack and coat with jam. While the sponges are setting, place a plastic tray under the wire to catch the excess fondant. Using a small ladle, nappe (cover) the sponge shapes with fondant, starting with the sponge furthest from you and working back towards yourself. This avoids dripping fondant on those already coated.

To improve the gloss, add a little piping jelly, gelatine, glucose, stock syrup or mallow. On no account add raw egg white: this is not recommended for reasons of possible contamination and/or fermentation, and because food regulations do not allow raw egg whites to be added to uncooked food. To add flavour and colour to the fondant, you can add spirit, liqueur, essence, chocolate couverture or bakers chocolate, coffee essence or powder, or essential oils. Flavour compounds should be used with care, as only a few drops are required. Always shake the bottle well before using them, as some compounds can settle and separate while standing.

Coating with chocolate or ganache

Ganache is a rich mixture of fresh cream and chocolate. The chocolate can be good quality couverture or cheaper bakers' compound chocolate. Compound chocolate made with vegetable fats and oils (rather than cocoa butter) should be melted to 49 °C (120 °F) for milk or plain coating chocolate. Boil the cream and then stir in the chopped chocolate to form a glossy, thick sauce. Use while warm as a coating for individual cakes, sponges and gateaux.

Sandwich the cakes or gateaux and place them onto a wire rack with a clean plasic tray below. Using a large ladle, nappe the product with two or three ladles of chocolate (for individual cakes one ladle will be sufficient). The rack can be agitated by shaking gently to ensure the coating covers the top and sides evenly. The excess coating will drip off onto the plastic tray. If possible, always use the force or gravity to allow the weight of the coating to flow over the product; the use of a palette knife will spoil the finish unless done quickly.

The ganache will have a thicker consistency as it cools, enabling it to give products a glass-like finish.

Coating with royal icing

See page 234.

Topping

Top dressings are added both prior to baking and after baking to finish the decoration of desserts, cakes, pastries, sponges and biscuit products.

The topping added to products before baking can easily over-colour. To prevent this, reduce the oven setting and check the goods while baking to determine the required finish. Toppings added to sponges, cakes or pastries should be balanced in proportion to the size of the base, and complementary in colour and flavour. If toppings such as toasted coconut or vermicelli are used, they should be added carefully. Use a plain ring cutter pressed through a card and positioned over the gateau as a useful frame through which to apply toppings without spilling them over the whole surface.

Moderation is the rule: in design, colour and flavour. The aim is always to enhance the overall appearance for your clients and customers. Your reputation and standard are constantly judged by the quality of finish, flavour and eating quality of your work.

Water icing

This is used to decorate or glaze a thin layer of opaque icing onto cakes and biscuits.

Boil some water, then stir in the icing sugar to form a smooth icing free from lumps. Small sponges can be dipped into the icing which forms a crust and sets within minutes. Cherries or other decorating media should be added before the icing sets, as this holds them in place. Water icing can be coloured and flavoured, but note that if chocolate is added, it can sometimes thicken the icing and you may need to add a little stock syrup to soften it to the required consistency for dipping or topping.

Royal icing

This is made from egg whites and icing sugar to decorate wedding, birthday or celebration cakes. Powdered egg white called *albumen* or *albumen substitute* is widely used in place of fresh egg whites to make royal icing.

Traditional royal icing is made by mixing 2 egg whites with 400 g (14 oz) of sieved fine icing sugar then beating this for 20 minutes to form a stiff, light icing mixture. If using albumen powder, mix this with water at the ratio of 75 g (3 oz) per 400 ml (14 fl oz) of water, and then leave overnight to extract maximum strength from the albumen. Strain the solution and stir in the sieved sugar, then mix on a medium speed for 20 minutes. You can add a little blue colour to the royal icing to improve the whiteness. When mixed, cover with a clean, damp jay cloth or place in an airtight container.

Piping royal icing

Always use fresh royal icing for piping work. Mix the fresh icing to work out the air and make a smooth mix which will pipe to a pearl finish free from bubbles. The icing can be worked on a sheet of finished glass to remove any air or grains of sugar. It can be hardened and strengthened by the addition of a little lemon juice, especially useful when you are using deep colours such as Christmas red or black.

For competition work use distilled water for the albumen solution and sieve all sugar though muslin. For strength with fine runout, use large newly laid hen eggs, where the whites will have very strong albumen. Note that this mixture will require a good beating to make a light royal icing.

Coating with royal icing

Coating a cake well with royal icing requires skill developed by constant practice. The icing should be at least 24 hours old before being worked to remove the air and form a smooth icing mixture. Cakes should be coated with three thin coats, using a plastic scraper for the sides and an icing ruler for the top. Cake dummies can be used to practise on. A good, flat surface on a cake will improve piping that is not quite accurate, but good piping will not enhance a poor surface. Remember that the icing is the most critical decoration in cake artistry, gateau and torten production.

Buttercream

Buttercream is used for many gateau and cake products. It is made from icing sugar and butter or another fat, and is only strictly a buttercream if it has a 22.5 per cent fat content. The mixture is susceptible to contamination and should be stored in the refrigerator. It also picks up odours easily from strong smelling foods, so always keep the buttercream covered with cling film.

Many recipes are available for the production of these creams and all can be flavoured and coloured according to the product and dish requirements. Many commercial recipes are available that use marshmallow, margarine and shortening rather than butter, and some are stabilised, while yet others are made with fondant or pectin. Never use too much buttercream as this can make the product sickly; use it to sandwich or decorate with moderation.

Method: Traditional buttercream
Beat fresh unsalted butter and sieved icing sugar together until light and free from any lumps, then spread or pipe the cream to fill, sandwich or decorate cakes and sponges.

Method: Boiled buttercream
This is less sweet than the beaten variety due to the sugar being boiled, and it produces a far better cream. Beat the egg yolks. Boil the sugar to 150 °C (300 °F), then slowly pour the boiled sugar onto the yolks, adding softened butter in small portions while beating. Mix until cool.

Italian meringue can be blended with softened butter to produce an equally useful cream that is not as sweet as traditional buttercream but takes up flavours well to produce a continental type buttercream.

Whipped cream
The term *cream* can only be applied to products filled, finished, topped or decorated with fresh dairy cream.

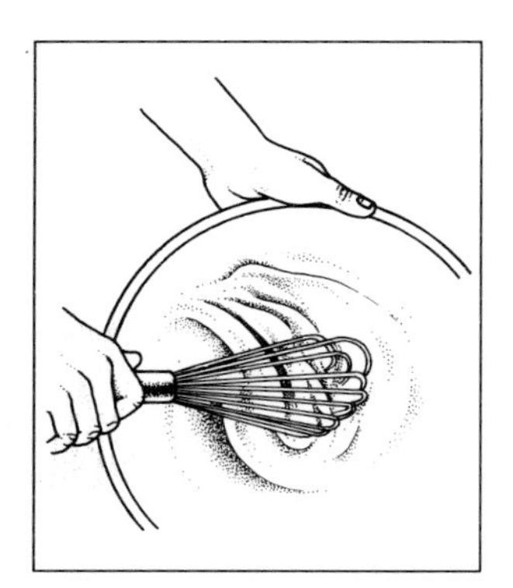

Whipping cream

Cream used for whipping should be either double or whipping cream. Double cream is more expensive and contains a higher butterfat content than whipping cream, but is quicker to whip to a piping consistency. It is also more stable in cases where liqueur or liquid flavours are added because of its high butterfat content. Double cream has a butterfat content of 48 per cent while whipping cream has a 3 per cent content. Whipping cream will give more volume when fully whipped than double or other fresh creams.

Remember the following points when whipping cream:
- whipping can be done by hand or machine
- if using a machine to whip cream, thoroughly clean the machine and preparation areas after mixing
- never whip cream in aluminium bowls or with aluminium whisks
- all equipment should be very clean
- over-beating will turn the cream to butter and buttermilk
- cream should be stored at 4 °C (40 °F)
- always use cream in strict date rotation. Never use cream that has turned sour for sweet products
- always return cream containers to the refrigerator; do not leave them standing in a warm environment
- whipped cream should be piped using a clean sterilised nylon piping bag rather than the older (less hygienic) cotton variety
- read the *1993 Food Hygiene Regulations* concerning the use and storage of fresh cream products.

Using fillings and decorations

- Always keep buttercream covered and chilled until required for use. Never store it near or with strong smelling foods.
- When you have finished working with fillings or icings, store excess

mixtures in a clean bowl covered with cling film or in an airtight container.

- Royal icing should never be stored in the fridge: this will make the icing heavy.
- Fruit preserved in sugar (e.g. cherries and angelica) should be rinsed in water to remove the sugar coating and then dried with a clean cloth.
- Coconut should be stored in a dry, cool, airtight container to prevent it from becoming rancid. Toasting coconut improves its flavour and extends its shelf life.
- Fresh cream should not be over-whipped or the buttermilk will separate and contaminate or spoil products being piped. Always place it back in the refrigerator after use, and only whip enough for each product type.
- Trim sponges and cakes carefully with serrated knives, following a safe technique that will cut or slice the sponges or cakes and take into account the safety of both yourself and your colleagues.

Decorating media

Decoration is an art form and there is a wide range of materials available. Essentially, the media used must complement and balance the overall design in order to be attractive and enhance the saleability of the product.

The following items are often used as decorating media:

- confiture fruits (glacé cherries, pineapple and ginger, angelica)
- crystallized flowers (e.g. rose petals, violets, mimosa, lilac)
- nuts (grilled or roasted, flaked, nibbed or split, especially almonds)
- dessicated coconut when toasted
- chocolate vermicelli.

Essential knowledge

The main contamination threats when preparing and using cake decorations are as follows:

- food poisoning bacteria can be transferred through to the cake decorations through use of unhygienic preparation areas, equipment and utensils
- food poisoning bacteria may be transferred from yourself to the food. Open cuts, sores, sneezing, colds, sore throats or dirty hands are all possible sources
- contamination will occur if flour or any foods are allowed to come into contact with rodents (such as mice or rats), or insects (house flies, cockroaches, silver fish, beetles). Fly screens should be fitted to all windows
- contamination can occur if products are stored at incorrect temperatures
- contamination can occur through products being left opened or uncovered. Foreign bodies (e.g. sack tape, metal objects, machine parts, string, cigarette ends, etc.) can fall into food, open containers and flour bins
- cross-contamination can occur if equipment is not cleaned correctly between operations
- incorrect waste disposal can lead to contamination.

To do

- Make a list of decorating media suitable for cakes and sponges.
- Draw a range of designs suitable for piping with chocolate.
- Find out how chocolate cigarettes are made.
- Ask your chef to demonstrate how small marzipan fruits are produced and finished as decorating media.
- Write out a list of all the items you are aware of that are suitable for use as decorating media.

Finishing, presenting and storing cooked cakes, biscuits and sponges

- Ensure the cooked products are cool before finishing (i.e. trimming, filling, spreading and smoothing, piping, dusting or dredging, coating, topping) according to product requirements.
- Never use cream to decorate products that are still warm.
- Biscuit products should be stored in airtight containers to prevent them from softening.
- Do not over decorate with buttercream or fresh whipped cream.
- Use decorating media with care to produce goods that have visual appeal.
- Decorate products so they are the same size, shape and finish.
- Present and store the finished cooked products according to the *Food Hygiene (Amendments) Regulations 1993.*

Health, safety and hygiene

Note all points given in Units G1 and G2 of the Core Units book concerning general attention to health, safety and hygiene. When preparing cake, biscuit and sponge products the following points are particularly important:

- Use all machines according to the manufacturer's instructions.
- Carry out all trimming and cutting with care and attention following safe techniques.
- Make sure that all moving parts of any mixing machines have stopped moving before removing the mixtures from the mixing bowl. Check that the correct machine attachment is in place for the production method you are going to use.
- Handle all products carefully, especially products with a greater percentage of sugar, e.g. brandy snaps and rich fruit cakes.
- Wipe up any spillages as soon as they occur.
- Make sure that there is adequate space on the work bench before removing full trays of cakes, biscuits or sponge products from the oven.
- Transport boiling liquids such as hot jam glaze with care and attention.

Disposing of waste

Waste materials should be disposed of cleanly and efficiently, to prevent contamination of uncooked and cooked cake and biscuit mixtures, products and ingredients and to prevent health hazards.

Planning your time

- Work in a clean and organised manner attending to any priorities and laid down procedures.
- Know what your responsibilities are. If in doubt, ask your supervisor.
- Clean up after yourself as you work. Wipe down work areas between each task or job; do not use glass or oven cloths for this purpose as this causes contamination.
- Refer to *Planning your time* (pages 1–4) concerning the general points that you should be aware of whenever you are working in a kitchen or bakery.
- Use any spare time to make sure trays, tins, hoops, moulds or frames are prepared ready for the next stage of production.
- While products are baking you should clean and clear work surfaces ready for the next stage of work and prepare any fillings, coatings, icing or toppings and decorating media such as cherries, chocolate, angelica or fresh fruits.

Unexpected situations

An unexpected situation might be a rush order, a customer request, an over-cooked batch of products or an emergency. Make sure you know how to deal with all of these, according to any laid down procedures.

What have you learned?

1 What are the main contamination threats when you are using and preparing cake decorations?
2 Why is it important to keep preparation and storage areas and equipment hygienic?
3 What are the main decoration methods?
4 What is *ganache* and which type of cake would you be most likely to use it with?

Extend your knowledge

1 The area of cakes, sponges and biscuits is a large one. Look up recipes for tea pastries and French fancies and find out how they are produced and finished, and when they are served. Find out about types of petit four glacé, sweetmeats and pastilles.
2 Look at the popular range of gateau, torten, layer and sponge cake products.
3 Investigate the world of cake artistry, looking at the types of decorative icings and pastes used, different shapes produced and styles of decoration, especially for celebration cakes.
4 Ask your library or colleagues if they have a copy of *Complete Pâtissiere* by E.J. Kollist. Read up on the many types of gateau, spice and petit four sec recipes.
5 Read *The Complete Pastry Work Techniques* by I. Nicolello, which contains many useful recipes, techniques and methods covering many of the products covered in this unit.
6 Discuss competition work with your chef. This can improve your skills, widen your interest and give you the opportunity to compete with fellow trainee chefs, pâtissiers and bakers from all over Europe.
7 Talk to other catering companies and businesses in your area. Find out how they organise production, what types of commercial and convenience products they use that you do not, and share ideas.

UNIT 2D13

Preparing food for cold presentation

ELEMENTS 1 AND 2: **Preparing and presenting cold canapés, open sandwiches and cooked, cured and prepared foods**

What do you have to do?

- Prepare ingredients of the correct type, quality and quantity for canapés and sandwiches.
- Prepare the correct cooked, cured and prepared foods and garnishes in the quantity and of the quality required.
- Prepare and present food products according to customer and dish requirements.
- Prepare garnishes according to customer and dish requirements.
- Prepare and finish canapés and sandwiches according to customer and product requirements.
- Store canapés, sandwiches and prepared dishes in accordance with food hygiene regulations.

What do you need to know?

- What the main contamination threats are when preparing and storing canapés, sandwiches or any foods for cold presentation.
- How to satisfy health, safety and hygiene regulations concerning preparation areas and equipment both before and after use.
- Why it is important to keep preparation and storage areas and equipment hygienic.
- How to plan your work to meet daily schedules.
- How to deal with unexpected situations.

Introduction

Preparing foods for cold presentation covers a wide area of food preparation. From an hors d'oeuvre through to a decorated sweet. Nearly every course on a menu could include food of some description which could be served cold.

This unit is aimed specifically at illustrating the preparation, storage and service of cold canapés and open sandwiches, and the presentation of savoury cooked, cured and pre-prepared foods. It will illustrate the importance of careful and safe production and presentation to ensure the customer is served the best possible product.

ELEMENT 1: Preparing and presenting cold canapés and open sandwiches

Planning your time

Producing sandwiches and (especially) cold canapés can be time consuming and requires a great deal of patience and care. Always work in a methodical manner.

Take the following points into consideration:

1. What types of bases, fillings and toppings are to be prepared?
2. What garnishes are to be used?
3. How many of each variety do you need to produce?
4. What equipment will you need to produce and store them?
5. What service dishes will you need for presenting the finished canapés?

You also need to prioritise your order of work, by deciding which jobs you need to do first. Establish the appropriate sequence, then produce the canapés in a methodical manner, beginning with the most time consuming jobs.

Preparation areas and equipment

When producing canapés or sandwiches you should have all the equipment that you will need close at hand and ready before you start work.

Follow the guidelines given below.

- Make sure that any chopping boards, knives and any other equipment needed is thoroughly clean.
- As you finish each canapé you will need to place it onto a tray; work out how many trays you will need and place these near your work bench, ready for use.
- If you are making canapés coated in aspic jelly, make sure you have racks clean and ready to hand for storing the canapés before coating them with aspic.
- When producing sandwiches, ensure that the bread is kept covered to prevent it from becoming dry.
- Make sure that all of the preparation work is completed before starting to make sandwiches. This means checking that the butter is softened, any purées are made, and any tomatoes, cucumber, meats, etc. are sliced ready for production. This will help your work flow and ensure that the sandwiches are made as quickly as possible.
- Remember that foods can be easily contaminated, especially through handling. In order to produce canapés you will need to handle them during production; keep this to a minimum and always work hygienically, using disposable gloves.
- Keep all cooked meat, fish, eggs and vegetables refrigerated until required at a temperature of not more than 4–5 °C (39–41 °F).
- Canapés require delicately cut garnishes, so make sure that knives are clean, sharp and organised tidily next to the cutting board before you start.
- Canapés can be served dressed on a clear set aspic jelly, a dish paper or a doily.

Health, safety and hygiene

Cold canapés and sandwiches are produced from raw, cooked or preserved commodities and once they are made there is no further cooking process involved. For this reason it is essential that they are produced in a clean, safe and hygienic manner. Canapés and open sandwiches are high risk foods that are easily contaminated by incorrect handling.

Make sure that you are familiar with the general points given in Units G1, G2 and 2D11 of the Core Units book. Pay special attention to the section on cross-contamination and the storage of cooked and raw food items.

Remember that lettuce can contain live food poisoning bacteria: always wash it well before use to prevent any soil from contaminating sandwiches or your work area. Watercress must also be washed well, as the water in which it grew may have been contaminated by waste products.

Essential knowledge

The main contamination threats when preparing and storing canapés and sandwiches are as follows:

- bacteria may be passed to the food from the nose, mouth, cuts, sores or unclean hands
- cross-contamination can occur through the use of unhygienic equipment, utensils and preparation areas. Always use colour coded boards for chopping, slicing and cutting foods
- foods stored at incorrect temperatures can become contaminated through the development of micro-organisms. Keep items refrigerated as much as possible
- frozen foods can become contaminated during the thawing process. Make sure that any frozen foods you need to use are thoroughly and correctly defrosted
- uncovered finished or partly prepared items can easily become contaminated
- bacteria may be transferred to food, utensils or preparation areas unless waste is disposed of correctly.

Preparing cold canapés

Canapés are attractive, bite-sized delicacies that can be served in a number of ways and on many occasions. They can be served as an appetiser before a meal, as accompaniments to drinks at receptions or cocktail parties or as part of a buffet display. Their delicate nature means that they often need to be prepared in advance and stored carefully to ensure that they remain attractive and appealing.

When producing cold canapés, you need to allow for many individual tastes and preferences, so you should aim to offer as wide a variety as possible. This applies both to fillings and toppings and the bases on which they are dressed. In this section we will look at how to produce canapés of different tastes, textures, finishes and garnishes, enabling you to produce an interesting and varied selection.

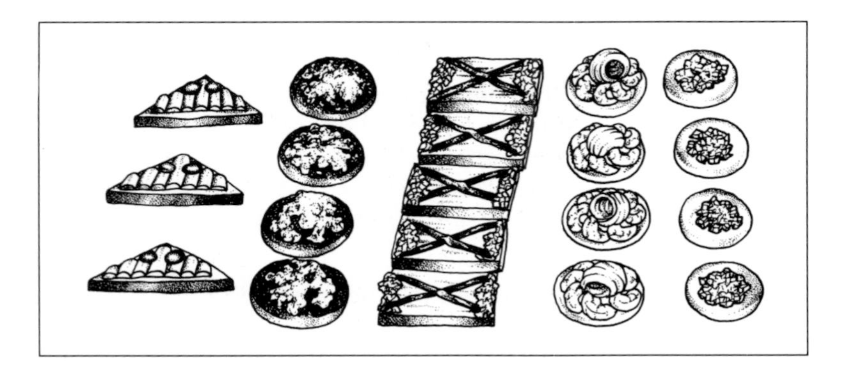
Dressed canapés

When making canapés we refer to *bases, toppings or fillings* and *garnishes. Bases* refers to the items onto or into which you place the *topping* or *filling. Garnishes* refers to the decorations you use to enhance appearance, vary texture and balance flavour.

Canapé bases

Bread

When bread is used as a base for canapés it is usually toasted, and the slice of toast may be referred to as the croûte. When preparing croûtes, remember the following points:

- once toasted, the bread must be allowed to cool before it is spread with butter. This prevents the butter from soaking into the toast and making it soft
- the butter will act as a seal to prevent any moisture from the topping soaking into the toast
- the croûte may be cut into shapes either before or after the topping is added; check which method you will need to use. Canapés topped with soft pâtés or cream cheeses may require the toast to be cut first, then the topping piped onto the different shapes. Smoked salmon canapés should be made by placing the salmon over the entire piece of toast, which is then cut to ensure that the croûte is totally covered.
- different types of bread can be used to produce croûtes; different breads give different flavours to the canapés. Likewise flavoured butters can be used to enhance the flavours of the toppings used.

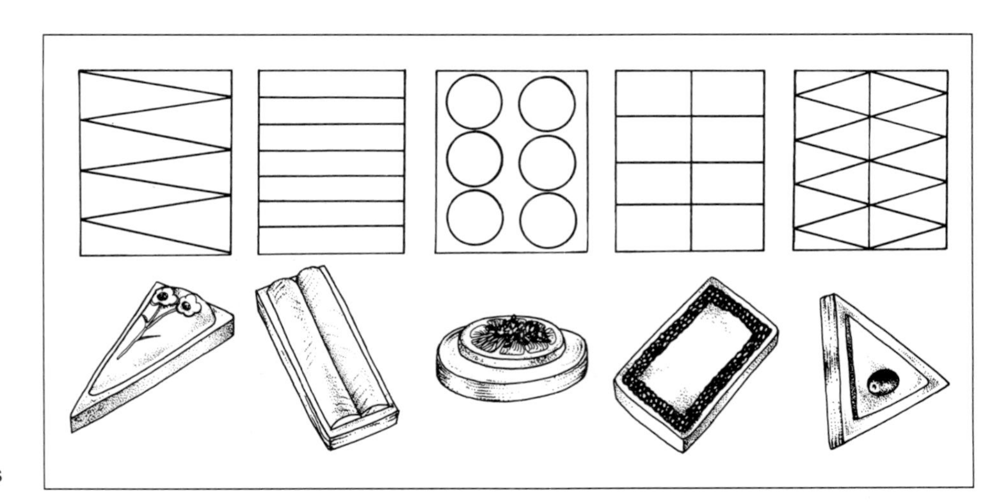
Cuts of toast and their resulting canapé shapes

Puff pastry

Puff pastry can be made in the kitchen (see page 183) or bought in either chilled or frozen form. When purchased made up, the pastry may come either in a thin roll or as a thick block to be shaped and rolled as required. Specific shapes can also be purchased, such as bouchées and vol-au-vent cases.

Puff pastry is used to make bite-sized canapés in the form of *bouchées.* These use small vol-au-vent cases of puff pastry as a base for a variety of fillings. The vol-au-vent cases are usually filled with mixtures that have been puréed or diced and then bound with a suitable sauce or dressing. The fillings may be piped or spooned into the puff pastry base.

A puff pastry base, unlike toast, cannot be sealed with butter, so some fillings may cause the bouchées to become soft. Bearing this in mind, always add the filling just before service to keep the pastry as crisp as possible.

Short pastry

Short pastry can be made in the kitchen (see page 179) or puchased ready made in frozen or chilled forms. Short paste is normally purchased in a block which must be rolled and cut to the shape and size required. Frozen pastry should be thawed at room temperature before use. Several pre-cooked items are available, such as small tartlettes and barquettes, which are ready for immediate use.

A savoury short paste is used for canapés such as small *tartlettes* or *barquettes.* Tartlettes are round pastry cases while barquettes are 'boat shaped'. For both of these types of canapé the pastry is rolled out thinly and used to line the moulds which are then *baked blind* (i.e. without the filling). Once cooked, the pastry cases can be removed from the moulds, cooled and sealed with butter (which may be flavoured if required). The filling can then be puréed and piped into them; chopped and bound with sauce/dressing and spooned into them; or thinly sliced, rolled or folded and then placed into them.

Barquettes and bouchées

Canapé fillings and toppings

Canapés are small, delicate and eaten with the fingers, which must be kept in mind when you are producing them. Sliced items should be thinly sliced and diced items should be finely diced, making them easier to eat.

Any sauces and dressings that you use in the production of your canapés should be used to complement the foods they are being used with; choose and use them carefully to avoid masking or hiding the flavour of the main ingredient.

Cooked and cured meats and poultry

Meats can be thinly sliced and either placed onto the canapé or rolled or folded onto one for attractive presentation. Poultry may be sliced, diced or puréed and then bound with an appropriate sauce to improve the overall texture. Meats and poultry can easily dry out, making them look unappetising: produce them as close to service as possible.

Examples of suitable meats and poultry: salami, pastrami, mortadella, liverwurst, Parma ham, liver sausage, smoked chicken, turkey, duck.

Cured fish and shellfish

As for meats and poultry, these can be sliced, diced or puréed and appropriate sauces used to enhance their appeal. Prawns bound with a cocktail sauce and placed in a bouchée or tartlette are a common feature in canapé presentations.

Examples of suitable fish and shellfish: rollmop herrings, bucklings, smoked oysters, soused herrings, smoked trout, prawns, smoked salmon, smoked fish roe, mussels.

Fresh vegetables and fruits

Fruits and vegetables can provide a variety of textures when used on cold canapés. When using cooked vegetables, it is essential that you ensure that they are well drained; if any liquid is left on the vegetables it will quickly make the canapé bases soft. Salad vegetables should be clean and crisp when used. They quickly become limp once prepared and stored and should therefore be added just before service.

When using fruits, remember that some will discolour very quickly and lose their appeal. Many fruits will also need to have skins, stones and pips removed where they are bitter and inedible.

Eggs

Eggs can be used hard boiled and either sliced, cut into sections or diced and bound with a sauce such as mayonnaise. Scrambled eggs are also occasionally used for canapés. When using eggs, take care not to overcook and discolour them and remember that they dry out very quickly, so should be stored carefully and produced as close to service as possible.

Hens' eggs are generally used, but quails', plovers' and gulls' eggs are also acceptable and provide interest.

Cheeses

Cheeses, depending on the variety used, can be sliced, diced, grated or even piped onto or into canapé bases. Different types of cheese will provide you with different textures, flavours and colours.

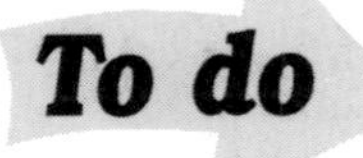

- Look at the variety of food items used in your establishment and decide which are acceptable for producing canapés and how they can be used.
- Find out the correct temperatures for storing the food items that you will be using for the production of your canapés.

Garnishes

Any garnish or decoration used on a canapé should not only enhance the appearance, but should also complement the taste. Taste and flavours must be considered as well as visual appeal.

These decorations should be neat, small, attractive and obviously fresh at all times. Remember that they are to be eaten.

Flavoured butters and mayonnaise can also be used as garnishes: simply pipe them over the canapé.

Suggested combinations for cold canapés

Bases	*Filling*	*Garnish*
Toast	Smoked salmon*	Lemon and picked parsley
Toast	Creamed cheese*	Caper and cayenne pepper
Toast	Pâté*	Black olive
Tartlette	Prawn and mayonnaise	Sieved hard boiled egg and chopped parsley
Tartlette	Asparagus	Julienne red pepper
Tartlette	Diced chicken and mayonnaise	Fan of cocktail gherkin
Bouchée	Salmon mousse	Slice of stuffed olive
Bouchée	Creamed tuna	Red fish roe

* These canapés may be glazed with aspic jelly

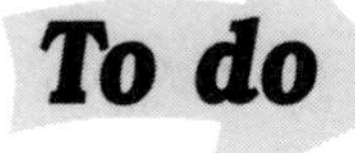

- Find out what canapés are commonly produced in your establishment.
- What bases are used for these canapés?
- Watch your supervisor or chef preparing canapés and notice the types of toppings and fillings they use.
- Familiarise yourself with the correct storage and final presentations of these canapés.

Preparing open sandwiches

Open sandwiches are a speciality of Scandinavian countries, and make up part of a traditional Smorgasbord buffet.

Today they are served at many different occasions, from light snack, brunch type meals to cocktail receptions. Open sandwiches are usually fairly substantial: they are made from a thick slice of bread covered with a variety of toppings and decorations. The end result should be colourful and appetising. There is no set size for open sandwiches, as this depends on the occasions for which they are being used and the style of service.

Open sandwiches are traditionally produced in the larder or cold section of a kitchen. It is very important to pay particular attention to the food hygiene regulations during the production of an open sandwich as the ingredients are highly susceptible to bacterial growth both during and after production.

It is also important not to produce open sandwiches too far in advance of service, as they can quickly lose their fresh, appetising appearance and become limp and unappealing.

A selection of open sandwiches

Types of bread

The list of breads that you can use is virtually endless, but these are a few of the common breads or bases that can be used in the production of an open sandwich:

- wholemeal bread
- wholegrain bread
- rye bread
- Vienna bread
- French bread
- pumpernickel
- brioche
- olive bread.

The bread base for an open sandwich should be cut fairly thickly to enable it to support the topping: approximately 0.75 cm ($^1/_4$ in) thick.

The size of the slice will depend on the occasion it is required for. The shape of the loaf obviously has a part to play in the shape of the sandwich, and the slices do not usually need to be trimmed square or cut into any specific shape. You would not normally cut off the crusts.

Butters

The type of butter you use on an open sandwich is a personal choice of either unsalted, lightly salted or salted. Alternatively the spread used could be a low-fat variety, should the client require it.

As in the production of a closed sandwich, one of the reasons butter is spread onto the bread is to help prevent any moisture from the topping soaking into the base, making it go limp and in turn making it difficult to pick up and eat.

You may wish to flavour the butter you use for spreading, such as using horseradish butter for a roast beef open sandwich. You may also wish to colour the butter. Butters that have been coloured and flavoured can be piped onto the open sandwich as part of the decoration.

To do

- Find out how many different type of bread bases you could use for an open sandwich from the types of breads delivered to your establishment.
- Make a list of 10 different ingredients you could use for flavouring butter. Which ones might you use on an open sandwich?
- Make a second column in your list. Write down any food items that are appropriate for accompanying each flavoured butter in an open sandwich.
- Find examples of ingredients that you could use to both colour and flavour butter.

Toppings

As with the production of any sandwich, the combination of the toppings is endless and items can be used on their own or in combination.

Ask yourself the following questions:
- do the combinations complement each other?
- are the textures contrasting?
- do the colours of the ingredients look appetising together?
- am I working in a safe and hygienic manner?

Here is a list of some of the commodities you may wish to consider when producing open sandwiches.

Meat and poultry
Salami, garlic sausage, mortadella, boiled ham, roast or boiled beef, chicken, turkey, tongue, bacon, pâté, foie gras, any type of smoked meats.

Fish and shellfish
Smoked or unsmoked eel, salmon, trout, mackerel, halibut, tuna, prawns, crab, lobster; pickled fish such as anchovies and herrings.

Eggs
Hens' and quails' eggs (hard boiled or scrambled), egg mayonnaise.

Vegetables and fruits
Asparagus, cucumber, all edible leaves, mushrooms, tomatoes, pimento, spring onions, radishes, etc.

Cheese
All types of sliced, grated or cream cheese.

Garnishes

All open sandwiches should be dressed then garnished or decorated before being served. The garnishes or decorations can be complex or simple, from a carved vegetable to a segment of lemon.

The following food items are examples of possible garnishes:
- *fruits:* segments of orange, lemon, apple, grapefruit and pineapple
- *pickles:* silver skin onions, capers, olives, gherkins and cauliflower
- *vegetables:* carrots, spring onions, radishes, cabbage, onions, celery. These can be grated, sliced, carved or shaped
- *butters:* piped flavoured and coloured butters
- *mayonnaise* and mayonnaise-based sauces.

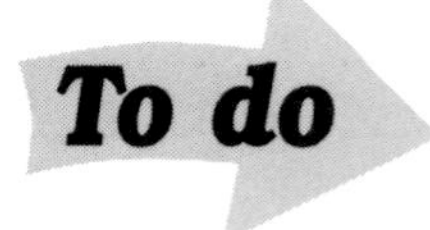

- Make a list of the different types of smoked meat suitable for use in open sandwiches.
- Produce a selection of open sandwiches for a coffee shop menu, giving a brief description of each sandwich.
- Ask your supervisor to discuss the correct storage conditions for cooked ingredients to be used in open sandwiches. Make sure you know the correct storage temperatures.

Producing open sandwiches

Method

1. Mise en place:
   - collect all ingredients, making sure that they are of the correct quality and quantity
   - ensure that you have the correct equipment and that it is clean and any knives are sharp
   - check that the work area is clean.
2. Cut a slice of bread 0.75 cm ($^1/_4$ in) thick and spread it with butter.
3. Cover the bread with a base of lettuce or similar ingredient, then cover with the chosen topping making sure that the bread is completely covered. (If you are using hot ingredients do not cover the bread with lettuce.)
4. Season the sandwich with condiments and seasonings.
5. Finish with a suitable decoration.
6. Arrange a dishpaper on a suitable service dish and place the open sandwich onto the dishpaper. You may also want to place some decoration onto the flat. Serve.

Storing canapés and open sandwiches

If at all possible, canapés and open sandwiches should be made and served immediately. If you do need to store them for a short time, remember that these products are decoratively garnished and so the way you store them must prevent the finished appearance from being damaged.

1. Present the canapés or sandwiches on a dish and vacuum pack, making sure that the vacuum is not too great.
2. Cover the dish with cling film, making sure the film does not disturb the garnish.
3. Store in a chilled environment.

In the past, canapés and sandwiches may have been covered with damp greaseproof paper or cloths while stored for a short time. This practice should be avoided, as it causes the food product to become damp and then to dry out as the covering dries.

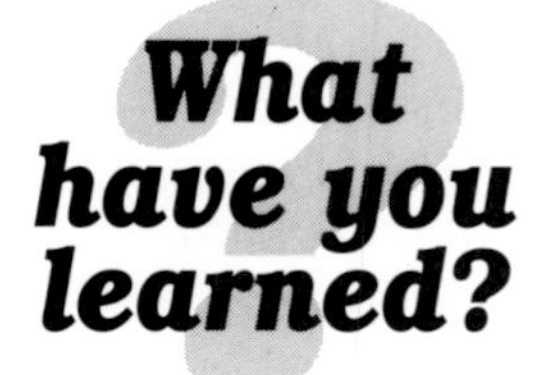

1 What are canapés? When might they be served?
2 What are the main contamination threats when preparing and storing canapés and sandwiches?
3 List five toppings or fillings that may be used in preparing canapés.
4 Why is it important to keep preparation and storage areas and equipment hygienic?
6 Suggest four sandwich toppings that might be used at a buffet.

ELEMENT 2: Presenting cooked, cured and prepared foods

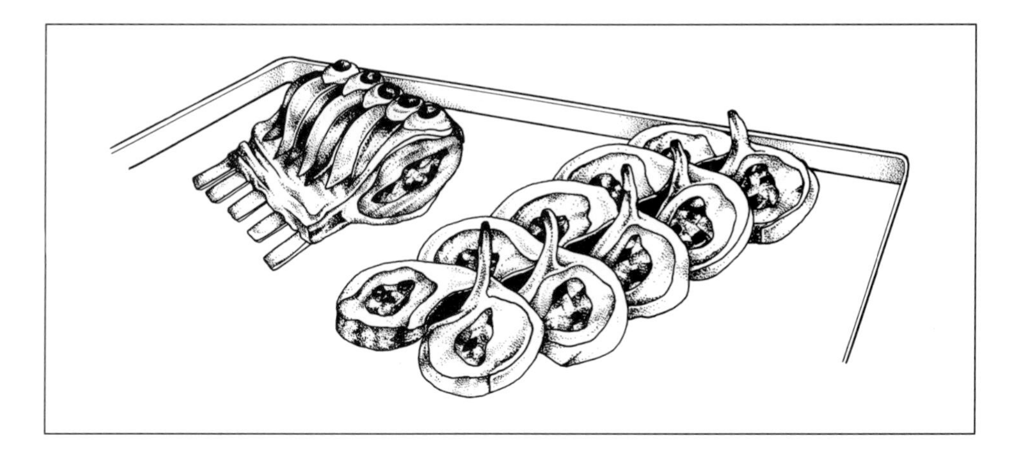

Stuffed loin of pork

Any presentation of cold foods should offer a wide range of choice, including fish, meat and poultry and prepared items. The dishes should all be attractively displayed and neatly garnished.

Most meat, poultry, game and fish items can be cooked and presented as part of a cold buffet presentation. Many of these can be cooked by traditional methods (such as roasting or boiling) or *cured.* By this we mean that foods are preserved by methods such as smoking, drying and salting. Smoked fish is often seen on the menu, but smoked meat and poultry items are becoming more and more common.

Also included in this element are the many prepared items available to you as the caterer. These would normally include a wide selection of pâtés, terrines, sausages, salamis and pressed meats.

Planning your time

- Ensure your work area is clean, tidy and clear of all unwanted equipment.
- Collect dishes and any other items you will need for presenting food before you start work.
- Assemble all the required ingredients in a methodical manner.
- Prepare the ingredients according to the recipe you are using.
- Present the items on the service dishes as required.
- Garnish each dish and then store in the appropriate way or serve as required.

Health and safety

As with any cold food there is always the possibility of contamination occurring. Always:

- ensure work surfaces and areas are kept clean
- handle food items as little as possible
- use plastic gloves when handling food
- keep food refrigerated for as long as you can
- display food under refrigerated conditions whenever possible
- remember and follow good personal hygiene practices and the health and safety regulations.

Essential knowledge

The main contamination threats when preparing and storing foods for cold presentation are as follows:
- bacteria may be passed to the food from the nose, mouth, cuts, sores or unclean hands
- cross-contamination can occur through the use of unhygienic equipment, utensils and preparation areas. Always use colour coded boards for chopping, slicing and cutting foods
- foods stored at incorrect temperatures can become contaminated through the development of micro-organisms. Keep items refrigerated as much as possible
- frozen foods can become contaminated during the thawing process. Make sure that any frozen foods you need to use are thoroughly and correctly defrosted
- uncovered finished or partly prepared items can easily become contaminated
- bacteria may be transferred to food, utensils or preparation areas unless waste is disposed of correctly.

Refrigeration management

- Raw and cooked foods must be stored in separate areas to prevent cross-contamination and at a temperature not exceeding 4–5 °C (39–41 °F).
- Food should always be covered.
- Bacteria is present in all foods; remember that refrigeration does not kill bacteria, but does help to prevent its growth. It is therefore essential to refrigerate cold items for as long as possible before service.
- When displaying food for service, monitor the temperature to check that a safe temperature is being maintained.
- Ideally any food displayed in a restaurant should be placed behind a *sneeze screen* to protect it while customers make their choice.

Food items

Cooked and cured meats

These items include all butcher meats which are usually cooked by roasting or boiling or prepared by curing.

Roasted and boiled joints can be presented as a whole joint (such as a fore rib of beef or whole decorated ham) and carved on the buffet as and when needed. They can also be pre-sliced before service when a quicker service time is required. Cured meats are generally pre-sliced for display.

Cooked poultry

Poultry is normally roasted or poached for cold presentation. Traditionally, poached poultry was presented coated in chaud-froid sauce, attractively decorated and glazed in aspic jelly. This style of presentation is now thought to be unhygienic due to the amount of handling required and the need to keep moving the item in and out of the refrigerator whilst it is being coated and decorated. This means that today poultry is more often roasted and then cut into joints (for chicken or duck), or sliced (for turkey).

Some poultry is poached, removed from the bone, diced and then presented as a salad.

Fish and shellfish

Several fish and shellfish items are eaten cold. These are normally cooked by poaching in fish stock or court-bouillon or eaten in a preserved form, such as smoked, canned or pickled. Fish is a delicate type of food and so is often cooked and presented whole (e.g. salmon or trout). It can also be pre-portioned before or after cooking or prepared into a variety of mousses and terrines.

Preserved fish is generally served as a single hors d'oeuvre or as part of a selection of hors d'oeuvres.

Crustacean items (e.g. crab) are cooked by traditional methods such as poaching and can be served plain or bound with a sauce or dressing.

Some molluscs (e.g. oysters) can be served raw or cooked. For instance, oysters are eaten raw or cooked depending on the recipe you are using; some are even smoked.

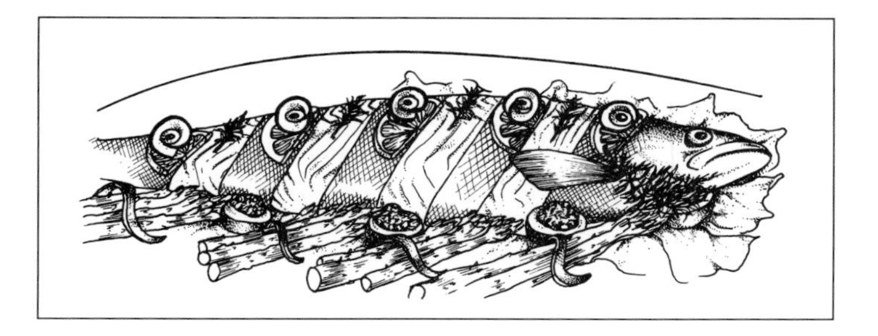

Decorated cold salmon

Prepared pâtés and terrines

Today there is a wide range of pâtés and terrines available, from traditional liver pâté to vegetable and fish terrines. These differ in both flavour and texture. You are able to choose from a very coarse mixture to a fine texture which can be easily spread on toast or biscuits.

For presentation these items can be left whole and garnished, or portioned by slicing or cutting into wedges and displayed on trays or dishes. Care must be taken when portioning to prevent the mixture from breaking and losing its appeal. The surface of pâtés and terrines will also become dry and unappetising if they are incorrectly stored.

All composite items like these are susceptible to contamination. It is essential to store, handle and present them in a safe and hygienic manner.

To do

- Visit a local delicatessen and list the variety of cooked and prepared items on offer.
- List joints of meat suitable for cold presentation and how each would be cooked and served.
- Find out which types of preserved fish and shellfish are suitable for cold displays.
- Look through recipe books to find examples of decoration used when displaying cold foods.
- Find out where cooked, prepared items are stored in your kitchen. Check the temperature of foods and the environment in which they are stored.

Dressings and garnishes

When presenting cold items of food you must ensure that they always look:

- attractive
- appealing
- fresh.

This may mean that if food is on display over a period of time it will need re-dressing and garnishing to maintain the appeal. You should bear certain factors in mind when displaying cold food.

Displaying food

- Look at the colour of the food items being displayed and the garnish being used. It is easy to get carried away and add garnishes that are so colourful the main item is lost.
- Provide a centre point to your display; even if you are only presenting one tray of food there will still be a focal point.
- Ensure your cutting and slicing is even. This makes a more attractive presentation.
- Think about how the customer is going to eat the food being displayed. Will they have a knife and fork? Or is it a fork buffet? This will influence the size of portions.
- When decorating the dish ask yourself whether the decoration will be destroyed once the first portion is removed.
- Remember that often the simplest design and decoration is the most attractive and successful.

Types of garnish

Vegetables, salad items and fruits can be prepared and presented in many ways to garnish and decorate food items. They are used to make the presentation attractive but they are also part of the dish and should always be edible. Some vegetables and fruits can be shaped into flowers for an attractive presentation, such as radishes, tomatoes and certain root vegetables. This type of decoration helps to enhance the appearance and appeal of the dish they accompany. You should learn how to prepare some of these items as shown below.

Decorative garnish

Storing prepared dishes

These items must be stored in a way that helps to prevent them from becoming dry and unattractive. Store them at a temperature of 1–3 °C (34–37 °F) and keep them covered with cling film or in a vacuum pack.

Essential knowledge

It is important to keep preparation and storage areas and equipment hygienic, in order to:

- prevent the transfer of food poisoning bacteria to food
- prevent pest infestation in storage areas
- ensure that standards of cleanliness are maintained
- comply with the law.

Dealing with unexpected situations

- Check that you know what to do in the case of accidents and emergencies. Note all the points given in Unit G1 of the Core Units book: *Maintaining a safe and secure working environment*.
- Be aware of your personal responsibilities within the kitchen. If in doubt, ask your supervisor.

What have you learned?

1. What do you need to consider when displaying cold items of food?
2. Why is it necessary to keep cold food items refrigerated?
3. What are the main contamination threats when preparing and storing foods for cold presentation?
4. Why do you garnish cold foods in an attractive manner?
5. At what temperature should you store cold foods?
6. Why do you need to plan your work?

Extend your knowledge

1. Look through the following books for details of particular types of cold canapes, open sandwiches and items that might be included in a cold presentation: *The Larder Chef* (M J Leto and W K H Bode); *Pâtés and Terrines* (F Ehlere, E Longe, M Raftael, F Wessel); *Chinese Appetisers and Garnishes* (Huang Su-Huei).
2. Look through supplier catalogues to discover the range of convenience items that are available for using as canapé bases.
3. Research the number of slices/pieces you can obtain from various types of breads. This will help you with purchasing and portioning control.
4. Look at the different types of hot canapés that can be produced. How do the production and storage methods differ from cold canapés?

UNIT 2D14

Preparing and cooking shellfish dishes

ELEMENT 1: Preparing and cooking shellfish dishes

What do you have to do?

- Prepare shellfish correctly and as appropriate for individual dishes.
- Combine prepared shellfish with other ingredients ready for cooking where appropriate.
- Prepare preparation and cooking areas and equipment ready for use, then clean correctly after use.
- Cook and finish shellfish dishes according to customer and dish requirements.
- Correctly store prepared shellfish not required for immediate consumption.
- Plan your work, allocating your time to fit schedules, and carry out the work within the required time.
- Carry out your work in an organised and efficient manner, taking account of priorities and any laid down procedures or establishment policy.

What do you need to know?

- The type, quality and quantity of shellfish required for each dish.
- What equipment you will need to use in preparing shellfish.
- Why it is important to keep preparation, cooking, storage areas and equipment hygienic.
- Why time and temperature are important when preparing and cooking shellfish.
- The main contamination threats when preparing, cooking and storing shellfish and shellfish dishes.
- How to satisfy health, safety and hygiene regulations, concerning preparation and cooking areas and equipment, and how to deal with emergencies.

Introduction

The coastal waters of Great Britain are rich in quantities of highly regarded shellfish: oysters from Colchester, Dublin Bay prawns, lobsters from Cornwall and Morecambe Bay shrimps. They offer a wealth of opportunity for caterers to express themselves through exotic dishes to a simple snack. Shellfish are available throughout the year and are prized by gourmets and domestic cooks alike as they may be cooked simply with a minimum of fuss, or with great skill and technique to provide complex dishes; with both types giving maximum flavour and pleasure. Certain shellfish are even consumed in their raw state.

Shellfish may contribute to a well-balanced diet by providing a nutritious food rich in protein and trace elements. They are a consistent source of iodine and also provide calcium, iron and sodium.

Their diversity of use has led to them to appear on menus as appetisers and hors d'oeuvres, soups, fish courses, main courses and as garnishes for many other dishes. They also lend themselves well as delicate items for decorative techniques.

Classification

Shellfish are divided into two main groups for culinary purposes:
- crustaceans
- molluscs.

Crustaceans

Crustaceans have jointed legs and a tough outer layer or shell covering the body known as an *exoskeleton*. The category includes: lobsters, crabs, prawns, shrimps, crawfish and crayfish.

Molluscs

Molluscs are soft-bodied creatures with protective shells which form their habitat. There are three main groups: *gastropods, bivalves* and *cephalopods.*

Gastropods have only one shell; this category includes whelks, cockles, winkles, and snails. *Bivalves* have two shells joined together; this category includes mussels, scallops, clams and oysters. The third group, *cephalopods,* are grouped as molluscs even though they have an internal transparent shell or bone (quill) in stead of a hard, external shell. This category includes squid, octopus and cuttlefish.

Quality points

Shellfish deteriorate quickly after death, making correct handling, storage and use essential if quality is to be maintained and the risk of contamination reduced.

They are best purchased live from a reputable supplier or direct from a wholesale market or port where the catch is landed daily, as this is the only way to ensure freshness. However, most shellfish can be purchased in a frozen state, and these are widely used in the catering industry.

Note that all shellfish must be *thoroughly washed* prior to preparation to remove any surface contamination from their natural habitat

Live crustaceans

Look for the following points in live crustaceans:
- a fresh salty sea-smell
- a lustrous and fresh looking shell
- black and glossy eyes (not dull or pale)
- no missing claws (on lobsters or crabs)

- fish that are heavy in proportion to their size
- fish that react strongly when handled
- no signs of limpness. Limp-tailed animals should not be purchased as a limp tail is a sign of deterioration
- lobster tails that spring back into place after being uncurled.

Note that the hen or female lobster is distinguished from the cock or male lobster by the breadth of the tail. Hen lobsters have a wider tail to facilitate the carrying of eggs or coral.

Cooked crustaceans

Look for the following points in cooked crustaceans:
- fish that are heavy in proportion to their size
- intact shells, with no visible damage
- no signs of discoloration or unpleasant smells (especially ammonia).
- undamaged packaging (if applicable).

Molluscs

In general, when shellfish with two shells are purchased they must be live. Look for the following points:
- tightly closed shells (this indicates whether they are live or not). If shells are open and do not close when sharply tapped the mollusc is dead and *must be discarded.* This is because it is impossible to know how long the mollusc has been dead and whether or not it is contaminated
- a fresh sea-smell
- an absence of excessive barnacles.

Storing shellfish

Shellfish are highly perishable food commodities requiring immediate storage upon delivery. It is essential to maintain high standards of storage procedures in order to minimise wastage and the risk of spoiling food. Spoilage can lead to the contamination of other foodstuffs and increase the risk of possible food poisoning. The fine eating quality of shellfish is dependant on *absolute freshness.*

Prolonged storage of cooked and raw shellfish must be avoided because of the danger of spoilage and contamination. This could lead to severe food poisoning if eaten, as a long storage period increases the risk of bacterial growth.

Key points

- Store live shellfish at a temperature of 1–5°C (34–41 °F).
- Keep the live shellfish in its packaging to avoid moisture loss.
- Keep it covered with wet cloths or sacking to retain moisture.
- Do not over-purchase shellfish as the keeping quality is poor.
- Check that shellfish are alive prior to cooking; *reject any dead or dying specimens.*
- Store cooked and uncooked shellfish separately *at all times* to avoid any risk of cross-contamination. Ideally they should be stored in separate refrigerators.
- Cooked shellfish should be covered and stored at 0–3 °C (32–37 °F).
- Frozen shellfish should be defrosted in a refrigerator overnight and

used within a short period of time (approximately 12 hours).
- *Never freeze shellfish:* this is a dangerous practice which could lead to severe food poisoning.
- Live lobsters can be kept for short periods of time in a specially aerated salt water tank. The water must be kept at a constant temperature suitable to the lobster.
- All live shellfish should be cooked and served as soon as possible after purcnase.

Health, safety and hygiene

The observation of the basic rules of food safety and hygiene as outlined in Units G1 and G2 of the Core Units book are critical when handling shellfish as you will be handling high risk foods.

Note the following points before you start to prepare or cook any shellfish dishes:
- always use the correct colour-coded board for preparing raw shellfish and make sure that you use a different board for cooked items. Do not work directly on the table
- keep your chopping board clean using fresh disposable wipes
- use equipment reserved for the preparation of raw shellfish. Where this is not possible, wash and sanitise equipment before use and immediately after use
- work with separate bowls for shellfish offal, bones and usable shellfish: never mix these as the risk of contamination from the offal is high
- always follow the correct storage procedures and temperature controls (see *Storing shellfish,* page 256)
- work away from areas where other cooked foodstuffs are being handled
- keep your preparation area clean. Wash your equipment, knives and hands regularly and employ the use of a bactericidal detergent or sanitising agent to kill bacteria
- dispose of all swabs immediately after use to prevent contamination via soiled wipes.

Essential knowledge

It is important to keep preparation and storage areas and equipment hygienic in order to:
- prevent the transfer of food poisoning bacteria to food
- prevent pest infestation in storage areas
- ensure that standards of cleanliness are maintained
- comply with the law.

Planning your time

When preparing shellfish, remember that some preparations and cooking methods are more time consuming than others. Some procedures may require only minutes and will be undertaken at a specific time (i.e. cooked to order *à la carte* style), while others may require lengthy preparation before cooking and may employ the use of fish stocks or court bouillons which must be prepared in advance.

Before starting any procedure, identify the basic culinary preparations required to complete the dish to the required establishment standard. Plan your approach to the dish carefully, addressing the longest and most time consuming jobs first. Familiarise yourself with all ingredients, cooking and storage methods involved. Make sure that you have everything in place before you start cooking the shellfish dish.

To do

- List the shellfish used in your establishment and group them under:
  - crustaceans
  - molluscs.
- Check with your supervisor how and where cooked shellfish and live shellfish are stored in your establishment.
- Identify those currently stored as live, cooked or frozen.
- Select a shellfish dish that appears on your establishment menu. Prepare a written time plan. indicating recipe requirements, mise en place and service requirements.

Equipment

Before starting to prepare or cook shellfish, decide what equipment and utensils you will need to complete the process. Arrange them, making sure they are clean and ready to use. You may need to use mechanical equipment such as mincers or food processors: do not use this equipment unless you have received instruction from your supervisor and you are also familiar with the safety procedures outlined in Units G1 and G2 in the Core Units book.

Knives

You will need a variety of cook's knives for efficient fish preparations. Read Unit 2D10: *Handling and maintaining knives* in the Core Units book, noting especially the safety points on handling knives.

You will need to be familiar with the following knives:

- filleting knife: a thin, flexible blade ideal for following bones closely
- fish scissors: used for trimming fins and tails in fish preparation
- oyster knife:
  a) a round-nosed, short-handled knife with a rigid blade and guard, designed for separating the flat shell from the bowl shell of the oyster
  b) a sharp pointed rigid short-bladed knife, designed for prising apart the shells of bivalve molluscs
- heavy duty cook's knife: a heavy bladed 25–30 cm (10–12 in) cook's knife for cracking and cutting through shells of crustaceans
- carving knife: a long thin bladed knife used for portioning raw and cooked shellfish.

Types of shellfish

Crab (*Crabe*)

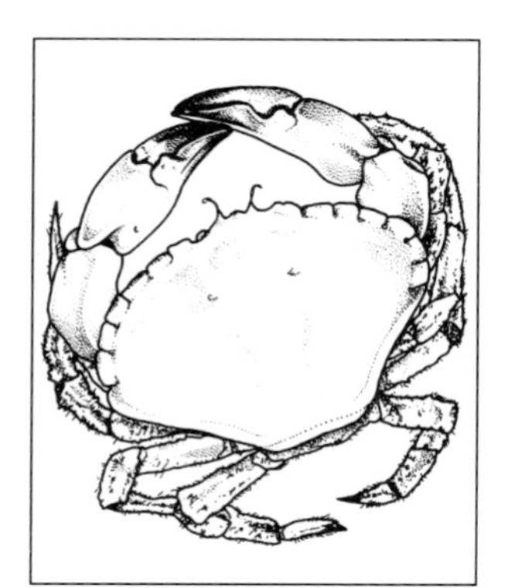
Edible brown crab

The brown edible crab is a large, powerful crab measuring over 20 cm (8 in) across the shell, although smaller ones are sometimes used. The large claws crack open to provide white meat, while brown meat is found in the shell.

Edible crabs are cooked in boiling salted water or *court-bouillon* (see page 75). Only the *dead man's fingers* (gills), the mouth and stomach bag cannot be eaten and must be discarded.

Crab is available all year round in fresh, tinned and frozen forms. Fresh crabs are best in summer.

Culinary uses

- Dressed crab: here both the brown and the white meat are arranged in the cleaned shell and usually decorated with sieved hard-boiled egg, chopped parsley, paprika, anchovy fillets and capers.
- Soup: bisque, bouillabaisse.
- Pâtés/mousses.
- Hot soufflés.

Lobster (*Homard*)

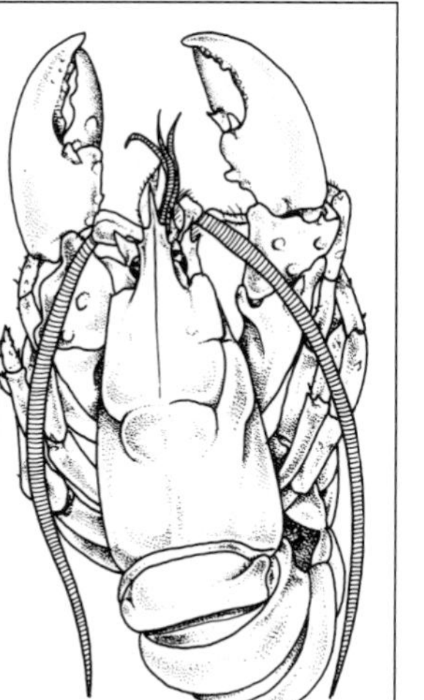
Lobster

Lobsters are a dark, bluish-black colour which changes to red when cooked. They may grow up to 60 cm (24 in) in length and have two powerful claws. The claws contain well-flavoured white meat, as does the tail/abdomen.

Hen (female) lobsters can contain eggs in the form of red roe or coral. Males and females contain creamy parts and greenish liver known as *tomalley,* which is used in soups and sauces. Raw lobster coral can be used for lobster butter to enrich sauces, while cooked lobster coral is often used for decoration and garnish.

The only part of the lobster which cannot be eaten are the intestinal tract and the gelatinous sac behind the eye, known as the *queen.* These parts are discarded because they impart a bitter flavour to the item being cooked.

Lobster is available all year round in fresh, tinned and frozen forms. Fresh lobsters are best in summer.

Culinary uses

- Soup: bisque d'homard.
- Sauce: sauce americaine.
- Grilled lobster: here the lobster is split lengthways before cooking.
- Boiled lobster: this is cooked in court-bouillon and served cold for salads or decorated for presentation on a cold buffet.
- Mouses: hot or cold.
- Soufflés: hot or cold.

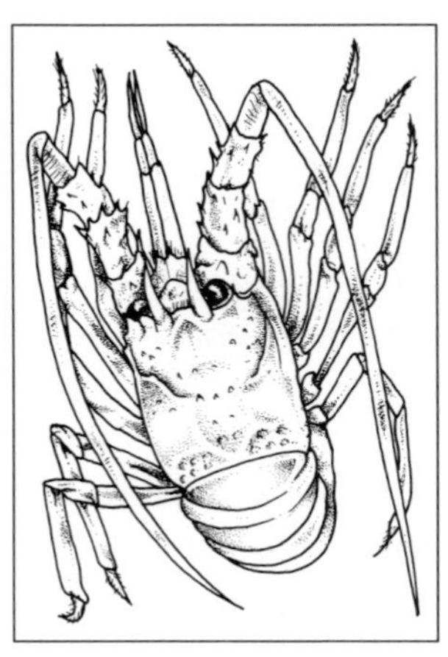

Crawfish

Crawfish/spiny lobster (*Langouste*)

Crawfish are reddish-brown in colour, with tiny spines dotted over the shell and long antennae. They resemble lobsters except that they have no claws, and unlike lobsters, they do not change colour when cooked. Crawfish may grow to 4.5–5.5 kg (10–12 lb) in weight. Most of the flesh is contained in the tail.

They are available fresh (cooked or raw) and frozen. Fresh crawfish are best in summer.

Culinary uses
As for lobster (page 259). They are particularly attractive in cold buffet displays.

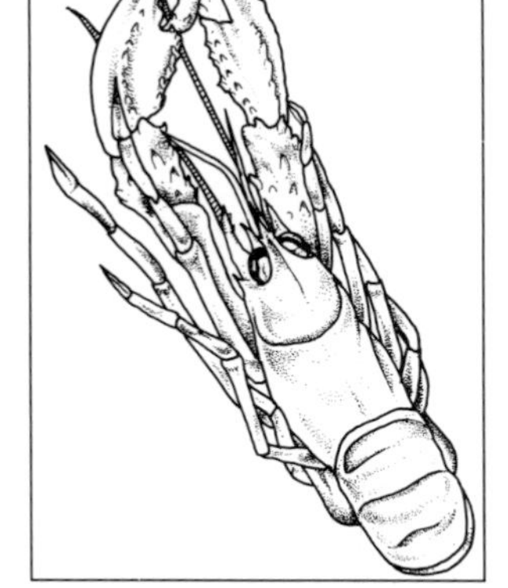

Dublin bay prawn

Dublin bay prawn/scampi (*Langoustine*)

These are rose-grey to pink in colour, and resemble a miniature lobster, growing to 18 cm (7 in) in length. Note that only the tail is used. They are available in the shell or shelled, fresh or frozen raw (encased in an ice glacé), or egg-and-crumbed ready for deep-frying.

Culinary uses
- Deep-frying: egg-and-crumbed.
- Shallow-frying: meunière style.
- Poached: with an accompanying sauce.
- Cooked as for classical lobster dishes.

Common prawn (*Crevette rose*)

These prawns are a pale, semi-transparent pink colour when raw, turning an opaque pinkish-red when cooked. Caught from shallow inshore waters, they are 9–10 cm (4 in) in length. They are usually only available cooked (whole or the tail), shelled, frozen or tinned.

Culinary uses
- Cocktails: usually accompanied by a mayonnaise-based sauce.
- Salads.
- Soup: bisque de crevettes.
- Garnish: for fish, egg, chicken, veal, avocado and rice dishes.

Jumbo prawn

Jumbo prawn/Mediterranean prawn (*Crevette rouge*)

These large prawns grow up to 20 cm (8 in) long and are light-pinkish to yellowish-grey in colour, turning pink-red when cooked.

Culinary uses
- Hors d'oeuvre: whole, unpeeled on a bed of crushed ice.
- Salads.
- Grilled or shallow-fried (raw tails only).
- Barbecued.

Shrimp

Shrimps are smaller than prawns, semi-transparent grey in colour with dark spots, changing to reddish-brown when cooked. They are usually only available cooked, either shelled, smoked, frozen, tinned or dried.

Culinary uses

- Soup: bisque de crevettes.
- Potted in butter.
- Snack item.

Crayfish (*Ecrevisse*)

This is a freshwater crustacean that lives in the muddy banks of rivers, streams and lakes. In appearance it is like a miniature lobster. Crayfish are pale pink in colour, changing to a dark reddish-brown when cooked. Only the tails are eaten, after removing the intestinal tract (failure to do this will render the flesh bitter). This is done by depressing the centre of the tail carapace and removing the middle part of the tail, extracting the gut in one piece. This process should be undertaken while the crayfish is live, although an alternative is to plunge the crayfish in boiling water for two minutes to kill humanely prior to cleaning.

Crayfish are available whole and fresh (cooked and raw) and frozen.

Freshwater crayfish

Culinary uses

- Soup: bisque d'ecrevisses.
- Boiled or stewed.
- Mousse: hot or cold.
- Sauce: Nantua sauce.
- Garnish: hot and cold for fish and chicken.

Scallop (*Coquille St Jacque*)

Scallops have a fan-shaped, ribbed, pinkish-red convex upper shell, and a white and flat under shell. The edible parts of a scallop consist of the large, round, white muscle and the orange-red tongue or coral. The frill is discarded. Open as illustrated or purchase opened in half shell.

They are available whole and uncleaned in their shell; freshly cleaned in a half shell or cleaned and frozen as king or queen scallops.

Culinary uses

- Ingredient for *Fruits de mer* mixture.
- Mousses: hot or cold.
- Stuffing for fish, supremes of chicken, escalopes of turkey.
- Poached: deep-poached in court-bouillon.
- Deep-fried and served with a suitable sauce or garnish.

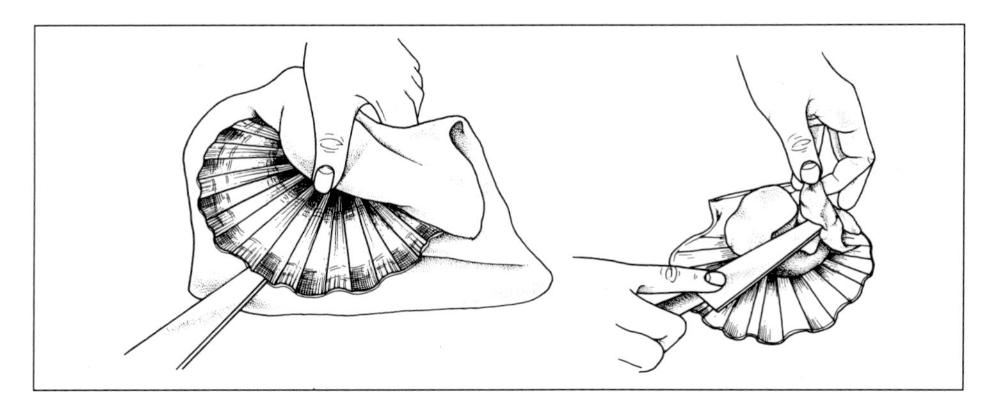

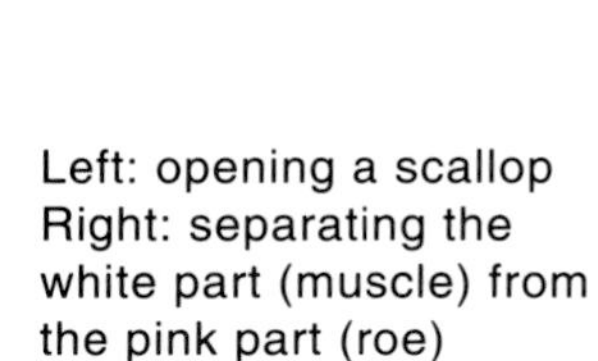

Left: opening a scallop
Right: separating the white part (muscle) from the pink part (roe)

Mussels (*Moules*)

Mussels are bivalves with a bluish-black shell, which has a pointed hinged end.

To prepare mussels, scrape the shells with a knife or brush to remove the barnacles, then place them in clean salted water. They will stay alive for several hours expelling sand and grit. Mussels have a collection of fine hair-like strands, known as the beard, along the hinge of the shell; this must be removed with a knife prior to cooking.

Mussels are available fresh in the autumn and winter, and frozen and pickled all year.

Culinary uses

- Soup: as the main ingredient or as a garnish item.
- Garnish: for rice, pasta and fish dishes.
- Popular mussel dishes include stuffed, curried, grilled, poulette and marinière.

Oyster (*Huître*)

Oysters are the most highly valued of all the molluscs. They are greyish in colour with a rough ridged shell. A number of varieties are cultivated, including the native, Portuguese and Marenne. They are eaten raw when there is an *R* in the month (i.e. they are not eaten from May to August). Oysters are opened at the hinged end, turned, bearded, and placed into the convex upper shell. Use an oyster knife to open the hinged shell: this has a short broad blade and a large guard to protect the hand.

Culinary uses

- Raw: with lemon juice.
- Garnish: for steak-and-kidney pudding, carpet bagger steak, and fish dishes.
- Oysters can be steamed, poached, grilled and baked, and served with various sauces or garnishes, e.g. Huître florentine.

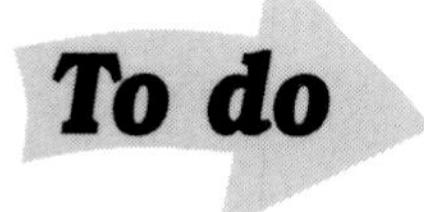

- Find examples of the following types of prawn: Dublin bay prawn, common prawn, jumbo (Mediterranean) prawn. Notice the difference in shapes and colour.
- Watch your supervisor preparing mussels.
- Under supervision, prepare some oysters to be eaten raw, using an oyster knife to open the shells.

Cooking methods

Shellfish can be cooked by a variety of methods, including: poaching, grilling, barbecuing, boiling, steaming and frying; they are a very versatile food item.

Key points

- Shellfish cooks very quickly and can easily be over-cooked, resulting in loss of flavour and a tough chewy texture.
- If the shellfish is to be served hot, it must be very hot and not held in a cooked state for prolonged periods of time. This would result in dry flesh and poor eating quality.
- Shellfish are highly perishable; stock must be rotated frequently and kept at the correct chilled temperature prior to cooking
- Shellfish has a delicate flavour and lends itself to simple preparations and garnishes, giving excellent results quite simply.
- Poached shellfish may be poached in fish stock or wine (see Unit 2D2: *Preparing and cooking fish,* page 75).

Cooking crustaceans

Boiling crustaceans

Lobster, crawfish, crayfish, Dublin Bay prawns and crabs are crustaceans that may be cooked by boiling. Many of the crustaceans purchased are already cooked by this method (or by steaming).

Crustaceans may be boiled in:

- boiling salted water with a little vinegar, or
- boiling pre-prepared court-bouillon (see Unit 2D2: *Preparing and cooking fish,* page 75).

Key points

- Allow approximately 15 minutes per lb for the first pound and 5 minutes per pound thereafter for lobster, crawfish and crabs in the simmering liquid.
- Crayfish require cooking times relative to their sizes; approximately 5–7 minutes will ensure thorough but not over-cooking. Make sure that you have removed the intestinal tract before cooking.
- Crustaceans may be removed from the boiling liquor and allowed to cool, or may be allowed to cool with the cooking liquor. If cooling in the liquor: reduce the overall cooking time by 20 per cent.
- After cooking the crustaceans may be served hot or cold, although further preparation is required for dressing cold lobsters, crawfish or crabs.
- For application and processes of poaching, grilling and shallow frying see Unit 2D2: *Preparing and cooking fish,* page 69.

Essential knowledge

Time and temperature are important when cooking fresh shellfish in order to:

- ensure a correctly cooked dish. Over-cooking of crustaceans or molluscs will render the flesh tough and dry, and affect the eating quality. Hot shellfish should also be cooked and served immediately: if held for service the quality of the dish will be badly affected
- prevent food poisoning.

Preparing a cooked lobster for cold presentation

1 Twist and remove the claws, legs and pincers from the body.
2 Crack open the legs and claws with a mallet or heavy knife. Remove the flesh and set aside. Discard the blade cartilage in the centre of the claw meat.
3 Split the lobster in half using a large, heavy knife. Draw the knife through the head towards the eye; then reinsert the knife and draw in the opposite direction, cutting completely through the tail or carapace. Pull into two distinct halves.
4 Discard the gelatinous sac in the top of the head directly behind the eyes.
5 Remove and discard the intestinal tract or canal which runs through the middle of the tail flesh.
6 Scrape out the soft head meat and the greenish liver or *tomalley,* and reserve this for other uses (butters, soups or sauces).
7 Remove and discard the gills situated in the cavities to either side of the body.
8 Remove the tail flesh in one piece keeping the red skin-like coating intact. Wash the shells thoroughly, then replace the claw meat into the head cavity and the tail meat into the carapace (this may be sliced or left whole).
9 Garnish to establishment requirements: normally a light salad garnish and lemon, accompanied by a sauce mayonnaise. Chill and serve very cold.

Preparing crab for cold presentation

Wash the live crab then plunge it into a boiling court-bouillon for 15–20 minutes according to size. Allow to cool in the liquor.

Dressed crab (illustrated overleaf)
1 Remove the large claws and sever at the joints.
2 Remove the flexible pincer from the claw.
3 Crack carefully and remove all flesh.
4 Remove the flesh from two remaining joints with the back of a spoon. (This is known as *white meat*).
5 Carefully remove and discard the soft under-shell, mouth and stomach bag.
6 Discard the gills (*dead man's fingers*) and the sac behind the eyes.
7 Scrape out the whole of the inside of the shell and pass through a sieve.
8 Season with salt, pepper, Worcester sauce and a little mayonnaise sauce, then thicken lightly with fresh white breadcrumbs. This mixture forms the *brown meat.*
9 Trim the shell by tapping carefully along the natural line.
10 Scrub the shell thoroughly and leave to dry.
11 Dress the brown meat down the centre of the shell.
12 Shred the white meat, taking care to remove any small pieces of shell and then season with salt, pepper and a little vinaigrette sauce.
13 Dress neatly on either side of the brown meat.
14 Decorate as required, using any of the following: chopped parsley, hard-boiled white and yolk of egg, anchovies, capers, olives.
15 Serve the crab on a flat dish, garnished with lettuce leaves, quarters of tomato and the legs. Serve a vinaigrette or mayonnaise sauce separately.

Preparing a cooked crab

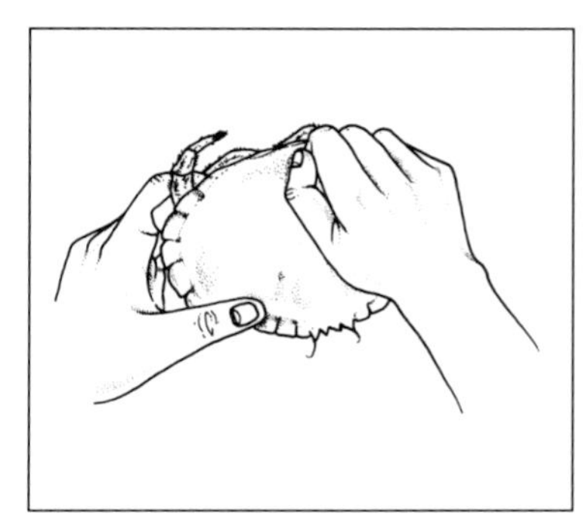
Loosen the shell by hitting the crab firmly on the back

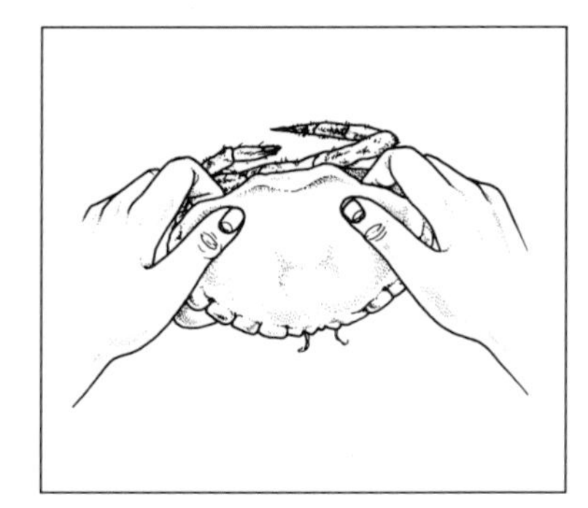
Stand the crab on one side with the shell facing you then force the body from the shell with your thumbs

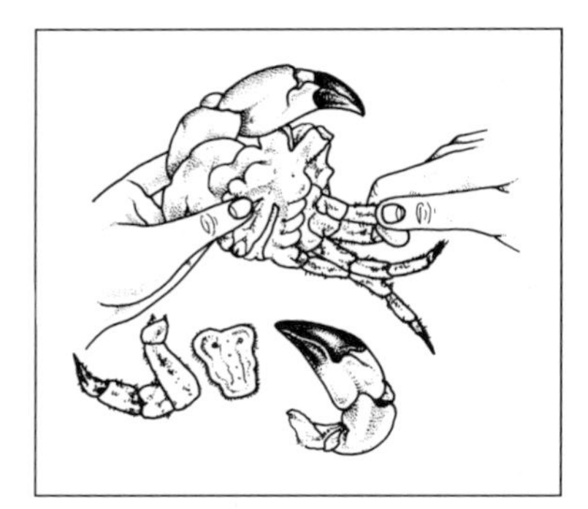
Twist off the tail, legs and claws. Crack open the legs and claws to remove the meat

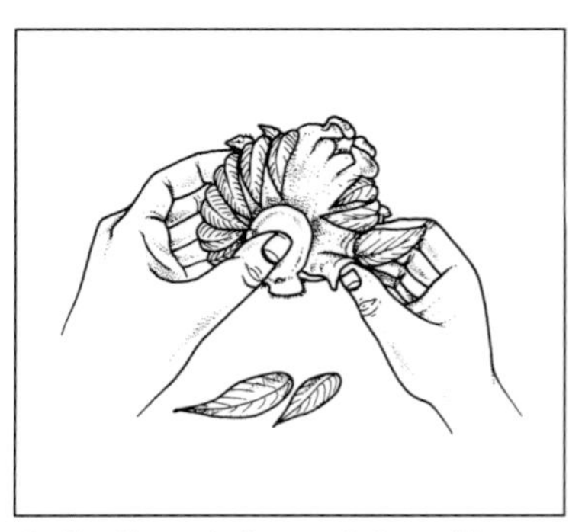
Pull off and discard the gills

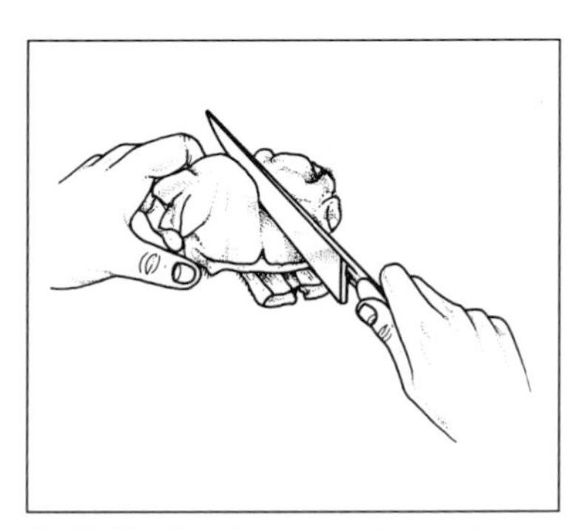
Split the body open down the centre and prise out the meat

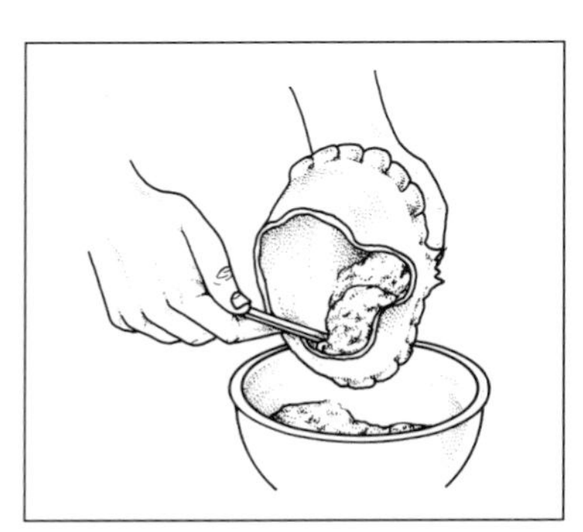
Scrape the brown meat from the shell

Preparing lobster for a complex hot presentation

Lobster thermidor (*Homard thermidor*)

1. Plunge the live lobster into a simmering court-bouillon, reboil and set to simmer gently for approximately 15 minutes per pound, depending on the size of the lobster.
2. Remove and allow to cool.
3. Place the cooled lobster onto a board and remove the claws, cracking and removing the flesh intact.
4. Split the lobster through the centre into two halves, then carefully remove the tail meat and creamy parts. Trim the fingers back to the shell and thoroughly clean.
5. Place on a baking tray and insert in a moderate oven to dry and heat lightly.
6. Slice the lobster tail and claw and reserve.
7. In a suitable pan, sweat some chopped shallots then add white wine, fish stock and bruised parsley stalks. Reduce to a syrupy consistency, remove the parsley stalks and add chopped parsley and double cream.
8. Reduce the mixture futher, then add 2 or 3 dessertspoons of high quality Mornay sauce to the required consistency without allowing the mixture to boil.

9 Put the pan to the side of the stove and cool slightly, *monter au beurre* (work up with butter) and add diluted English mustard to taste. Correct the seasoning.

10 In a minimum amount of butter, carefully sauté slices of lobster, gently binding off the heat with the prepared sauce.

11 Remove the shells from the oven and coat the base with the sauce. Arrange the lobster neatly in the shell or carapace, coat with sauce and place in a hot oven to glaze. Alternatively, finish under the salamander.

12 Remove the lobster from the tray and serve on a serviette on a polished silver flat.

Notes

- At Step 8, a spoonful of hollandaise is often used in conjunction with the sauce Mornay.
- A slice of truffle may be placed on top of the finished lobster; or interspersed with carefully arranged slices of lobster before glazing.

Grilling lobsters and crawfish

Lobsters and crawfish may be blanched in boiling salted water, split and then grilled, or split from raw, seasoned and grilled. If the lobster is split live the intestinal tract and stomach must be removed prior to seasoning and cooking. If the lobster has been blanched to kill and then split the cord must be removed at this stage.

Serve grilled lobster with a warm butter sauce or savoury butter, lemon and parsley.

Essential knowledge

The main contamination threats when preparing and cooking fresh shellfish are as follows:

- contamination from the food handler due to inadequate personal hygiene; i.e. from the nose, mouth, open cuts and sores and unclean hands to the food
- cross-contamination from use of unhygienic equipment, utensils and preparation areas
- contamination through products being stored at incorrect temperatures. Fresh, live shellfish are at risk if stored outside the range of 0–3 °C (32–37 °F). They will deteriorate quickly above that temperature and increase the risk of bacterial growth. Live crustaceans should be cooked and cooled quickly in court-bouillon and stored below 5 °C (41 °F) in a separate container from any raw commodities to avoid cross-contamination
- contamination through incorrect thawing procedures when shellfish is prepared from frozen. Frozen shellfish must be defrosted in a refrigerator and never re-frozen; this is dangerous to health
- use of dead or dying shellfish. These are very likely to be contaminated and must never be used
- contamination through incorrect disposal of waste
- contamination through products being left uncovered.

Cooking shrimps and prawns

Most shrimps and prawns are purchased pre-cooked, the exception being king or *jumbo* prawns which are purchased fresh or frozen uncooked in the shell. Remember that in shrimps and prawns only the tails are edible.

As for other crustaceans, a court-bouillon may be used for king prawns although grilling, shallow-frying, deep-frying and barbecuing are excellent methods of cooking this delicacy.

Menu examples: Grilled king prawns with garlic and parsley butter; Shallow-fried king prawns with bacon and garlic; Deep-fried butterfly king prawns with tartare sauce; Marinaded barbecued king prawns.

The smaller shrimps or prawns are used in an enormous variety of ways, such as: stir-fried with rice and vegetables; cooked in a combination for paella, soups, mousses, soufflés; potted; or simply prepared as a chilled shrimp or prawn cocktail.

Scampi/Dublin Bay prawns

Scampi can be cooked in a number of ways:

- boiled in a court-bouillon and served hot or cold
- shallow-fried à la meunière
- grilled with a savoury butter,
- incorporated into a stir-fry mixture with rice and vegetables
- poached with a sauce
- (most popularly) coated with flour, eggwash, breadcrumbs and deep fried, eaten with a tartare sauce.

When cooking frozen scampi, ensure that the flesh is defrosted *thoroughly* to permit even cooking, otherwise only surface bacteria may be destroyed.

In their shell, scampi make an excellent garnish and are most handsome on a cold buffet display. The peeled tails are used for kebabs, mousses, mousselines, terrines of shellfish and fish stews. The shells are reserved for use in bisque style soups or shellfish sauces.

Cooking molluscs

Cooking mussels

Mussels produce a large quantity of moisture as they cook, so you need only add a small proportion of liquid to assist cooking.

They are cooked or blanched in a pan over a fast direct heat. The mussels are cooked when the shell opens. Take care not to over-cook the mussels as this will affect the texture, rendering them tough and chewy.

The resulting cooking liquor may form the basis of an accompanying sauce, as in the case of the classic Moules marinière, or the cooked mussels may form a garnish for another dish.

Health and safety
Mussels feed by filtering out nutrients from the water. This makes them subject to contamination by disease organisms and bacteria. It is essential to purchase mussels only from a reputable supplier and to discard any open shells before cooking.

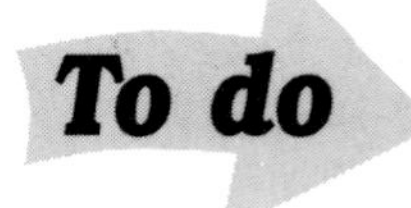

- Read Unit 2D2: *Preparing and cooking fish,* (pages 57–80).
- Watch your chef or supervisor preparing a cooked lobster for cold presentation. Notice the equipment and techniques used.
- Find out whether grilled lobsters and crawfish are usually blanched before grilling in your establishment. Ask your supervisor why this decision has been made.
- Check the temperature of stored live shellfish. Does it fall within the correct range?

Cooking scallops (*Coquilles St Jacques*)

Scallops may be shallow-fried, lightly grilled or poached and served with a wide variety of sauces. Note that if over-cooked, scallops become tough and chewy and completely spoiled. When poaching it is advisable to detach the orange roe to reduce the cooking time. Owing to of the delicate flavour of the scallop, care must be taken when selecting the appropriate sauce or garnish as the natural delicacy can be lost by the use of strong, over-powering flavours.

Cooking oysters (*Huîtres*)

Oysters are considered by gourmets to be at their best when consumed raw with a little lemon juice. When oysters are to be eaten raw they should not be prepared until they are required for service in order to retain their flavour and natural liquids.

Preparing oysters

1. Hold the oyster in one hand in a cloth for protection.
2. Using a special oyster knife, prise the shells are apart through the natural ligament that joins the shells (this is at the narrow part of the oyster).
3. Discard the flat shell, without spilling the juices, and loosen the oyster in the shell. Serve on a bed of crushed ice.

Oysters may be lightly grilled from raw, wrapped in bacon and used for a savoury (e.g. Angels on horseback), served on toast, or they may be lightly poached and served with an accompanying sauce, (e.g. Oysters florentine). Care is needed in cooking oysters as they become tough if over-cooked or from being stored for a long period at a high temperature.

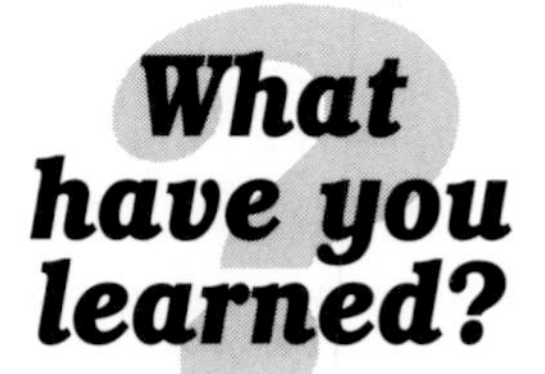

1. What are the main contamination threats when preparing and cooking fresh shellfish?
2. Why is it important to keep preparation and storage areas and equipment hygienic?
3. Why are time and temperature important when cooking fresh shellfish?
4. Why must dead or dying molluscs be discarded? How can you tell if they are dead?
5. How should live shellfish be stored?
6. When preparing cooked lobster and cooked crab, which parts are discarded?
7. What effect does over-cooking have on shellfish?
8. Why is it essential to wash shellfish prior to cooking?

Extend your knowledge

1. Find out about shellfish not covered by this unit. What types of shellfish are they? How might you prepare and cook them? What part of the world are they from?
2. Find out how long shellfish take to mature in their natural habitat until they are ready for eating. List the different types of shellfish and their growing time.
3. Some shellfish are farmed. Find out which ones are and which ones are not. Why is this?
4. Find out about preserving shellfish. Many shellfish may be frozen but some may be preserved by other methods. What are these methods and which types of shellfish are they used for? Find out how preserved shellfish may be used in menus.

UNIT 2D15

Preparing cook-chill food

ELEMENTS 1 AND 2: **Portioning, packing, blast-chilling and storing cook-chill food**

What do you have to do?

- Portion and pack food in accordance with laid down procedures.
- Blast-chill and store food in accordance with laid down procedures.
- Seal and label food containers correctly.
- Monitor and record food temperatures.
- Follow stock rotation procedures and maintain accurate records.
- Store cook-chill items under the correct conditions in accordance with laid down procedures.
- Keep accurate records of received, stored and issued food items.
- Carry out your work in an organised and efficient manner in accordance with current regulations.

What do you need to know?

- The required type, quality and quantity of food for cook-chilling.
- How and when food temperatures are monitored and recorded.
- Why food containers must be sealed and labelled correctly.
- Why food items must be handled carefully and remain undamaged at all stages.
- Why storage areas must be kept clean and tidy, and secure from unauthorised access.
- Why time and food temperatures are important when preparing cook-chill foods.
- How cook-chill food (in food containers) can be transported undamaged to storage areas within a required time.
- How to satisfy health, safety and hygiene regulations concerning preparation areas and equipment both before and after use.
- How to plan your work to meet daily schedules.
- How to deal with unexpected situations.

Introduction

A cook-chill system is one where food is prepared before it is required, quickly chilled to 0–3 °C (32–37 °F) and stored at that temperature for up to five days until required for sale or consumption. It is a system that is widely used in institutional catering such as hospitals, schools and industrial catering. An increasing number of small commercial outlets within the hotel and restaurant, in-flight and catering-on-the-move sectors also pre-prepare and store finished dishes, and are therefore using the cook-chill system although this is not always recognised as such.

Advantages of the system

Cook-chill production and service offers a number of advantages to the caterer:

- *a cost-effective use of labour.* Food may be prepared within normal working hours when there is the most available labour, and served at times when labour is more expensive; e.g. very early mornings, evenings, weekends and holiday periods
- *flexibility in the menu.* A variety of stored food allows the caterer to offer a greater number of dishes without suffering the potential wastage of conventional cooking and serving systems
- *consistent standards.* A standard quality of dishes is more easily attained because the dishes are produced in bulk before being individually packed. This applies particularly where the production occurs in a large kitchen or Central Production Unit (CPU)
- *fresh-tasting food.* Cook-chilling does not allow ice crystals to form in the foods, so the regenerated (re-heated) dishes are closer to their original taste, colour and texture than is the case with some frozen foods or foods that have been kept hot for a long time
- *fast service times.* The regeneration (reheating) of chilled food is very quick in comparison with frozen products; dishes may be removed from chill cabinets only minutes before being served to the client
- *a lower equipment requirement* in the regenerating kitchen. The kitchens where regeneration takes place (known as satellite kitchens) may be some distance from the Central Production Unit (CPU), and these need less equipment than for conventional cooking and, provided there are strictly enforced guidelines, fewer skilled staff.

Disadvantages of the system

As in any system, there are also a number of disadvantages that have to be considered before adopting a cook-chill operation:

- *high equipment cost at production stage.* The cost of buying the quick chiller, packaging material, general equipment, chilled storage facilities and temperature control and monitoring equipment is considerable
- *the need for a large working space.* Cook-chill usually demands more space than conventional production, both for storage and packing
- *higher risk of food poisoning.* The potential for food poisoning on an epidemic scale is greater, since food is usually cooked in greater bulk than conventional cooking, temperature controls are more critical and there is a possibility of poor stock control. These factors can all cause contamination to take place in large quantities of food
- *risk of nutritional loss.* In addition to the food poisoning potential, if the highest standards of cooking and storage are not employed there is an added danger of nutritional and microbiological decay, reducing the flavour, texture, taste and food value of the dishes
- *detailed re-training* of management and production staff. This is needed to obtain maximum benefit from the system and to avoid the new range of hazards. Such training should include the adoption of Hazard Analysis Critical Control Point procedures in which each step of the process is analysed. This pinpoints potential hazards, allowing management to develop any procedures necessary to minimise risk.

ELEMENT 1: Portioning, packing and blast-chilling food

Suitable types of food

Meat and poultry

All meat, poultry, game and offal can be successfully chilled. Meat dishes that need to be sliced, such as roast beef, are cooked, chilled, quickly sliced and then packed for storage. It is not possible to serve rare (acceptably under-done) roast, fried or grilled meat dishes using cook-chill, because the regenerated food must be heated to an internal temperature of 70 °C (158 °F) for 2 minutes, thoroughly cooking the food.

Vegetables and fruits

Most cooked potato, vegetable and fruit dishes are suitable for cook-chill, with the exception of dishes containing a mixture of cooked and uncooked ingredients, such as composite salads. Salad items such as lettuce, watercress, radiccio, endive and other leaf salad vegetables should not be incorporated into dishes prior to chilling but added fresh before service.

Vegetables reheated and garnished ready for service

Fish

All pre-cooked fish dishes are suitable for cook-chill processing. However, if modern combination regeneration ovens are not available, deep-fried fish dishes are best cooked completely in the finishing kitchen for direct service to the client.

Sauces and soups

Most sauces and soups are suitable for cook-chill, although those using egg as a thickening agent and those with a high fat content (e.g. hollandaise, béarnaise, mousseline sauces) need particular care when regenerating to avoid curdling. Soups thickened with a liaison (cream and egg yolk), must have this added immediately before service rather than before being chilled.

Egg dishes

Most egg dishes are easily prepared on demand and cook-chill offers no great advantage to the caterer in this area. However, omelettes, scrambled, poached and fried eggs provide a satisfactory end product. Some caterers, particularly airlines, use eggs as part of meals provided as breakfasts.

Desserts

Basic cooked desserts chill well, with the exception of those that rely on batters lifted by eggs, such as soufflés. Mousses, jellies, pastry goods, yeast goods and fruits are all suitable for chilling. Composite sweet dishes that require a hot base with a raw fruit or cream topping are not suitable unless the two components are packed separately.

Preparing, cooking and chilling dishes

Guidelines for these are given in the table on pages 274–6.

Health, safety and hygiene

Note all points given in the Core Units book concerning general attention to health, safety and hygiene. When chilling and storing foods, the following points are particularly important:

- *hygiene*: since the objective of chilling is to extend the life of cooked food, hygienic preparation and storage is of prime importance. The extended life of the product also extends the opportunity for dangerous bacteria (particularly salmonella, stapylococcus aureus, clostridium perfringens, escherichia coli and listeria) to contaminate and grow in the food
- *time and temperature*: the limits for these must be adhered to very strictly and accurate records of time and temperatures of stored goods must be kept
- *prevention of cross-contamination*: the possibility of cross-contamination between cooked and uncooked foods must be kept to an absolute minimum.

Time and temperature

Some bacteria leave dangerous toxins in food even after they have been killed and others can protect themselves from the effects of heat. This is why the strict time and temperature controls exist and why they must be adhered to by all concerned. All of the following points are essential when carrying out preparation within the cook-chill process.

- The temperature of uncooked ingredients delivered from stores must not exceed 5 °C/41 °F for high risk foods such as poultry, fish, dairy produce and cooked and uncooked meat. The temperature of food must not exceed this level during the preparation process.
- The temperature of cooked food must reach a minimum of 70 °C (158 °C) for at least 2 minutes and chilling should commence within 30 minutes.
  Reason: this temperature will kill the majority of harmful bacteria which cause food poisoning. By chilling food quickly to a temperature which prevents bacterial growth, you prevent any remaining bacteria from having time to multiply.
- Food must be chilled down to 3 °C (37 °F) within a period of 90 minutes. Whole joints above 2.5 kg ($5^1/_2$ lb) may not be used.
  Reason: 90 minutes is the shortest practical time that this temperature can be achieved in a controlled way. Large joints (being dense and thick) cannot meet this requirement and so cannot be safely reheated.

See *Temperature recording* in Element 2 on page 286 for temperature control during the storage and transportation stages.

Preparing, cooking and chilling dishes for the cook-chill process. *Source*: Regethermic UK Ltd.

Food item	Preparation	Cooking	Chilling/chilling times	Portioning plated & bulk
Fish:				
Fried, meunière, fish cakes and fingers	Preparation of batter: make a thicker mixture or prepare mixture using an oil batter. By using this type of mixture the batter will not break away from the surface of the fish and it ensures a crisp result	In usual way. As fish is a delicate product do not overcook or flesh becomes flaky, difficult to handle and even dry	Place fish onto cooling racks, lined with damp greasproof paper (to prevent sticking). Ensure bottom side of fish uppermost. *Do not overlap,* but ensure even circulation of air around product. Time: 20 mins.	Turn fish to correct way up and ensure that in both bulk/plated systems, battered goods do not overlap to prevent sticking. This is not so important with fish cakes, etc. Brush fried fillets with butter for attractive glaze. Garnish with lemon wedges after regeneration. Do not place parsley on lemon, or parsley will turn yellow
Poached		In usual way, making sure fish does not break up during cooking process to spoil presentation	Chill in cooking liquor to avoid surface drying. Cover once chilled to 3 °C (37 °F). Time: 30 mins.	Place onto plate or in dish as usual. Garnish. If product is to be served with a sauce, chill fish and sauce separately and put together when cold
Red/white meats:				
Roasted	Butcher joints into 2–3 kg (5–7 lb) units if chilling as whole joints	In usual way ensuring correct seasoning. Red meat should be cooked rare/medium rare	Allow meat to cool and place on clean dishes to chill. By chilling whole, meat juices are retained and there is little flavour loss. If chilling slices, chilling times will be reduced but meat liquor will need to be doubled. Time: whole 90 mins sliced 30 mins.	Cut medium slices. Bulk portion: place a few vegetables or meat trimming along one corner of dish. Lay first slice on top of this layer and place each successive slice overlapping the one below. *Never* put thick gravy on meat: use a stock or jus lié. Plated: arrange slices of meat to the side of vegetables, preventing direct contact with the plate. Chicken joints: ensure a large surface area is in contact with the dish and that height of joint is not too great and so touching underside of lid
Boiled	Small joints 2–3 kg (5–7 lb) each	In usual way. Can be immersed in cold water to partly cool. Drain meat prior to chilling	Chill in trays until core temperature 2 °C (35 °F). Best results obtained from joints chilled whole. Time: whole 90 mins. sliced 30 mins.	As for sliced roast meats. If required, coat with cold sauce when meat is cold

Food item	Preparation	Cooking	Chilling/chilling times	Portioning plated & bulk
Red/white meats:				
Grilled (chops, steaks, hamburgers, cutlets)	Ensure chops, etc. are not too thick: average 1–2 cm ($^1/_2$–$^3/_4$ in)	Seal under very hot grill to seal in the juices. Cook rare/medium rare. Hamburger mixture is best served the day after cooking	Chill on trays in single thickness to ensure fat drains to the bottom of the dish. Ensure product chilled immediately after grilling. Time: 30 mins.	As for roast meat avoid direct contact with the plate/dish. Arrange overlapping items as for roast sliced meat. No food should touch the underside of the lid. Garnish: tomato, water-cress, etc. Do not add liquid
Fried (bacon, burgers, sausages)		In normal way ensuring product is fully cooked. Drain off fat	Chill in single layers to ensure fat drains to bottom of dish. Once chilled, place into clean dishes to leave excess fat behind. Time: 30 mins.	Bacon: as for sliced meat. Sausages: single thicknesses in contact with the dish. Burgers: overlapped as for chops, steaks, etc.
Stewed/braised (casseroles, stews)	Cut meat into small portions and serve two small pieces instead of one large one to avoid undue thickness	In usual way. Check that flour based sauces are *fully* cooked out. If not, sauce will thicken during regeneration	Chill in bulk maximum thickness of 6 cm (2$^1/_2$ in). If a finer presentation is required, meat/sauce can be reportioned after chilling. Time: 60 mins.	Bulk: already portioned prior to chilling. Plated: remove meat from sauce, re-plate then pour sauce over product after skin is stirred in
Offal: Liver, kidneys, etc.	Slice into 1 cm ($^1/_2$–$^3/_4$ in) units	Liver: with or without flour. Cook as normal. Drain	Chill on cooling racks to allow blood to drain away from product and thus prevent discolouration and tainting. Time: 30 mins.	As for roast, sliced meats. Coat with cold sauce or onion gravy if required
Binding sauces:	Make sauce thinner than usual (by approx. 15%)	Check that the sauce is fully cooked out. If not flour will absorb more liquid	Chill immediately in shallow bulk dishes. Agitate during chilling to ensure product is chilled to the centre. Time: 30 mins.	Never portion sauce into bulk containers with no solid content. When sauce is cold, eradicate skin and nappe over item to be coated. Check none is in contact with lid
Eggs: Scrambled	Season egg/milk mixture well	Cook until egg begins to scramble, remove from heat and allow heat of product to complete process to a soft/runny consistency	Transfer immediately to shallow dishes and chill. Stir during chilling. Consistency will thicken during chilling as temperature drops. Time: 20 mins.	Place cold egg mixture onto cold buttered toast, in pastry case or on fried bread. Procedure is the same for bulk and plated. Depth 2.5 cm (1 in) or less in bulk dishes. If insulating layer not used, regeneration time must be reduced

Food item	Preparation	Cooking	Chilling/chilling times	Portioning plated & bulk
Potatoes:				
Mashed/creamed		Add more liquid to make product looser and less dense, thereby assisting the chilling and regeneration stages, as the potato absorbs more liquid when chilled. Season well	Bulk: 50–60 minutes. Plated: (individual piped potato, etc.): 30 mins. e.g. by piping or fluting to expose maximum surface area	Prior to chilling either using a piping bag or by filling dish to maximum 2.5 cm (1 in) depth and fluting surface with palette knife or fish slice. Portion to ensure even heat penetration,
Roast/sauté/ parmentier	Cut potatoes into quite small pieces ensuring adequate heat penetration in specified time	Blanch, finishing off in oven/ frying pan/fryer until potatoes are golden brown and crisp. Season	Chill as quickly as possible, avoiding undue heaping of product in dish, to extend chilling time. Time: 45 mins.	Ensure seasoning is correct and there is a single layer of potato in dish to ensure adequate heat penetration.
Baked jacket	When selecting potatoes for baking make sure they are not too large: 100–15 g (4–5 oz). This assists regeneration and ensures they are within recommended food depth	Wash/dry. Rub salt into skins as usual. Bake. Rub off excess salt potato is to be filled, chill filling separately. Time: 45 mins.	As quickly as possible avoiding undue heaping of vegetables to ensure product chills fast. If are large, cut neatly in half	In normal fashion ensuring potato is not in contact with underside of lid. If potatoes
Vegetables, rice, pasta, pulses and boiled potatoes:				
Green and root vegetables, boiled potatoes, boiled rice, pasta, pulses	In normal way. For boiled potatoes ensure potatoes are cut into small pieces (about the size of new potatoes)	In normal way using plenty of salt. Best cooked *en croquant* (with a bite). Once cooked, refresh under cold running water and glaze with a little melted butter/margarine (optional). Re-season	Product can be chilled following portioning. Ensure rice and pasta are chilled in shallow layers to prevent bottom layers becoming starchy and to minimise chilling times. Time: 15 mins.	When portioning the heavier products, e.g. rice, swede, pasta, etc. remember to ensure depth in dish is within limits to ensure heat penetration. Any vegetable touching underside of lid will dehydrate and discolour
Tinned	Open tin and drain away any unwanted juice or liquid	No need for further cooking in service dishes after portioning. Although vegetable is already cold it is very important that the item is chilled to 3 °C (37 °F). Time: 15 mins.	Once drained, re-season and chill	As above, reheated as a freshly prepared vegetable
Pastry goods:				
Savoury, fruit	Best results are obtained from pastry goods cooked in bulk and portioned or enclosed as individual items, e.g. Cornish pasties	In normal way. Maximum thickness 4 cm ($1\frac{1}{2}$ in). Check all pastry including crimped edges are contained within depth of dish (i.e. not standing proud)	Chill in bulk before cutting or portioning. Time: 1–2 hours (depending on density of content)	Cut when cold or shrinkage away from cut line will occur, exposing contents. Portion onto plates cold, so contents do not run over plate (they gel when cold). Do not pour sauce over surface of pastry or it will be soggy.

Essential knowledge

Time and temperature are important when preparing cook-chill food in order to:

- prevent contamination from food poisoning bacteria
- maintain the food value of the products
- retain the taste and texture of the dishes
- allow standard regeneration (reheating) processes to be used.

Planning your time

The time between cooking food, packing it and chilling it is very short. For this reason it is very important that every stage of the portioning-packing-chilling-storing cycle of the operation is planned with precision. Problems will arise if dishes are produced more quickly than they can be packed, or packed when there is not enough chilling or storage space available.

The Operations Manager is responsible for scheduling production so that none of the sections of the operation is required to work above its maximum capacity. Within the schedule, the staff responsible for each operation have to ensure that they have the necessary equipment to perform the task successfully.

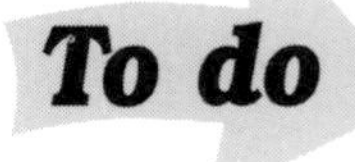

- Find out the colour coding system for preparation equipment in your kitchen. List the colours used and the food items they refer to.
- Check and note the temperatures of your chiller and holding refrigerators. Do they comply with the regulations?
- Look at temperature records and note any changes that have occurred.

Production section

The manager will produce written schedules for each area. An example of a production section schedule is given at the top of the next page.

The volume of the production unit will vary according to the scale of the operation and will be controlled by the Unit Manager.

The production schedule will go to the stores so that the correct quantities of ingredients will be available to the cooks at the time required. The schedule will allow enough time between the preparation of dishes and the arrival of ingredients for the next batch of dishes for cleaning down and sterilisation of equipment and work areas. This is a necessary precaution against cross-contamination.

Selecting containers

When selecting the containers for cook-chill, you need to ensure that they are capable of withstanding both the chilling and reheating process without changing their structure or contaminating the food.

PRODUCTION SECTION: Schedule

DATE..5 June 1993

SHIFT LEADER ....John Hurst

DISH PRODUCTION (1): Braised beef

RECIPE NUMBER..26

BATCH NUMBER..15131E3

QUANTITY ....50 × 5 portion packs = 250

STORES DELIVERY ....8.00 a.m.

TIME REQUIRED FOR PACKING ....12.15 p.m.

DISH PRODUCTION (2): Beef stew and dumplings

RECIPE NUMBER..29

BATCH NUMBER..25132E5

QUANTITY ....50 × 5 portion packs = 250

STORES DELIVERY ....9.30 a.m.

TIME REQUIRED FOR PACKING ....13.45 a.m.

In order to be acceptable:

- containers should be a suitable size for the number of portions that they are required to contain
- the depth of the container should not exceed 5 cm (2 in). Deeper containers will not cool quickly enough
- containers should be stackable without compressing the food they contain in order to utilise space fully
- container lids should have an airtight fit or be capable of being sealed with a non-contaminating seal.

The principal packing materials are aluminium foil, paperboard cartons and plastics.

Aluminium foil containers

Aluminium foil containers are ideal for the majority of purposes, as they can withstand high and low temperatures, have excellent moisture retention properties, and are non-toxic, odourless and good conductors of heat. Modern plastic-coated foil is protected against acids and alkalines present in some foods, as the coating prevents the metal in the foil from coming into direct contact with the food. The coating also means that foil containers may be used in microwave ovens.

Paperboard cartons
Paperboard cartons are usually coated with wax, polyester or polythene, none of which are able to withstand the temperatures required for regeneration very effectively. Since this is the cheapest form of packaging it is often used for dishes that are to be served cold.

Plastic containers
Plastics have many of the properties of aluminium but cannot withstand very high temperatures (above 120 °C/248 °F). One particular advantage of plastic containers is that they are available in many attractive designs, enhancing the presentation of food.

Preparation for portioning and packing

Equipment and clothing

Assemble all the equipment and containers that you need in order to portion and pack the dishes being prepared. Never use the same equipment for portioning/packing and processing raw food or cooking food: the equipment may be contaminated with harmful bacteria that will transfer to the cooked food.

Most establishments using a cook-chill system have colour-coded equipment with highly visible markings that identify the area in which the items are to be used. Make sure that you are familiar with these. The operations management should be very strict in ensuring that this equipment is not moved from one area to another.

This colour-coding is often also used for protective clothing so that staff are confined to one area of a production unit by the colour of the uniform they are wearing.

Before you start

You must have:
- sufficient containers of the correct type
- clean service/portioning equipment
- prepared labels listing the dish name, time, date and batch number
- adequate chiller space for the food you are packing
- adequate cold storage space for the food once it has been chilled
- a probe thermometer in order to check that the dishes are at acceptable temperatures within the required time.

There will be a schedule for the packing section, as for the production section. It will contain details as given in the example schedule at the top of the next page.

Portioning, packing and chilling food

1 Place exactly the required quantity of food in each container in accordance with your establishment requirements. Scoops, ladles and spoons of the correct size will help you to control portion sizes and will speed up the process. If greater accuracy is required, you will need to use scales.

PACKING SECTION: Schedule

DATE .... 5 June 1993

SHIFT LEADER .... Peter Brewer

DISH TO PACK (1): Braised beef

BATCH NUMBER 15131E3

QUANTITY 250

PACKS .... 50 × 5 portions

AVAILABLE TIME .... 12.15 p.m.

PACKED FOR CHILLING BY .... 12.40 p.m.

PACKING MATERIAL .... Foil pack no. 14 (8 × 6 inches)

CHILLER .... No. 1

CHILL STORE .... East 3

DISH TO PACK (2): Beef stew and dumplings

BATCH NUMBER 25132E5

QUANTITY 250

PACKS .... 50 × 5 portions

AVAILABLE TIME .... 13.45 p.m.

PACKED FOR CHILLING BY .... 14.25 p.m.

PACKING MATERIAL .... Foil pack no. 10 (8 × 8 inches)

CHILLER .... No. 2

CHILL STORE .... East 5

Remember:

- all constituents of the dish to be packed should be ready at the same time
- you have only 30 minutes to portion and pack before the chilling process must commence
- composite dishes or meals should be presented as you have been instructed.

2 Completely seal all containers as soon as possible to prevent contamination from the air and deterioration while in cold storage.

3 Fix labels to each container.

4 Pack the containers onto racks for chilling (see *Blast-chilling* on page

283). The size of the racks will depend on the type and capacity of the chiller you are using, but they are usually either transportable racks (e.g. they can be placed onto a trolley) or are specialist trolleys designed to fit completely into the chiller.

5 Check the temperature of the food using a probe thermometer. This should be done twice: once immediately before chilling starts, and then again after the food has been in the chiller for 90 minutes. These times and temperatures should be recorded on a document that is kept for reference.

Checking and recording the internal temperature of a batch of chilled dishes using a probe thermometer

Portioning and packing different types of food

Sliced meats
Layer the meat slices into the tray so they are overlapping, and continue until the slices completely cover the base of the dish. Keep the layers to a constant level: do not allow some areas to become much higher than others. Thick layers will take longer to reheat. Use gravy sparingly, making sure that some gravy runs under the meat slices. Too much direct contact between the meat and the container will cause the meat to dry out.

Meat, poultry and fish dishes with sauces
Chill both the meat, poultry or fish and sauce before combining. Place the meat or fish in the container and then add the cold sauce. Do not use excessive amounts of sauce: the final dish will be hard to serve and may be unacceptable to the client.

Meat steaks, chops and cuts
Place the meat into the container so that only the heel of the meat comes into contact with the container, making sure that the eye of the meat is protected by any bone.

Meat stews/casseroles
Portion after chilling so that the portions of stew do not form a skin. Do not fill the container too deeply: the food will take longer to reheat. A depth of 4 cm ($1^1/_2$ in) is recommended.

Egg dishes
Never place cooked egg directly onto the base of a container. Eggs should be placed onto concassée, vegetables or toast to prevent them from drying out.

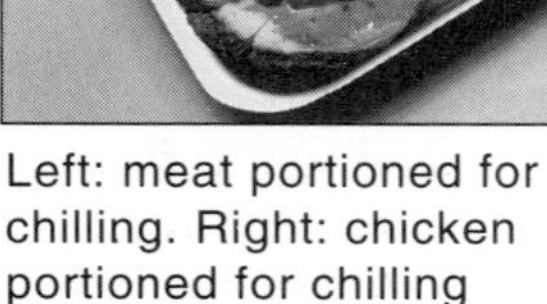

Left: meat portioned for chilling. Right: chicken portioned for chilling

Vegetables

Place vegetables evenly into the dish, to a depth of approximately 4 cm ($1\frac{1}{2}$ in). Vegetables benefit from being tossed in butter or packed with small pats of butter. Make sure that any vegetables are fully drained before packing: any remaining cooking liquor will make the dish difficult to serve and quality will be badly affected.

Potatoes

Whole potatoes should be even in size and shape so that they will all take the same amount of time to reheat. Place them into the container so that all of them have some direct contact with the container. Mashed potatoes should be spread evenly across the base of the container, then fluted on top to allow as much surface area as possible to become exposed.

Pastries

Large pies may be cooked and chilled in the same dish. Individually portioned pies should be removed from their cooking containers and placed flat on the chilling container. Do not allow them to overlap. Make sure that pastry lids are at least 1.5 cm ($\frac{1}{2}$ in) below the lid of the container to prevent them from browning during regeneration.

Essential knowledge

Portions must be controlled when filling packages in order to:
- ensure efficient stock control
- control costs
- ensure that sufficient food is delivered to regenerating kitchens
- ensure that the standard regeneration procedures are safe to use.

Essential knowledge

Food containers must be sealed correctly before storage in order to:
- protect the food from airborne contamination
- reduce the dehydration effects of chilling
- improve the presentation of the dishes
- prevent tampering with finished dishes
- prevent the evaporation of moisture when reheated.

Essential knowledge

Food containers must be labelled correctly before storage in order to:
- allow easy identification of the contents
- facilitate stock control
- ensure that correct stock rotation is carried out
- identify product and 'use by' information
- enable Environmental Health Officers to check that food safety laws are being complied with
- allow dishes and products to be tracked.

Blast-chilling

Blast-chilling is a widely used and efficient method of reducing the temperature in cooked and, in some cases, raw food at a very rapid rate. Correctly packed food is stacked onto racks or trolleys designed to fit into the chillers. Powerful fans then circulate the air in the chiller over refrigerated elements and the food within the cabinet. The movement of the air reduces the insulating properties of the air on the surface of the food and allows the temperature of the food to be reduced very rapidly.

Sensors placed in the food and in the cabinets stop the cooling process as soon as the desired temperature is reached, at which point the chiller becomes a holding cabinet. In large scale operations chillers normally have two doors. The first connects the kitchen to the chiller, allowing the containers to be loaded into the chiller. The second leads from the chiller directly into the chill storage room, allowing the products to be removed from the chiller without any possibility of temperature variations occurring.

Left: blast chillers
Right: racks being withdrawn from a blast chiller

Key points in the preparation of food for cook-chill production

- Raw materials must be of the highest quality and freshness and must be stored separately from the finished product.
- Cooking must take place as soon as possible after preparation to avoid spoilage, deterioration or bacterial growth.
- Any cook-chill food bought in must always be constantly evaluated and checked to see that it meets quality control standards. Food must be received at the right temperature, date coded and transferred to the right temperature/conditions.

- Where appropriate, cook-chill food may be removed from cartons before storing. All cook-chill food must be transferred after production/transportation into hygienic storage containers.
- Every effort must be made to avoid the risk of cross-contamination during preparation and storage.
- The highest standards of personal hygiene must be maintained at all times.

What have you learned?

1. What period of time is allowed between the completion of cooking and the start of the chilling process?
2. What are the critical temperatures for cooking and storing cook-chill food?
3. What are the particular hazards of the cook-chill system of catering?
4. What is the maximum length of time allowed between the cooking and the consumption of chilled foods?
5. Why must portions be controlled when filling packages?
6. Why are time and temperature important when preparing cook-chill food?
7. Why must food containers be sealed and labelled correctly before storage?

ELEMENT 2: Storing cook-chill food

Trolley racks in the chill store awaiting despatch

Transporting the chilled food

The cooked food needs to be transported from the chillers to the regenerating kitchens. As you learned earlier (see *Blast-chilling*, page 283), the food is packaged and packed into racks designed to fit the chillers. These racks are also used to move the stock from one location to another; i.e. into a holding store and then on to the satellite kitchen where the food will be finally prepared for service. This method reduces the handling of the actual containers to a minimum, which in turn reduces the possibility of damage occurring. When smaller quantities of prepared and chilled food have to be delivered to a store or kitchen, the batch is put together in the chill store on a suitable number of racks before being transported.

The speed of transportation is critical as every minute spent out of the cold store produces an increase in temperature in the food which can dramatically reduce the shelf life of the product (see *Temperature recording*, page 286). Refrigerated vans or lorries must be used for any transportation other than a very short internal distance. Note that the same regulations apply to refrigerated vehicles as apply to non-mobile refrigerators; temperatures have to be checked and recorded regularly.

Chilled food must be transported correctly in order to:
- minimise temperature variations in the product
- reduce the risk of contamination
- eliminate loss of stock due to damage
- comply with hygiene and food safety legislation.

Accepting food for storage

No food other than chilled and packed food should be stored in a chill store, as to do so would create a risk of cross-contamination. It would also result in more people needing access to the store which can cause problems with temperature control.

Chilled food must not be accepted for storage unless it is properly labelled. The information on the label must include:
1. the description of the contents (i.e. the name of the dish)
2. the date and time of chilling
3. the batch number
4. the storage temperature
5. the use-by-date.

Additional information may include nutritional information which may be necessary for special medical diets (e.g. 'Salt Free' or 'Gluten Free') or for religious purposes (e.g. 'Kosher Prepared' for those of the Jewish faith or 'Hallal Prepared' for Muslims).

Temperature control during storage, transportation and regeneration

The following points are essential during these stages:
- Food must be stored at 0–3 °C (32–37 °F) for no longer than five days. This includes the day that the food is re-heated.
  Reason: most, but not all bacterial activity stops at these temperatures. After five days there could still be some build up of harmful bacteria, which could dangerously contaminate the food when re-heated.
- When food is distributed from the cold store the temperature must be carefully monitored. If the temperature of the food exceeds 5 °C (41 °F) at any time, it must be consumed within 12 hours or discarded. If it exceeds 10 °C (50 °F) it must be either consumed immediately or discarded.
  Reason: food stored at 5–10 °C (41–50 °F) will allow bacteria to multiply slowly, but providing the preparation and packing is carried out under strict hygienic control the food will remain safe for a maximum of 12 hours. As soon as 10 °C (50 °F) is reached, the food rapidly

becomes dangerous if not consumed immediately.
- When removed from chilled storage the cooked food must be reheated to an internal temperature of at least 70 °C (158 °F) and held at this temperature for at least two minutes.
  Reason: this temperature will kill most dangerous bacteria present.

Temperature recording

The temperature of cook-chill food must be monitored throughout the production and storage processes in order to keep the risk of contamination by food poisoning bacteria to a minimum. It also makes it possible to trace any problems back to a particular process or department.

The importance of accurate and continuous temperature monitoring cannot be stressed enough. Temperatures should be taken at the core (the centre of the thickest part) of the product using an accurate probe thermometer which should be cleaned with a special 'once-only' bactericidal wipe after each use. It is also essential that the thermometers used are recalibrated regularly to ensure that they are accurate. Records must be kept of these recalibration tests.

The temperature of food should be taken with an accurate probe thermometer at the following times:
- on delivery (maximum: 5 °C/41 °F for high risk foods such as meats, poultry, fish, dairy produce, cooked meat)
- when moved from the stores to the cooking area (maximum: 5 °C/41 °F for high risk foods such as meats, poultry, fish, dairy produce, cooked meat)
- when cooked (minimum 70 °C/158 °F for two minutes)
- when chilled (maximum 3 °C/37 °F)
- at regular intervals whilst in storage (maximum 3 °C/37 °F; if higher than this temperature see page 285 for action)
- when moved from chill to service area (maximum 3 °C/37 °F)
- when re-generated (minimum 70 °C/158 °F for two minutes)
- whilst held for service (minimum 70 °C/158 °F).

Under current hygiene regulations these temperatures must be recorded with dates, times and the name of the person responsible. Each establishment will have standard procedures to ensure that the records are maintained and filed for future reference for every batch of food that is stored and moved or served. These records must be made available to the Environmental Health Inspector on demand.

Essential knowledge

It is important to monitor and record food temperatures regularly in order to:
- prevent contamination from incorrect storage conditions
- ensure flavour and texture is maintained.

Stock control system

All food must be booked into the cold store and the label details recorded. Items required for dispatch should be those that were first into the store (the oldest). An efficient stock rotation system should be in place to ensure this. No food should be held in the cold store longer than five days. Daily checks should be made both through records and the actual stocks held to ensure that no out-of-date items are held.

The temperature of the cold storage unit, irrespective of size should not vary from the range of 0–3 °C (32–37 °F). If there is no continuous recording system for the cold store, the temperature should be checked at least twice a day and the time and temperature recorded on a log. These records should be safely filed in accordance with the established procedures of the cook-chill unit.

Before sending food for consumption, the chilling date should be recorded on an issue sheet. Food dispatched for consumption may need to be transported in refrigerated containers in order to keep the temperature below 3 °C (37 °F).

Essential knowledge

Stock rotation procedures must be followed in order to:
- prevent damage or decay to stock
- ensure that older stock is used before newer stock.

Limiting access

Access to cold storage areas should be restricted to authorised people only. This can be achieved by using digital locks with special codes, personal keys or systems using a 'booking in and out' log. In large chill stores there must be some form of external indication that there is someone in the store and when they entered the store. This is essential as unnecessary opening of the cold store door would destabilise the temperature and could cause spoilage of the food. There is also, as with any store, the need to guard against the possibility of theft.

Remember that a cold store is a hazardous area and unauthorised people may become trapped inside with very serious consequences.

Essential knowledge

Storage areas must be secured from unauthorised access in order to:
- prevent pilferage or damage by unauthorised persons
- prevent injury to unauthorised persons
- prevent unnecessary opening of store doors, which would destabilise the temperature.

Legal requirements

A caterer or manufacturer should bear in mind the potentially dangerous (and expensive) outcomes that contravention of the Food Safety Act of 1990 can entail. The labelling of products, recording of temperatures, attention to hygiene and training given to staff would all form part of a legal defence known as *due diligence,* should any legal action be taken. For this to be a successful defence, the caterer must convince the court that all the requirements of the law have been complied with and that the 'accepted customs and practices' of the trade have been carried out. It is also necessary to have documentary evidence to prove these facts.

To do

- Find out how times and temperatures are recorded in your establishment, and how long the records are kept.
- Find out who is authorised to enter your cold store.
- What safety precautions apply to your unit?

Key points: chilling, storing and regenerating food

- The rate of cooling will depend on a number of factors including container size, shape or weight, and food density and moisture content.
- The product must be labelled with the date of production and a strict system of stock control must be in operation. Temperature control during food distribution should be very closely monitored.
- For reasons of safety and palatability the food must be reheated (regenerated) quickly. It must be heated to 70 °C (158 °F) for 2 minutes, maintained at a minimum temperature of 63 °C (145 °F) and consumed within 15 minutes.
- Food to be served cold should be consumed within 4 hours after removal from chilled storage.
- No food, once reheated, should be returned to the refrigerator. All uneaten, reheated food should be destroyed.

Dealing with unexpected situations

Establishments that operate cook-chill systems will have detailed operational procedures for dealing with every non-routine situation.

The most important of these will deal with refrigeration faults in chillers and storage, abnormal temperature variations in storage, below specification deliveries from suppliers, and incorrect processing times.

It is your responsibility to be aware of these procedures and to follow them exactly; there is no margin of error.

What have you learned?

1. What temperature should chilled food be stored at?
2. What does *First in –First out* mean when referring to controlling stocks of chilled food?
3. Why must stock rotation procedures always be followed?
4. Why is it important to monitor and record food temperatures regularly?
5. Why must storage areas be secured from unauthorised access at all times?

Extend your knowledge

1. Find out about the full legal requirements concerning the production of cook-chill food. You will find the relevant laws included in The Food Safety Act 1990 and The Food Hygiene (Amendment) Regulations 1990 and 1991. You should also read the following government guidelines: Guidelines on Cook-chill and Cook-freeze Catering Systems (Dept of Health 1989), Guidelines on the Food Hygiene (Amendment) Regulations (Dept of Health 1990 and 1991).
2. You can find out more about cook-chill generally by reading A Guide to Cook-chill Catering by Lewis Napleton (International Thomson Business Publishing, 1991).
3. To increase your understanding of contamination threats and how to prevent these, you may like to read Croners Food Hygiene Manual (Croner Publications, 1991).

Preparing cook-freeze food

ELEMENTS 1 AND 2: Portioning, packing, blast-freezing and storing blast-frozen food

What do you have to do?

- Portion, pack and cover food in accordance with laid down procedures.
- Blast-freeze and store food in accordance with laid down procedures.
- Seal and label food containers correctly.
- Rotate stock holdings and use stock according to date ordered.
- Monitor and record food temperatures in accordance with laid down procedures.
- Keep accurate records of received, stored and issued food items.
- Carry out your work in an organised and efficient manner in accordance with current regulations.

What do you need to know?

- The type, quality and quantity of food required.
- Why portions must be controlled when filling packages.
- Why time and temperature are important when preparing cook-freeze foods.
- Why food items are handled carefully and remain undamaged.
- Why it is important to monitor and record food temperatures regularly.
- How to transport food containers undamaged to storage areas within a required time.
- Why food containers must be sealed and labelled correctly.
- Why it is important to follow stock rotation procedures and maintain accurate records.
- Why storage areas must be kept secure from unauthorised access and kept clean and tidy.
- How to satisfy health, safety and hygiene regulations concerning preparation areas and equipment both before and after use.
- How to plan your work to meet daily schedules.
- How to deal with unexpected situations.

Introduction

A cook-freeze system is one where food is prepared well before it is required, quickly blast-frozen to –18 °C (0 °F) then stored at that temperature until required for sale or consumption (for up to 3 months). It is a system that has been adopted by many institutional caterers such as hospitals, schools and industrial catering facilities.

Advantages of the system

Cook-freeze production and service offers a number of advantages to the caterer:

- *a cost-effective use of labour.* Food may be prepared within normal working hours when there is the most available labour, and served at times when labour is more expensive; e.g. very early mornings, evenings, weekends and holiday periods
- *greater buying power.* Centralising production brings economies of scale and economies in buying supplies in larger bulk
- *flexibility in the menu.* A variety of stored food allows the caterer to offer a greater number of dishes without suffering the potential wastage of conventional cooking and serving systems
- *consistent standards.* A standard quality of dishes is more easily attained because the dishes are produced in bulk before being individually or multi-portion packed
- *fresh-tasting food.* Blast freezing does not allow large ice crystals to form in the foods, so the regenerated (re-heated) dishes are closer to their original taste, colour and texture than is the case with foods that have been kept hot for a long time. The large ice crystals that form when foods are frozen slowly (as in a domestic freezer) puncture the structures of foods and cause loss of texture, moisture and flavour. It is therefore essential to buy specifically designed fast-freezing equipment for this task
- *fast service times.* The regeneration (reheating) of frozen food is relatively quick using microwaves and combination ovens; dishes may be removed from freezer cabinets close to service times
- *a lower equipment requirement* in the regenerating kitchen. The kitchens where regeneration takes place (known as satellite kitchens) may be some distance from the Central Production Unit (CPU), and these need less equipment than for conventional cooking. Provided there are strictly enforced guidelines, they may also need staff with fewer culinary skills.

Disadvantages of the system

As in any system, there are a number of disadvantages that have to be considered before adopting a cook-freeze operation:

- *high equipment cost* at production stage. The cost of buying the quick freezer, packaging material, general equipment, frozen storage facilities and temperature control and monitoring equipment is considerable
- *the need for a large working space.* Cook-freeze usually demands more space than conventional production, both for storage and packing
- *higher risk of food poisoning.* The potential for food poisoning on an epidemic scale is greater, since food is usually cooked in greater bulk than conventional cooking, temperature controls are more critical and the possibility of poor stock control can all allow contamination to take place in large quantities of food
- *high energy consumption.* The consumption of energy (i.e. electricity) for freezing and storage is considerable
- *high cost to prevent damage to ecology.* The refrigerant gases are not environmentally friendly (they contain CFCs) and the alternative less-damaging refrigerants are considerably more expensive
- *detailed re-training* of management and production staff. This is needed to obtain maximum benefit and to avoid the new range of

hazards. Such training should include the adoption of Hazard Analysis Critical Control Point procedures in which each step of the process is analysed. This pinpoints potential hazards, allowing management procedures to be developed to minimise risk.

ELEMENT 1: Portioning, packing and blast-freezing food

Equipment

The equipment used in the production of food for the freezing process can be the same as that for a conventional kitchen. Ideally, however, in order to reduce the possibility of cross-contamination between cooked and raw food, the kitchen should be designed so that the stores delivery point, stores and preparation areas are separate from the cooking, packaging and freezing areas.

The packaging area can vary in size and complexity, from a full conveyor system for a large production unit to a table in a side room for a small system. Within the packaging area there must be a sealing system which will remove the air from the container holding food and then seal it. The sealers are usually manually operated, with the exception of large capacity cook-freeze operations. There are three methods of fast-freezing foods.

- The first method involves freezing food in a blast freezer, where sub-zero air is blown over the surface of the food to be frozen.
- The second method uses an immersion freezer, where the containers are immersed in a cabinet containing liquid refrigerant. Until recently the liquid refrigerant normally used was liquid nitrogen or carbon dioxide, but other special liquids have now been developed which are easier and cheaper to use.
- A third method which may be used for high capacity systems involves the use of tunnel freezers, which move the containers along a tunnel whilst they are sprayed with a very cold refrigerant.

The system used depends on the design capacity of the cook-freeze unit.

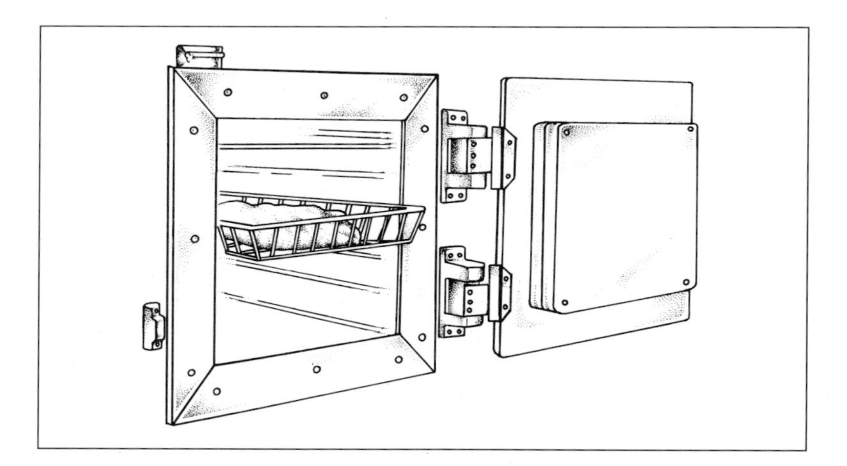

A blast freezer

In order to set up a cook-freeze unit on a commercial basis, the unit must be capable of producing and selling around 5,000 meals per day. This inevitably means that a specialist fleet of vehicles will be required to distribute the meals at sub-zero temperatures to a number of geographical locations not necessarily very close to the production unit.

Suitable types of food

Meat and poultry

Most meat, poultry, game and offal can be successfully frozen. Meat dishes that need to be sliced, e.g. Roast beef, are cooked, chilled, quickly sliced and then packed for freezing and storage. It is not possible to serve rare (acceptably under-done) roast, fried or grilled meat dishes using cook-freeze because the food must be cooked to an internal temperature of 70 °C (158 °F) and held at this temperature for two minutes.

Cooked pork can be frozen but pork tenderloin has a very high water and internal fat content and a relatively short storage life (a maximum of eight weeks). Ground pork (e.g. sausages) and cured pork (bacon and ham) have a very short life even at the frozen temperature of –18 °C/0 °F (a maximum of four weeks). The preservatives used in these products can oxidise during their storage life and develop a rancid flavour or smell which is not dangerous but tastes unpleasant.

Vegetables and fruits

Most cooked potato, vegetable and fruit dishes are suitable for freezing with the exception of composite salads containing cooked and raw ingredients. Salad items that are to be eaten raw such as lettuce, watercress, radiccio and endive should be added fresh to dishes just before service, not added to dishes prior to freezing.

The delicate cellular structures and high water content of these leafy foods causes changes in the food during the harsh freezing process which badly affects the flavour and textural quality (they become limp). The use of conventional garnishing material after regeneration (reheating) is recommended in order to prevent customers from becoming 'processed food fatigued'.

Fish dishes

All pre-cooked fish dishes are suitable for cook-freeze processing. Salmon, crab, lobster and shrimp/prawn dishes freeze well but have a shorter shelf life.

Sauces and soups

Most sauces and soups are suitable for cook-freeze although those with egg as a thickening agent and those with a high fat content (e.g. hollandaise, béarnaise, mousseline sauces) will need particular care when regenerating to avoid curdling. Soups thickened with a liaison (cream and egg yolk) must have this added immediately prior to service and not before being chilled.

There are specialist modified starches on the commercial market which can be used as thickeners for sauces and soups in place of a conventional roux (fat and flour). These solve the problem of curdling which can occur using a roux method. Waxy maize flour, cornflower and arrowroot are all acceptable thickeners of this type. This is one example of how conventional recipes have to be modified to allow for the hostile environment of the freeze/regeneration cycle.

Egg dishes

Most egg dishes are easily prepared on demand and cook-freeze offers no great advantage to the caterer.

Desserts

Basic cooked desserts freeze well, with the exception of those that rely on batters lifted by eggs, such as soufflés. Mousses, jellies, pastry goods, yeast goods, cooked fruits and uncooked fruit are all suitable for freezing, although in many cases the high sugar content prevents hard freezing.

This sometimes causes concern as it may lead you to think that the product is not cold enough or that the freezer is not operating at the correct temperature. Composite sweet dishes that require a hot base with a raw fruit or cream topping are not suitable unless the two components are packed separately.

Preparing and cooking dishes

Most conventional recipes may be frozen successfully provided suitable thickening agents are used for accompanying sauces and soups. Waxy maize flour should be used to replace wheat flour as a thickening agent as it does not curdle when regenerated. Sauces thickened with cornflour and arrowroot will also reheat satisfactorily.

Do not garnish finished dishes until the food has been reheated. This is particularly important for any fresh garnish, such as parsley sprigs, tomato or salad items. These do not freeze well as they have a high water content and become limp and unappetising when thawed.

The dishes outlined in the Unit 2D15: *Preparing cook-chill food* (pages (270–88) would all be suitable for freezing using the cook-freeze methods of production.

Food preparation in a cook-freeze operation

Health, safety and hygiene

Note all points given in the Core Units book concerning general attention to health, safety and hygiene. When freezing and storing foods, the following points are particularly important:

- *hygiene.* Since the objective of freezing is to extend the life of cooked food, hygienic preparation and storage is of prime importance. The

extended life of the product also extends the opportunity for dangerous bacteria (particularly salmonella, staphylococcus aureus, clostridium perfringens, escherichia coli and listeria) to contaminate and grow in the food. For more detailed information on this subject see the further reading list at the end of this chapter
- *time and temperature.* The limits for these must be adhered to very strictly and accurate records of time and temperatures of stored goods must be kept
- *prevention of cross-contamination.* The possibility of cross-contamination between cooked and uncooked foods must be kept to an absolute minimum.

Time and temperature

Some bacteria leave dangerous toxins in food even after they have been killed and others can protect themselves from the effects of heat. This is why the strict time and temperature controls exist and why they must be adhered to by all concerned. All of the following points are essential when carrying out preparation within the cook-freeze process.

- The temperature of uncooked ingredients delivered from stores must not exceed 5 °C/41 °F for high risk foods such as meats, poultry, fish, dairy produce and cooked meat. The temperature of food must not exceed this level during the preparation process.
- The temperature of cooked food must reach a minimum of 70 °C (158 °C) for at least 2 minutes and freezing should commence within 30 minutes.
  Reason: this temperature will kill the majority of harmful bacteria which cause food poisoning. By freezing quickly to a temperature which prevents bacterial growth, any remaining bacteria does not have time to multiply.
- Foods must be chilled down to a temperature of –5 °C (19 °F) within a period of 90 minutes. The food should then drop again to –18 °C (0 °F). Whole joints above 2.5 kg (5½ lb) may not be used.
  Reason: 90 minutes is the shortest practical time that this temperature can be achieved in a controlled way. Large joints (being dense and thick) cannot meet this requirement and so would not be safe when reheated.
- However food is distributed from deep-freeze storage, if the temperature of the food exceeds 5 °C (41 °F) it must be consumed within 12 hours or discarded. If it exceeds 10 °C (50 °F) it must be consumed immediately or discarded.
  Reason: food stored at 5–10 °C (41–50 °F) will allow bacteria present to multiply slowly. However, providing the preparation and packing was carried in out under strict hygienic control and in accordance with the regulations, the food will remain safe for a maximum of 12 hours. Once 10 °C (50 °F) is reached the danger from bacteria increases dramatically and the food would become hazardous if not consumed immediately.

Remember: some bacteria leave dangerous toxins in food even after they have been killed and others can protect themselves from the effects of heat. This is why the strict time and temperature controls exist and why they must be adhered to by all concerned.

See *Temperature control and recording* in Element 2 on page 301 for temperature control during the storage and transportation stages.

Essential knowledge

Time and temperature are important when preparing cook-freeze food in order to:

- prevent contamination from food poisoning bacteria
- prevent any bacteria present from multiplying
- kill bacteria susceptible to heat
- present the food to the consumer in the best possible condition.

To do

- Find out the colour coding system for your preparation equipment and list the colours used and the types of food they are used for.
- Check and note the temperatures of your freezer and holding freezer cabinets. Do they comply with the regulations?
- Look at temperature records and note any changes that have occurred.

Planning your time

Because of the short time that is required between the cooking of food and the packing and freezing it is very important that you plan every stage of the portion-packaging-freezing-storage cycle of operation precisely. You will create severe problems if you produce more dishes than you can pack, or pack more than you can freeze and/or store.

Production scheduling

The Operations Manager is responsible for scheduling production so that none of the sections of the operation is required to work above its maximum capacity. Within the schedule, the staff responsible for each operation have to ensure that they have the necessary equipment to perform the task successfully.

The manager will produce written schedules for each area. An example of a Production Section Schedule is given on page 278.

Packaging scheduling

The volume of the production unit will vary according to the scale of the operation and will be controlled by the Unit Manager.

The production schedule will go to the stores so that the correct quantities of ingredients will be available to the cooks at the time required. The schedule will allow enough time between the preparation of dishes and the arrival of ingredients for the next batch of dishes for cleaning down and sterilisation of equipment and work areas. This is a necessary precaution against cross-contamination.

The Packing Section Schedule may look like the one shown at the top of the next page.

Selection of containers

In selecting the containers for cook-freeze the person responsible needs to ensure that they are capable of withstanding both the freezing and reheating process without any change in their structure or contamination of the food. The one common factor for all containers used for cook-

PACKING SECTION: Schedule

DATE..5 June 1993

SHIFT LEADER ....Peter Brewer

DISH TO PACK (1): Braised beef

BATCH NUMBER..15131E3

QUANTITY 250

PACKS ....50 × 5 portions

AVAILABLE TIME..12.15 p.m.

PACKED FOR FREEZING BY ....12.40 p.m.

PACKING MATERIAL ....Foil pack no. 14 (8 × 6 inches)

BLAST FREEZER ....No. 1

FREEZER STORE ....East 3

DISH TO PACK (2): Beef stew and dumplings

BATCH NUMBER..25132E5

QUANTITY 250

PACKS ....50 × 5 portions

AVAILABLE TIME..13.45 p.m.

PACKED FOR CHILLING BY..................................14.25 p.m.

PACKING MATERIAL ....Foil pack no. 10 (8 × 8 inches)

BLAST FREEZER ....No. 2

freeze is that they must be able to maintain a near-vacuum (the exclusion of all air) in order to reduce the spoilage rate within the freezing storage.

In order to be acceptable:
- containers should be a suitable size for the number of portions that they are required to contain
- the depth of the container should not exceed 5 cm (2 in). Deeper containers will not cool quickly enough
- container lids should have an airtight fit or be capable of being sealed with a non-contaminating seal.

The principal packing materials are aluminium foil, paperboard cartons and plastics.

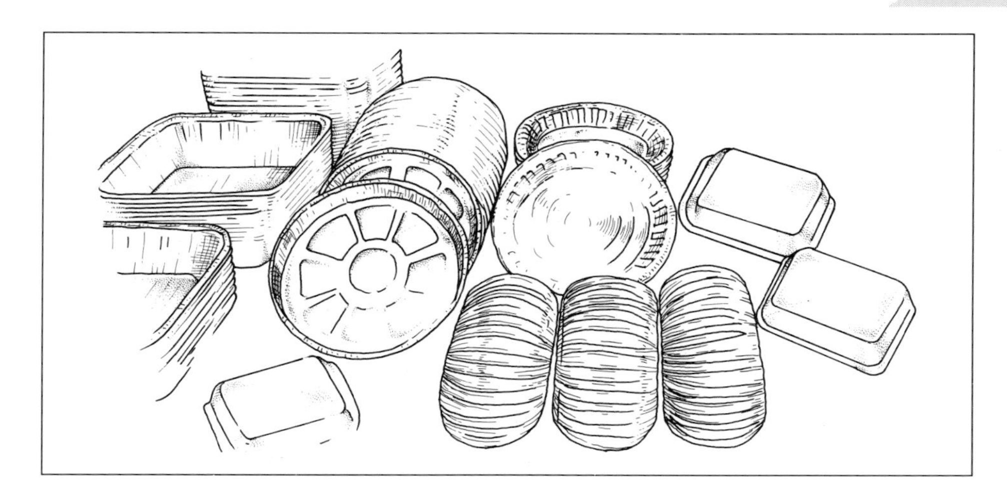

Containers for packaging cook-freeze food

Aluminium foil containers
Aluminium foil containers are ideal for the majority of purposes, as they can withstand high and low temperatures, have excellent moisture retention properties, and are non-toxic, odourless and good conductors of heat. Modern plastic-coated foil is protected against acids and alkalines present in some foods, as the coating prevents the metal in the foil from coming into direct contact with the food. The coating also means that foil containers may be used in microwave ovens.

Paperboard cartons
Paperboard cartons are usually coated with wax, polyester or polythene, none of which are able to withstand the temperatures required for regeneration very effectively. Since this is the cheapest form of packaging it is often used for dishes that are to be served cold.

Plastic containers
Plastics have many of the properties of aluminium but cannot withstand very high temperatures (above 120 °C/248 °F). One particular advantage of plastic containers is that they are available in many attractive designs, enhancing the presentation of food. Plastic can also be heat-sealed, allowing food to be packaged in heat-sealed bags, which is useful for small quantities of liquids and single portions of dishes that can be reheated in the bag.

Preparation for portioning and packing

Equipment and clothing

Assemble all the equipment and containers that you need in order to portion and pack the dishes being prepared. Never use the same equipment for portioning/packing and processing raw food or cooking food: the equipment may be contaminated with harmful bacteria that will transfer to the cooked food. Most establishments using a cook-freeze system have colour-coded equipment with highly visible markings that identify the area in which the items are to be used. The operations management should be very strict in ensuring that equipment is not moved from one area to another. This colour-coding is often also used for protective clothing so that staff are confined to one area of a production unit by the colour of the uniform they are wearing.

Before you start

You must have:
- sufficient containers of the correct type
- clean service/portioning equipment
- prepared labels with the dish, time, date and batch number
- adequate blast freezer space for the food you are packing
- adequate cold storage space for the food once it has been frozen
- a probe thermometer in order to check that the dishes are at acceptable temperatures within the required time.

Portioning, packing and freezing food

1. Place exactly the required quantity of food in each container in accordance with establishment requirements. Scoops, ladles and spoons of the correct size are an aid to portion control and will speed up this process. If greater accuracy is required, you will need to use scales. Remember:
   - all constituents of the dish to be packed should be ready at the same time
   - you have only 30 minutes to portion and pack before the freezing process must commence; i.e. to complete Steps 1, 2, 3 and 4
   - composite dishes or meals should be presented as instructed.
2. Completely seal all containers as soon as possible to prevent contamination from the air and deterioration while in cold storage.
3. Fix labels to each container.
4. Pack the containers onto racks for freezing. The size of the racks will depend on the type and capacity of the freezer you are using, but they are usually either transportable racks (e.g. they can be placed onto a trolley) or are specialist trolleys designed to fit completely into the freezer.
5. Check the temperature of the food using a probe thermometer. This should be done twice: once immediately before freezing starts, and then again after the food has been in the freezer for 90 minutes. These times and temperatures should be recorded on a document that is kept for reference.

Essential knowledge

Portions must be controlled when filling packages in order to:
- standardise the cost
- control costings
- facilitate stores control
- help service staff to serve the correct size portion
- standardise the thawing or reheating process
- allow the sealing to be properly accomplished.

Essential knowledge

Food containers must be labelled correctly before storage in order to:
- accurately identify the contents of a container
- enable stocktaking to be completed quickly and accurately
- indicate important information regarding the packing date and the 'use by' date
- inform service staff of the number of portions contained in the package.

Essential knowledge

Food containers must be sealed correctly before storage in order to:
- prevent spoilage due to contact with the cold air
- prevent spillage prior to freezing
- allow safe stacking without damage to containers.

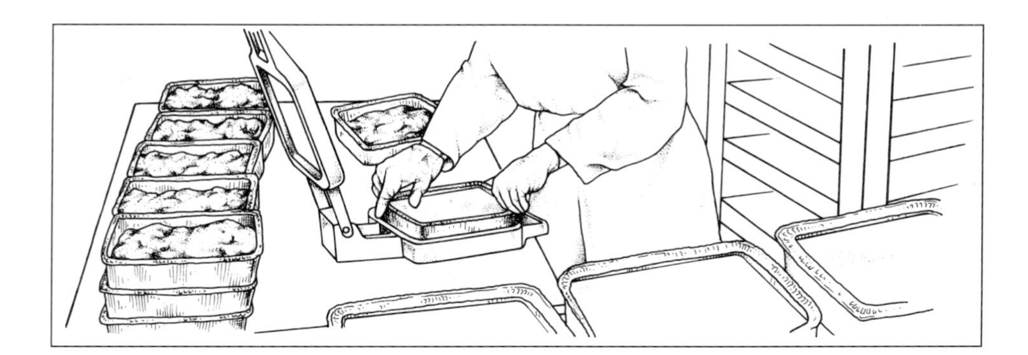

Packing food for cook-freeze

Portioning and packing different types of food

Follow the same procedures as for cook-chill food (see page 281).

Blast freezing

Blast freezing is the most widely used method for reducing the temperature of cooked food in medium to small scale operations. Freezers of this type are usually classified by the quantity of food that they can bring down to −5 °C (19 °F) within one hour.

Portioned and sealed dishes are loaded onto racks which allow air to circulate freely between the shelves. The racks are then pushed into the freezing chamber where sub-zero air is blown over the containers by powerful fans. The movement of the air reduces the insulating properties of the containers and lowers the temperature rapidly. Blast freezers are not efficient as storage and the power consumption is very high, so the products should be removed from the cabinets and stored in deep-freeze storage as soon as the desired temperature is reached. The deep-freeze storage will then reduce the temperature still further to −18 °C (0 °F).

Key points in the preparation of food for cook-freeze

- Raw materials must be of the highest quality and freshness and must be stored separately from the finished product.
- Cooking must take place as soon as possible after preparation to avoid spoilage, deterioration of quality or bacterial growth.
- Any cook-freeze food bought in must always be constantly evaluated and checked to see that it meets quality control standards. Food must be received at the right temperature, date coded and transferred to the right temperature/conditions.
- Where appropriate, cook-freeze food may be removed from cartons before storing. All cook-freeze food must be transferred after production/transportation into hygienic storage containers.
- Every effort must be made to avoid the risk of cross-contamination during preparation and storage.
- The highest standards of personal hygiene must be maintained at all times.
- Times and temperatures must be carefully controlled and recorded.

What have you learned?

1. How does cook-freeze differ from cook-chill production and storage?
2. Why is it absolutely essential to comply with all hygiene regulations when preparing food for cooking and freezing?
3. Why must portions be controlled when filling packages?
4. Why are time and temperature important when preparing cook-freeze foods?
5. Why must food containers be sealed and labelled correctly before storage?
6. What period of time is allowed between the completion of cooking and the start of the freezing process?
7. Why should equipment be colour-coded?
8. What is the best depth for food containers when packing?
9. What is the maximum length of time allowed between the regeneration and the consumption of frozen foods?

ELEMENT 2: Storing cook-freeze food

Personal safety

Make sure that you:

- always wear protective clothing when entering a deep freeze store. Temperatures of –18 °C (0 °F) and below can cause frostbite and respiratory illnesses very quickly
- follow your unit's safety procedures. Always notify your colleagues that you are working in the deep freeze store: they may accidentally lock you in if you forget
- know where the alarm switches are inside the freezers in case you get into difficulties
- know what to do if the alarm is sounded by a colleague.

Accepting food for storage

No food other than prepared frozen and packed food should be stored in a freeze store, as to do so would create a risk of cross-contamination. It would also result in more people needing access to the store which can cause problems with temperature control.

Frozen food must not be accepted for storage unless it is properly labelled. The information on the label must include:

1. the description of the contents (i.e. the name of the dish)
2. the date and time of freezing
3. the batch number
4. the storage temperature
5. the use-by-date.

Additional information may include nutritional information which may be necessary for special medical diets (e.g. 'Salt Free' or 'Gluten Free') or for religious purposes (e.g. 'Kosher Prepared' for those of the Jewish faith or 'Hallal Prepared' for Muslims).

Temperature control during storage, transportation and regeneration

The following points are essential during these stages:

- Food must be stored at –18 °C (0 °F) for no longer than three months. *Reason:* although bacterial activity ceases at these temperatures, there is a tendency for food to dehydrate and change colour after eight weeks due to the absorption of oxygen by the surface of the food. This is known as 'oxidisation' and shows up as freezer burn. Freezer burn can be reduced by using good packaging which excludes air from the product.
- Ideally food should only be stored in freezers for eight weeks. This is not a hard and fast rule, and food is sometimes kept up to a maximum of six months.
  Reason: the nutritional content will be seriously affected if food is kept for longer than this. Foods with a high fat content may go rancid and become unpalatable.
- When food is distributed from the freezer the temperature must be carefully monitored. If the temperature of the food exceeds 5 °C (41 °F) at any time, it must be consumed within 12 hours or discarded. If it exceeds 10 °C (50 °F) it must be either consumed immediately or discarded.
  Reason: food stored at 10 °C (50 °F) will allow bacteria present to multiply slowly, but providing the preparation and packing is carried out under strict hygienic control the food will remain safe for a maximum of 12 hours. As soon as 10 °C (50 °F) is reached, the food rapidly becomes dangerous if not consumed immediately.
- When removed from frozen storage the cooked food must be reheated to a core temperature of at least 70 °C (158 °F) and held at this temperature for at least two minutes (and served within 15 minutes). Some foods may be reheated straight from the freezer. However, most food will need to be defrosted in a thawing cabinet to chill temperature 5 °C (41 °F) before reheating begins.
  Reason: this temperature will kill most dangerous bacteria present.

Temperature recording

The temperature of cook-freeze food must be monitored throughout the production and storage processes in order to keep the risk of contamination by food poisoning bacteria to a minimum. It also makes it possible to trace any problems back to a particular process or department.

The importance of accurate and continuous temperature monitoring cannot be stressed enough. Temperatures should be taken at the core (the centre of the thickest part) of the product using an accurate probe thermometer which should be cleaned with a special 'once-only' bactericidal wipe after each use. It is also essential that the thermometers used are recalibrated regularly to ensure that they are accurate. (Records must be kept of these recalibration tests.)

The temperature of food should be taken with an accurate probe thermometer at the following times:

- on delivery (maximum: 5 °C/41 °F for high risk foods such as meats, poultry, fish, dairy produce, cooked meat)
- when moved from the stores to the cooking area (maximum: 5 °C/41 °F for high risk foods such as meats, poultry, fish, dairy produce, cooked meat)

- when cooked (minimum 70 °C/158 °F for two minutes)
- when frozen (maximum –5 °C/19 °F after 90 minutes)
- at regular intervals whilst in storage (maximum –18 °C/0 °F; if higher than this temperature see page 294 for action)
- when moved from freezer to service area (maximum –18 °C/0 °F)
- when re-generated (minimum 70 °C/158 °F for two minutes)
- whilst held for service (minimum 65 °C/149 °F).

Under current hygiene regulations these temperatures must be recorded with dates, times and the name of the person responsible. Each establishment will have standard procedures to ensure that the records are maintained and filed for future reference for every batch of food that is stored and moved or served. These records must be made available to the Environmental Health Inspector on demand.

Essential knowledge

It is important to monitor and record food temperatures regularly in order to:
- prevent contamination from incorrect storage conditions
- ensure flavour and texture is maintained.

Stock control system

All food must be booked into the cold store and the label details recorded. Items required for dispatch should be those that were first into the store (the oldest). An efficient stock rotation system should be in place to ensure this. No food should be held in the cold store longer than 12 weeks. Daily checks should be made both through records and the actual stocks held to ensure that no out-of-date items are held.

The temperature of the cold storage unit, irrespective of size, should not exceed –18 °C (0 °F). If there is no continuous recording system for the cold store, the temperature should be checked at least twice a day and the time and temperature recorded on a log. These records should be safely filed in accordance with the established procedures of the cook-freeze unit.

Before sending food for consumption, the freezing date should be recorded on an issue sheet. Food dispatched for consumption may need to be transported in refrigerated containers in order to keep the temperature below –18 °C (0 °F) if the food is to be held for some time on another site or kitchen.

Essential knowledge

Stock rotation procedures must be followed in order to:
- prevent damage or decay to stock
- ensure that older stock is used before newer stock.

Limiting access

Access to cold storage areas should be restricted to authorised people only. This can be achieved by using digital locks with special codes, personal keys, or systems using a 'booking in and out' log. In large freezers there must be some form of external indication that there is someone in the store and when they entered the store. This is essential as unnecessary opening of the frozen store door would destabilise the temperature and could cause spoilage of the food. There is also, as with any store, the need to guard against the possibility of theft.

Remember that a cold store is a hazardous area and unauthorised people may become trapped inside with very serious consequences.

Essential knowledge

Storage areas must be secured from unauthorised access in order to:
- prevent pilferage or damage by unauthorised persons
- prevent injury to unauthorised persons
- prevent unnecessary opening of store doors, which would destabilise the temperature.

Legal requirements

A caterer or manufacturer should bear in mind the potentially dangerous (and expensive) outcomes that contravention of the Food Safety Act of 1990 can entail. The labelling of products, recording of temperatures, attention to hygiene and training given to staff would all form part of a legal defence known as *due diligence*, should any legal action be taken. For 'due diligence' to be a successful defence, the caterer must convince the court that all the requirements of the law have been complied with and also that the 'accepted customs and practices' of the trade have been carried out. It is also necessary to have documentary evidence to prove these facts.

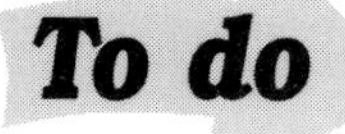

- Find out how times and temperatures are recorded in your establishment, and how long the records are kept.
- Find out who is authorised to enter your cold store.
- What safety precautions apply to your unit?

Key points: freezing, storing and regenerating food

- The rate of cooling will depend on a number of factors including container size, shape or weight, and food density and moisture content.
- The product must be labelled with the date of production and a strict system of stock control must be in operation. Temperature control during food distribution should be very closely monitored.
- For reasons of safety and palatability the food must be reheated (regenerated) quickly.
- Food to be served cold should be consumed within four hours after removal from chilled storage unless it is in a refrigerated display unit.
- *No food, once reheated, should be returned to the refrigerator.* All unconsumed, reheated food should be destroyed.

Dealing with unexpected situations

Establishments that operate cook-freeze systems will have detailed operational instructions for dealing with every non-routine situation.

The most important of these will deal with refrigeration faults in freezers and storage, abnormal temperature variations in storage, below specification deliveries from suppliers, and incorrect processing times.

It is your responsibility to be aware of these instructions and to follow them exactly; there is no margin of error.

What have you learned?

1. What temperature should frozen food be stored at?
2. Why must stock rotation procedures be followed?
3. Why is it important to monitor and record food temperatures regularly?
4. Why must storage areas be secured from unauthorised access?
5. What does *First in – First out* mean when referring to controlling stocks of frozen food?

Extend your knowledge

1. Find out about the full legal requirements concerning the production of cook-freeze food. You will find the relevant laws included in *The Food Safety Act 1990* and *The Food Hygiene (Amendment) Regulations 1990 and 1991.* You should also read the following government guidelines: *Guidelines on Cook-chill and Cook-freeze Catering Systems* (Dept of Health 1989), *Guidelines on the Food Hygiene (Amendment) Regulations* (Dept of Health 1990 and 1991).
2. To increase your understanding of contamination threats and how to prevent these, you may like to read *Croner's Food Hygiene Manual* (Croner Publications, 1991).

UNIT 2D18

Preparing and cooking vegetables and rice

ELEMENTS 1, 2 AND 3: Preparing and cooking vegetable and rice dishes

What do you have to do?

- Prepare cooking and preparation areas and equipment ready for use and clean after use.
- Prepare, cook and finish vegetables and vegetable and rice dishes according to customer and dish requirements.
- Store prepared and cooked vegetables and rice in accordance with food hygiene regulations.
- Satisfy health, safety and hygiene regulations concerning preparation and cooking areas and equipment.

What do you need to know?

- The type, quality and quantity of vegetables and rice required for each dish.
- How to plan your time efficiently, taking account of priorities and any laid down procedures.
- Why it is important to keep preparation, cooking and storage areas and equipment clean and hygienic.
- The main contamination threats when preparing, storing, cooking and finishing both vegetables, vegetable dishes and rice dishes.
- Why time and temperature are important when cooking vegetables.
- How to deal with unexpected situations.

ELEMENT 1: Preparing vegetable dishes

Introduction

Vegetables are an important part of our diet. Green vegetables are a good source of vitamins (especially Vitamin C) and mineral salts, while root vegetables provide starch and sugar for energy. When handling vegetables, always remember that they are *living organisms,* which means that they deteriorate very quickly unless stored and prepared carefully. Potatoes develop a greenish colour and produce shoots unless stored correctly and used quickly; for example, asparagus will continue to use its stored sugar to grow, developing tough, woody fibres.

There are nine different types of vegetables: roots, tubers, bulbs, leaves, flower heads, stems, fungi, vegetable fruits and legumes. You will need to think about type when storing, preparing or cooking vegetables.

Examples of all of these are given in the tables on the next three pages, together with information on how to check for freshness and quality, general preparation points and recipe examples. You may also be asked to prepare frozen vegetables and vegetable proteins for inclusion in menus; these are discussed at relevant points throughout this chapter.

Quality

Britain uses the *EEC Vegetable Quality Grading System.* There are four main classes:

- Extra class: top quality
- Class I: good quality
- Class II: reasonably good quality
- Class III: low marketable quality.

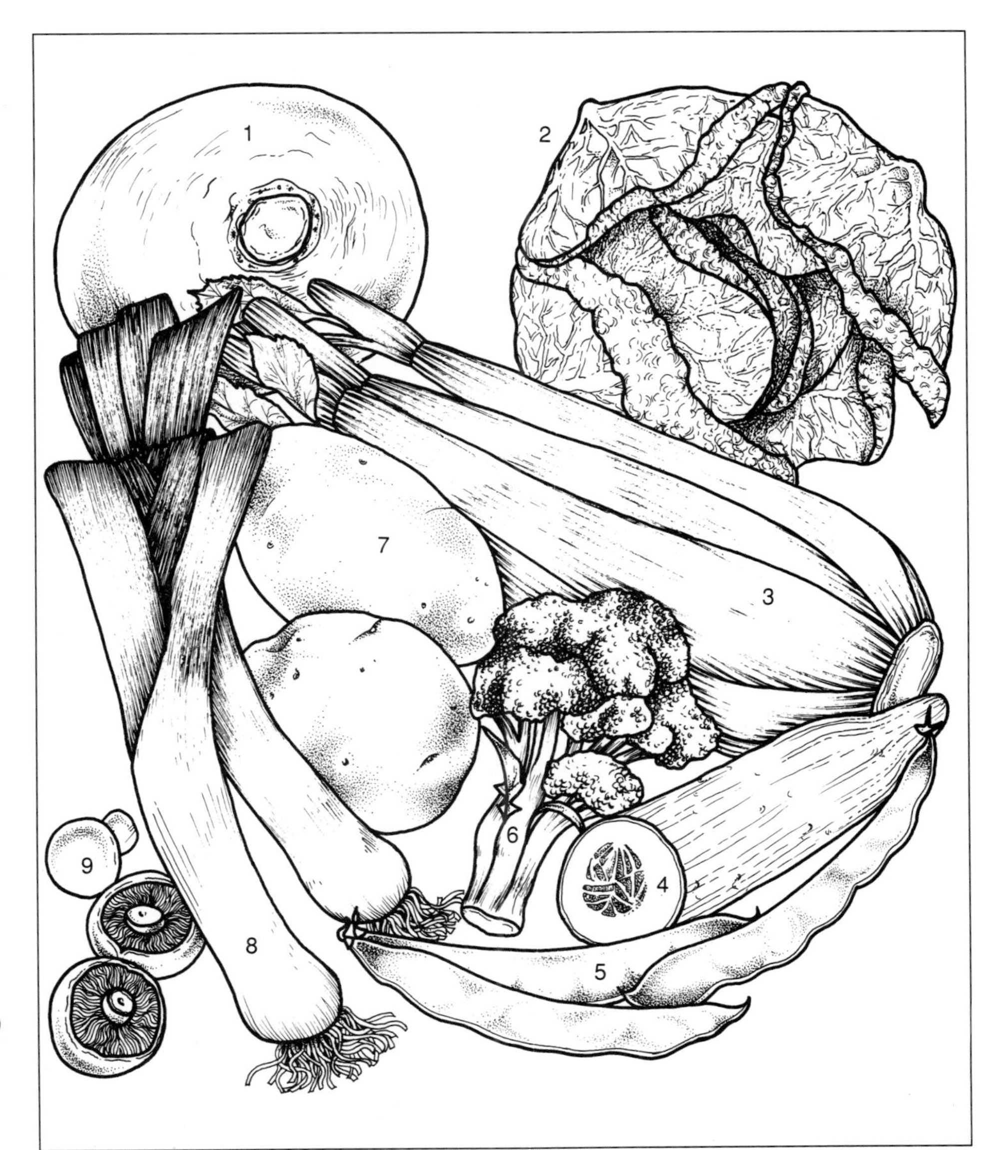

Examples of the nine vegetable types:
1 swede (root)
2 cabbage (leaf)
3 celery (stem)
4 cucumber (fruit)
5 runner beans (legume)
6 broccoli (flower)
7 potatoes (tuber)
8 leeks (bulb)
9 mushrooms (fungi)

Vegetable types

Vegetable	Type	Quality	General preparation	Examples of use and quantity
Artichoke (globe)	Flower	Stiff leaves, no dryness	Cut off the stalk and top third, trim off points of outer leaves. Rub cut surfaces with lemon to prevent discolouring; tie with string to hold shape. Remove choke before/after cooking: spread apart top leaves, remove furry choke	Globe artichokes *(Artichauts en branche*): 1 per portion
Artichoke (Jerusalem)	Tuber	Should not be misshapen or small	Wash, thinly peel and wash again. Keep in salted water ready for cooking	Jerusalem artichokes in cream sauce (*Topinambours à la crème*): 1 per portion
Asparagus	Stem	Tight, well-formed heads, stems not dry or woody. Graded by spear length and shoot width	Cut woody parts from base of stem. Remove tips (spurs) from leaves using back of a small knife. Thinly peel white part of stems downwards using small knife or French peeler. Wash well	Asparagus au gratin (*Asperges au gratin*): 6–8 pieces per portion
Aubergine	Fruit	Firm, no soft patches	Wash, peel, trim then slice as dish demands	Stuffed aubergine (*Aubergine farcies*) 5 for 10 portions
Beans, French	Legume	Crisp, medium size, break crisply under pressure	Wash, top and tail. Leave small beans whole: cut large beans into 7 cm ($2\frac{3}{4}$ in) pieces	Boiled French beans (*Haricots verts a l'anglaise*): 500 g (1 lb) for 3–4 portions
Beans, Broad	Legume	Young, tender pods of uniform size	Remove beans from pods	Buttered broad beans (*Fèves au beurre*): 500 g (1 lb) for 2 portions
Beans, Runner	Legume	Crisp, medium size, break crisply under pressure	Wash, top and tail, string. Cut into 4–7 cm ($1\frac{3}{4}$ – $2\frac{3}{4}$ in) strips	Runner beans nature: 500g (1 lb) for 3–4 portions
Beetroot	Root	Firm, sound, blemish-free	Screw off green leaves; do *not* cut tapering root (this causes vegetable to bleed). Wash well	Buttered beetroot (*Betterave au beurre*)
Broccoli	Flower	Small, fresh-looking heads, crisp stalks	Wash thoroughly and drain. Cut off stalk and any damaged outer leaves	Broccoli with hollandaise sauce (*Brocolis hollandaise*): 1 kg (2 lb) for 8 portions
Brussels sprouts	Leaf	No limp, yellowing leaves; tightly grown leaves; compact	Trim stalks, cut off any discoloured leaves. Cut an 'X' in stem base to ensure even cooking	*Brussels sprouts nature* (*Choux de bruxelles à nature*): 1 kg ($2\frac{3}{4}$ lb) for 10 portions

Vegetable	Type	Quality	General preparation	Examples of use and quantity
Cabbage	Leaf	No discoloured leaves, tightly-grown, compact	Wash, cut away outer leaves, quarter and remove hard centre core	Cabbage (*Chou nature*): 1 kg (2 lb) for 5 portions
Carrots	Root	Firm, not too large	Peel, wash and shape as for recipe	Buttered carrots (*Carottes au beurre*): 500 g (1 lb) for 4 portions
Cauliflower	Flower	No discoloured leaves or damaged, discoloured curds	Wash, remove outer leaves, trim stem and hollow out using peeler	Boiled cauliflower (*Chou-fleur nature*): 1 for 4 portions
Celery	Stem	Firm, tightly grown. Thick, plump at base, smooth sticks	Remove outer stalks, trim heads and root, peel to remove fibres, wash well	Braised celery (*Céleri Braisé*): 2 heads for 4 portions
Chicory	Stem	Firmly packed, crisp, no discoloured or curling leaves	Remove outer leaves if necessary, trim stem, wash	Braised chicory (*Endives brais-ées*): 500 g (1 lb) for 3 portions
Courgettes	Fruit	Firm, straight, light green, blemish-free	Wash, top and tail. Cut as recipe demands	Stuffed courgettes (*Courgettes farcies*): 1 per portion
Cucumber	Fruit	Straight, firm	Peel, shape as recipe demands	Cucumber in cream sauce (*Concombres à la crème*): 3 for 10 portions
Leeks	Bulb	Well-shaped, curling leaves no discoloured or slimy leaves	Cut off roots and trim tops; remove any discoloured leaves. Cut lengthways, wash to remove any soil inside leaves	Braised leeks (*Poireaux braisés*): 500 g (1 lb) for 2 portions
Lettuce	Leaf	Leaves bright in colour, fresh look. No brown, slimy leaves or brown patches	Trim base, remove outer leaves. Leave whole for cooking; separate leaves for salad. Wash	Braised lettuce (*Laitue braisée*): 2 for 4 portions
Marrow	Fruit	Firm, well-shaped with soft skin	Peel, seed, cut as recipe demands	Stuffed marrow (*Courge farcie*)
Mushrooms	Fungi	Not limp, broken or sweaty-looking	Trim base of stalks, wash and drain. Cut as recipe demands	Grilled mushrooms (*Champignons grillés*): 500 g (1 lb) for 4 portions
Onions	Bulb	Firm, regular shape, no soft patches at neck	Wash, peel, trim roots. Cut as recipe demands	French fried onions (*Oignons frits à la française*): 500 g (1 lb) for 4 portions
Parsnips	Root	Firm, sound, blemish-free. No splits up root	Wash, peel, rewash. Remove hard core if mature. Shape as for recipe	Buttered parsnips (*Panais au beurre*): 500 g (1 lb) for 3 portions

Vegetable quality	Type	Quality	General preparation	Examples of use and quantity
Peas	Legume	Plump, crisp, bright green pea pods	Shell and wash (*mange-tout* are cooked whole, i.e. within pods)	Peas French style (*Petits pois à la française*): 1 kg (2 lb) for 4 portions
Potatoes	Tuber	Firm, sound, no signs of damage	Wash, peel, rewash and shape as recipe demands. New potatoes may be peeled after cooking	Plain boiled potatoes (*Pommes nature*): 2 lb for 4 portions
Salsify	Root	Young roots with fresh grey-green leaves, tapering roots	Wash, peel, rewash; cut off top and tapering root. Keep in water with lemon juice if not cooked immediately	Buttered salsify (*Salsifis au beurre*): 500 g (1 lb) for 4 portions
Seakale	Stem	Crisp, not wilted; leaves bright in colour	Remove any discoloured leaves, trim roots, wash under running water	Seakale Mornay (*Chou de mer Mornay*): 500 g (1 lb) for 3 portions
Spinach	Leaf	Bright leaves, crisp	Remove stalks, wash several times in cold water to remove soil and grit	Leaf spinach (*Epinards en branches*): 1 kg (2 lb) for 4 portions
Swedes	Root	Firm, sound, free from spade marks	Trim stalk and root, peel thickly, wash. Cut as recipe demands	Buttered swedes (*Rutabaga au beurre*): 500 g (1 lb) for 2 portions
Sweet potatoes	Tuber	Firm, sound, no signs of damage	Scrub, peel if necessary	As for potatoes (above)
Tomatoes	Fruit	Firm, regular shape, bright colour, no blotches	Wash, remove eye using small knife	Grilled tomatoes (*Tomates farcies*): 1–2 per portion
Turnips	Root	Firm, sound, no signs of worm holes	Wash, trim stalk and root ends, peel thickly, rewash and shape as for recipe	Glazed turnips (*Navets glacés*): 500 g (1 lb) for 4 portions

TVP

TVP is a high protein vegetable product derived from soya beans, wheat, cotton-seed, oats and other vegetables. It is generally made from soya beans, which have a very high protein content and are readily available. Available in several forms, colours and flavours, it is a flexible type of food of particular interest to vegetarians.

Caterers use TVP to replace part of the meat in traditional dishes, as this lowers the cost without severely affecting the nutritional value. As a meat extender it can be used to replace anything from 10 to 50 per cent of the meat, especially in dishes such as stews, casseroles, pasties, pies and curries. When TVP is used in this way the percentage of meat replaced must be stated on the menu. It can be bought in chunks, strips, flakes, granules or as fine mince and comes in brown or pink shades. The blander tasting varieties absorb the flavours of the cooking liquor and other ingredients used in the dish, while the types flavoured to taste such as beef, pork, etc. add to the overall flavour of the dish.

It is an invaluable product for vegetarians, providing a high level of protein from a vegetable source, which enables them to enjoy dishes traditionally based around meat.

Storing vegetables

- Store all vegetables in a cool, dry room at an even temperature of approximately 4–8 °C (39–46 °F). Low temperatures will keep spoilage to a minimum, but note that the temperature should not fall below 3 °C (37 °F) as this will cause the vegetables to deteriorate. Air should be allowed to circulate around the vegetables to prevent alcohol building up in them and causing skin damage.
- Store green vegetables on well-ventilated racks. Root vegetables should be removed from their delivery sacks and stored in bins or on racks. Salad vegetables may be left in their containers.
- Remember that vegetables have the best flavour and the most food value when fresh. They start to deteriorate as soon as they are picked. Cut off any leaves on root vegetables before storing to prevent the sap rising from the roots.
- If vegetables are allowed to dry out or wilt, harmful substances will become concentrated in some areas, destroying parts of the cell walls and valuable Vitamin C. The vegetables will start to discolour and develop undesirable flavours. Green vegetables lose Vitamin C quickly if stored for a long time, or if they become bruised or damaged.
- Remember that mould spreads quickly, especially to damaged vegetables. Unpack all vegetables from their delivery packaging (bags, sacks, etc.) to keep disease and damage to a minimum, then place them on well-ventilated racks.
- Do not store vegetables close to fruit or other porous products; the flavours will be affected, sometimes surprisingly: carrots stored next to apples will develop a bitter taste.
- Potatoes must be stored in a cool, dry, dark place; warmth and light will cause them to turn green. The colour changes as the potatoes begin to produce poisonous alkaloids. Try to avoid using potatoes in this state. If it is unavoidable, peel them very thickly. Always cut out any eyes or shoots; these are also harmful.
- Frozen vegetables should be stored at −18 °C (0 °F) or below. Rotate stock so that you always use the longest-held vegetables first. Check the *use by* date and if necessary, throw away any vegetables still held after this date (most frozen vegetables can be stored in a freezer for 3–12 months; check the packaging). Never use vegetables from a torn or crushed packet as the product may be damaged or suffer from freezer burn.

Preparing vegetables

Planning your time

Be aware of what dishes you will be preparing, so that you know what equipment you will need to use, which dishes need to be prepared first, cooking times and so on. Preparing a production time plan will help you to think about priorities. When cooking vegetables remember that they deteriorate fast when held for service, so should be cooked or finished last, and cooked in batches. However, bear in mind that you cannot leave too many items close to service, so make sure that your time plan spreads work evenly throughout your time in the kitchen.

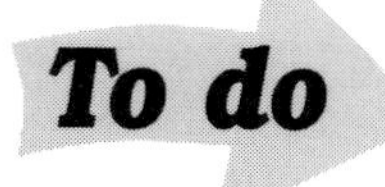

- Find out where the fresh vegetables are stored in your kitchen. Which types are stored where?
- Try to find at least one vegetable of each type (e.g. root, stem, bulb, etc.) in your kitchen. Look for the features indicating quality, by referring back to the table on pages 307–9.
- Watch a delivery of fresh vegetables being made to your kitchen. How are the vegetables unpacked and stored? How is rotation of stock managed?
- Check the temperatures of areas in which vegetables are stored in your kitchen (including any deep-freezers).

Health, safety and hygiene

Make sure you are familiar with the general points given in Units G1 and G2 of the Core Units book. Remember that soil can contain live food-poisoning bacteria, so always scrub vegetables before preparation to prevent any soil from contaminating your work area. Vegetables grown in water (e.g. watercress) must be thoroughly washed as the water may have been contaminated by animals.

Essential knowledge

The main contamination threats when preparing and storing raw vegetables are as follows:
- bacteria may be transferred from the nose, mouth, open cuts and sores or unclean hands
- cross-contamination may occur between unhygienic equipment, utensils and preparation areas and the raw vegetables
- micro-organisms may develop in vegetables stored at incorrect temperatures
- moulds may develop if vegetables are stored under moist conditions rather than in dry, airy storage spaces
- bacteria from raw vegetables can pass to cooked vegetables if they are stored in the same area
- incorrect thawing procedures carried out on frozen vegetables may lead to contamination
- bacteria from waste may be transferred to vegetables, utensils or preparation areas unless disposed of correctly
- uncovered prepared vegetables may easily become contaminated.

Preparation areas, cooking areas and equipment

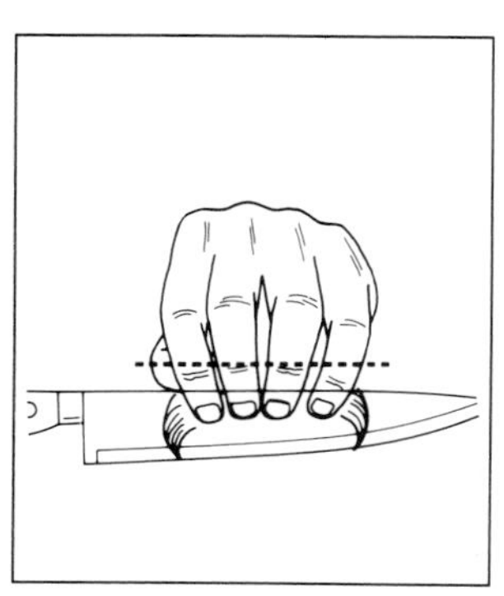

The correct way to hold a vegetable while cutting

When chopping vegetables, keep the following points in mind:
- use the correct chopping boards for preparing vegetables: check the colour coding system used in your kitchen. Do not use unhygienic wooden boards
- keep cooking areas clean. Do not leave pan handles jutting over the edge of the stove where they could cause accidents
- only put out the tools and equipment you need; keep your work area uncluttered
- for preparing vegetables you will probably need to use a vegetable peeler, and one or more cooks' knives: usually those with blades of 10 cm (4 in), 20 cm (8 in) or 25 cm (10 in)
- remember that many pieces of equipment are difficult to clean properly because they feature small holes,e.g. mandolins and graters. They

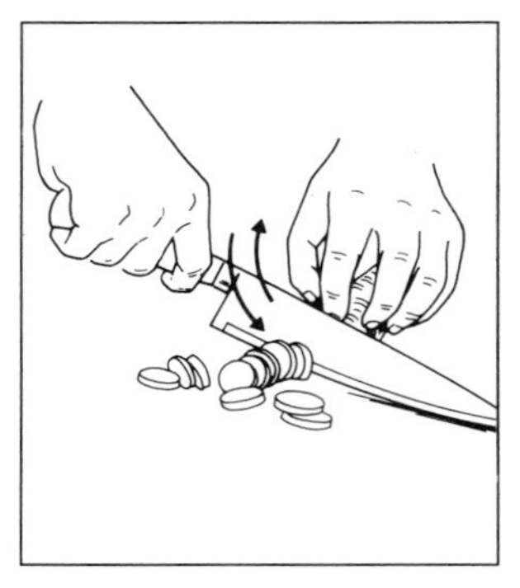

The correct way to slice a vegetable

can be breeding grounds for bacteria
- never use a blunt knife. The knife is more liable to slip and cut you, and the blunt edge will bruise the vegetable more than a sharp edge would, causing loss of Vitamin C
- remember that you will have less control when cutting sideways *across* a vegetable (e.g. when slicing an onion) than when cutting downwards
- when chopping or slicing vegetables, use the correct techniques. For chopping, keep the fingers of the hand not holding the handle on top of the blade at all times, and use a pivotal action. For slicing, hold the item to be cut by placing your fingers on top of the vegetable (finger-tips curled inwards) and parallel to the knife blade; keep your thumb tucked behind
- when chopping vegetables, always work from left to right; i.e. keep the uncut vegetables to your left and the chopped ones to your right. Put the chopped vegetables into clean bowls as soon as you finish chopping them
- try not to carry knives around the kitchen. When you need to do this, always carry them with the point facing downwards and the sharp edge facing behind you (see Unit 2D10: *Handling and maintaining knives* in the Core Units book).

Essential knowledge

It is important to keep preparation and storage areas hygienic at all times in order to:
- prevent the transfer of food poisoning bacteria to food
- prevent pest infection in storage areas
- ensure that standards of cleanliness are maintained
- comply with the legal health and hygiene regulations for the UK.

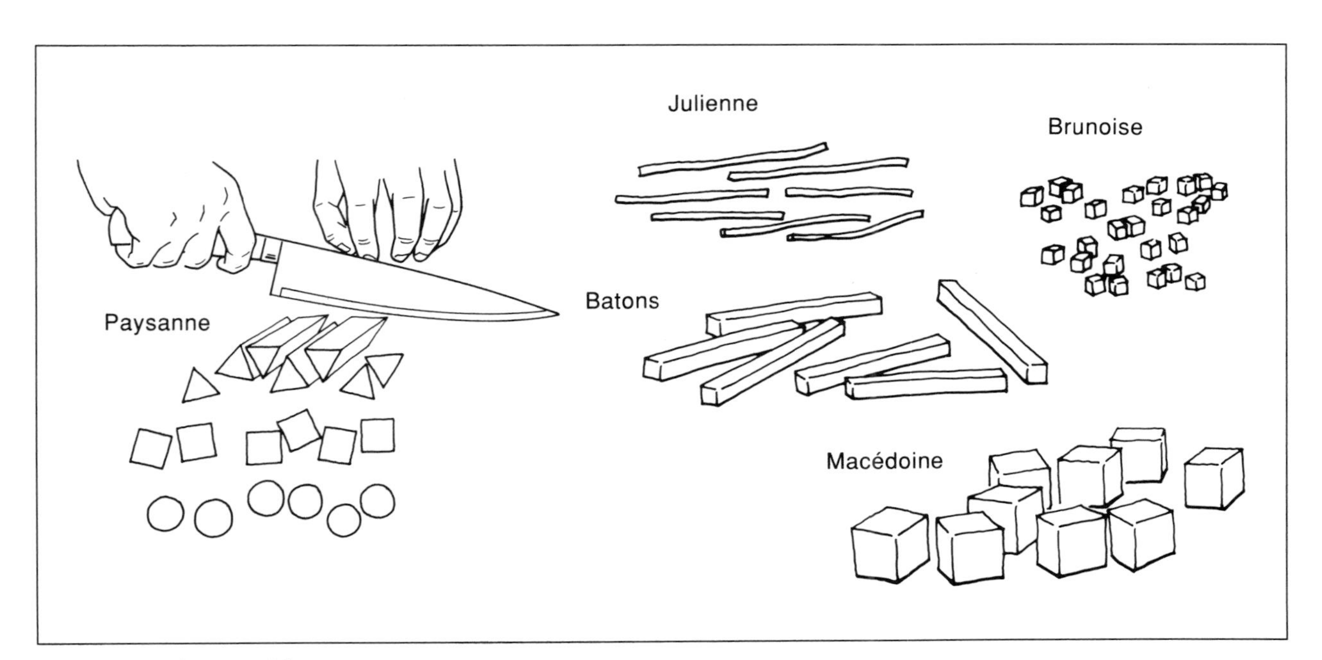

Basic cuts of vegetables

Cuts of vegetables

Certain dishes will require you to shape (or *turn*) vegetables. There are six types of cuts commonly used when preparing vegetables:

- Macédoine: small cubes
- Jardinière: batons
- Paysanne: either triangles, squares or circles
- Julienne: thin strips
- Brunoise: fine dice
- Barrel shapes.

Preparation methods

Puréeing

This is a common method of preparing vegetables. The finished purée may be served as a vegetable, a garnish to a main dish (e.g. braised ham may be served with a garnish of spinach purée), or as a base for a mousse/terrine. *Coulis* is the name given to a light purée used as a sauce accompaniment to a dish.

Method

1. Simmer the vegetables gently, using only enough liquid to cover them.
2. When cooked, refresh them under cold running water and then drain.
3. Pass the cooked vegetables through a sieve or mouli, or liquidise them.
4. Reheat the purée in a saucepan to draw out any excess water.
5. Add any butter, milk or cream to correct consistency and check seasoning. Serve in a vegetable dish, shaping the purée to a dome.

Optional: When using vegetables with a high water content, such as French beans or cauliflower, add an amount of starchy vegetable (creamed potato or rice) to achieve the correct texture. Make sure that the flavour of the thickening agent does not affect the main ingredient.

Stuffing

Many vegetables can be stuffed: especially marrow, aubergine and mushrooms. The main vegetable may be part-cooked before stuffing (e.g. Stuffed aubergine) and finished under a grill or salamander, or uncooked (Stuffed marrow) and then oven-cooked. Small vegetables like mushrooms cook quickly enough to be stuffed raw and then cooked either in the oven or under a salamander.

Stuffings vary; they usually include cooked, finely chopped vegetables, and herbs and seasonings. They may also include sausage meat or cooked meat, and rice. *Duxelle* is a popular stuffing, consisting of lightly cooked, chopped shallots and mushrooms combined with breadcrumbs (where necessary, to achieve the required consistency).

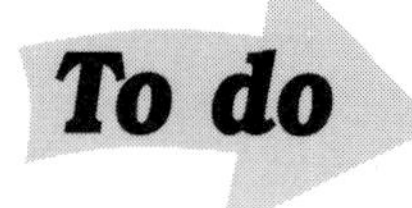

- Watch your supervisor carry out general preparation on at least three types of vegetables. Notice the equipment and techniques used.
- Find out what colour coding system is used in your kitchen. Make sure you know which boards to use for chopping vegetables.
- Watch your supervisor cutting vegetables into some of the types of cut given above. Notice knife techniques and steps followed.

Use of moulds

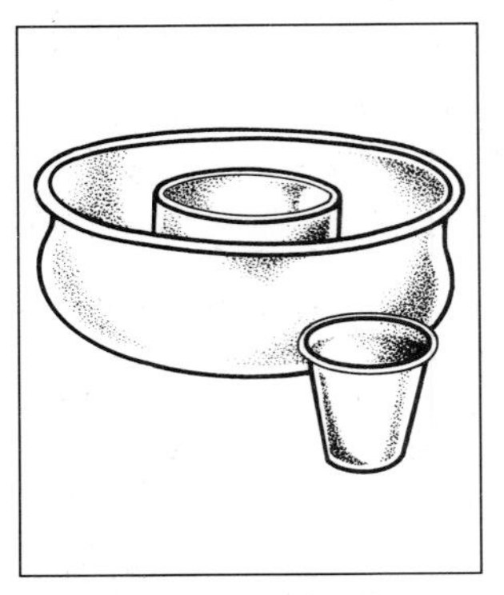

Moulds that might be used for vegetable mousses: savarin (top); dariole (bottom)

Vegetable mousses and mousselines may involve the use of moulds to shape the finished dish (*mousselines* is the name given to mousses prepared in individual moulds). Check with your supervisor before preparing mousses to find out how many and which moulds you need to use.

Mousses may be served either hot or cold.

Hot mousses

1 Purée the vegetables then place in the refrigerator to chill.
2 Add cream, eggs and seasoning.
3 Pass through a sieve then pour into moulds.
4 Place the moulds into a bain-marie or steamer to cook gently.
5 When cooked, turn out the moulds onto service plates. To do this: place the service dish upside-down over the mould, hold the mould and dish tightly together, then gently turn both over, allowing the mousse to slide from the mould to the dish. Serve immediately.

Cold mousses

1 Place the clean empty moulds in the refrigerator to cool.
2 Prepare and cook ingredients as necessary, then liquidise ingredients to a smooth paste and add any cream.
3 Line the mould with aspic if required.
4 Pour the mousse into small, individual moulds or one large mould and place in the refrigerator to set.
5 When set and required for service, the moulds can be turned on to service dishes. It may be necessary to warm the moulds slightly (to loosen the mousse from the side of the mould) before turning out; if so, either dip the mould into tepid water or wrap a hot cloth around the outside of the mould. Garnish as required.

Types of moulds

These include: darioles, Charlottes, savarins, tartlets, barquettes and many other types used both for sweet and savoury preparations. Note that spoons (particularly teaspoons and dessert spoons) may also be used for moulding vegetable purées or potato dishes (e.g. Pommes Dauphine). To do this: dip a pair of spoons into hot water, shape the purée or potato mixture between the two spoons and then gently transfer the moulded mixture onto the service dish or cooking tray as required.

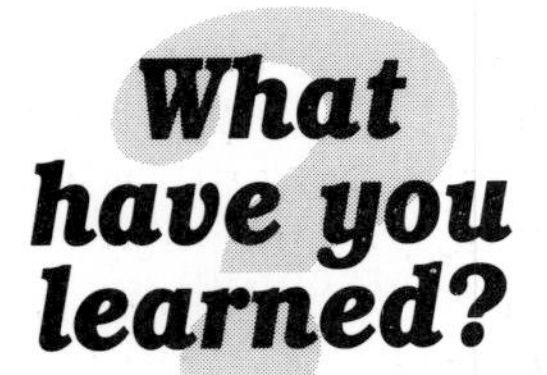

1 What are the main contamination threats when preparing and cooking vegetables?
2 Why is it important to keep preparation, cooking and storage areas hygienic?
3 Name two vegetables from each of the following types: tuber, root, stem, fruit.
4 What considerations should you make when storing tuber, stem and flower vegetables?
5 What temperature should frozen vegetables be stored at? Why should you never use frozen vegetables from damaged packets?

ELEMENT 2: Cooking and finishing vegetable dishes

Keep the following points in mind when cooking vegetables:
- young, fresh vegetables will cook more quickly than fully-grown vegetables or those that have been stored for some time
- when boiling vegetables, remember that as a general rule, root vegetables should be placed in cold water and then brought to the boil, while vegetables grown above ground should be placed into boiling water. New potatoes are an exception to the rule: always start them in boiling water. However, many chefs today start *all* vegetables in boiling water to preserve the Vitamin C content
- be aware of how vitamins can be lost, and what you can do to prevent this: do not prepare vegetables too far ahead of cooking time; always use a sharp knife during preparation to prevent bruising; do not immerse them in water before cooking (unless they would discolour if they were not immersed, e.g. potatoes); cook them as quickly as possible; serve them immediately when cooked
- always cook vegetables as quickly as possible. By keeping cooking time to a minimum, you will allow vegetables to retain maximum flavour, nutritional value and colour. Cooking vegetables in high speed steam cookers is one way of achieving this.

Essential knowledge

Time and temperature are important when cooking vegetables in order to:
- achieve the required food textures and flavours
- ensure that the finished dish is cooked correctly
- control the amount of fat absorbed by the food during frying processes
- prevent food poisoning.

Cooking methods

Roasting

Roasting is a dry cooking process where food is cooked in an oven or on a spit with a small amount of fat or oil (see pages 13-15). It is not suitable for most vegetables because of their high water content. Those that can be successfully roasted include: potatoes, sweet potatoes, parsnips, onions and Jerusalem artichokes.

Potatoes

These roast well, because the hot fat combines with the starch in the potatoes to give a crisp surface. Cut them to an even size, and place into cold water to prevent discolouration. Parboil them if necessary (to reduce the roasting time), then place them into a large roasting tray brushed with oil. Make sure the tray is large enough to expose all the potatoes to heat from both the tray and the hot air from the oven. Brush over the potatoes with oil, and place them in a hot oven (230 °C/445 °F) until cooked. Do not turn them while cooking. Parsnips are roasted following the same method, but note that they take less time than potatoes (about 30 minutes) and the finished vegetables are not as crisp.

Other vegetables
Where possible, cut the vegetables into long, flat shapes, as these will cook faster and therefore dry out less. Place the prepared vegetables into a hot roasting tray containing hot fat or oil. *Never* bring vegetables into contact with cold fat or oil: they will absorb too much of the fat and become greasy, which is undesirable for taste and health reasons. Place them in a hot oven (230 °C/ 445 °F) and baste them as necessary during cooking.

Recipe examples: Roast potatoes (*Pommes rôtis*), Roast parsnips (*Panais rôtis*).

Baking

Baking is another dry cooking process, where food is cooked by the dry, convected heat of an oven (see pages 22–3). As with roasting (above), the dryness of the process makes it unsuitable for most vegetables, although this effect can be lessened by leaving the skin on the vegetables for protection. Potatoes are the most commonly baked vegetable, but others, such as tomatoes, marrow and pimento can be successfully stuffed and baked.

Potatoes
Baking potatoes in their skins allows them to keep much of their Vitamin C content. Use the more floury types of potato for baking, to achieve the best texture (e.g. King Edwards, Désirée). Wash thoroughly, then make small slits in the skin to prevent them from bursting during baking. Brush the base of a tray with fat or oil, sprinkle on a fine layer of salt and place the potatoes on top. Place the tray into a moderate oven (200 °C/395 °F) for 1–1½ hours. Test for cooking by squeezing one potato gently between your hands (protected by a clean dishcloth). When cooked, cut a cross into the top of each potato and press the sides to open.

Baked potatoes

Other vegetables
Some vegetables can be stuffed and baked (such as marrow). These are usually covered while cooking, and are therefore partly cooked by the moist steam from within the pot, which helps to prevent the vegetable from drying out. The prepared, stuffed vegetable is placed in a moderate oven (180–200 °C/355–395 °F) often with some stock, and cooked until tender. Stuffed tomatoes are cooked uncovered, although the top of the tomato is often placed over the stuffed portion to act as a lid.

Recipe examples: Baked potatoes with sour cream, Stuffed tomatoes (*Tomates farcies*).

Grilling

This is another of the dry cooking processes, cooking food by radiated heat generated by infra-red waves (see pages 15–17). As with roasting and baking, it is not suitable for most vegetables, causing them to shrivel and dry out.

The most commonly grilled vegetables are mushrooms, tomatoes, aubergines and courgettes. Clean or wash the vegetables, place them on a grilling tray and brush with a little fat or oil. Cook them under a grill or salamander until tender.

Vegetable kebabs are often cooked by grilling. Place the prepared, chopped vegetables onto a skewer, brush with oil and place under a grill or salamander. Turn during grilling to ensure even cooking and serve immediately when cooked.

Recipe examples: Vegetable kebabs, Grilled mushrooms (*Champignons grillés*).

Shallow-frying, griddling or frying

This is also a dry cooking process, where the food is heated through contact with a hot cooking surface and hot fat or oil (see pages 17–19). The process is suitable for many vegetables in a variety of ways. Some are cooked entirely by this method, some are partly cooked by shallow-frying and then completed by, for example, braising, while others are shallow-fried to finish (having already been partly cooked by, for example, boiling).

Method

1. Heat the pan until the oil is at the correct temperature (175–195 °C/ 350–400 °F approximately). This is important because if the fat is too cool the vegetables will absorb too much and become greasy, while if it is too hot they will burn.
2. Add the vegetables, which may be cooked or partly-cooked (see below). Be careful not to shallow-fry too many at once: they will break up during cooking and may release so much liquid that they will start to simmer in the liquid rather than fry in the oil. If this happens, drain off the liquid during cooking.
3. Toss the vegetables as necessary during cooking using a fish slice. Try to avoid turning them too often or too vigorously, or they will start to break up.
4. Drain carefully when cooked.

Suitable vegetables

Cooked completely by shallow-frying: mushrooms, onions, chicory, aubergines, courgettes, peppers, tomatoes.
As a first step before stewing, braising: carrots, leeks, celery.
As a method of finishing after boiling, steaming: Brussels sprouts, French beans, cauliflower, potatoes.

Recipe examples: Peperonata, Sauté potatotes (*Pommes sautés*).

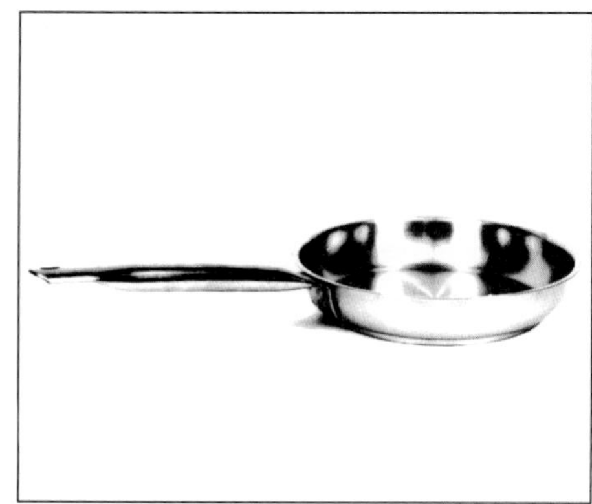

Equipment for shallow-frying: a sauteuse (left); a frying pan (right)

Boiling vegetables

Known as *l'Anglaise* or *nature,* this is a wet cooking process, where food is cooked in boiling, lightly salted water (see pages 5–7). It is suitable for almost all vegetables for several reasons. The method is fast, softening the vegetable while retaining most of the colour and vitamins. It is healthy, as no fat is added during cooking, and salt may be omitted. Fungi are not normally boiled.

Frozen vegetables
When boiling frozen vegetables, immerse them in boiling water while still frozen. Never thaw them first, as this causes vitamin loss and can unpleasantly affect colour and texture. Always read any instructions on the packaging thoroughly.

Considerations and method

- The cooking time depends on the vegetable, but also on the age, quality and size. Young, fresh vegetables will cook more quickly, as will those cut into smaller pieces (although this increases the loss of Vitamin C).
- Choose a pan that will hold the vegetables easily but requires the minimum amount of water to cover the vegetables; this also cuts down on Vitamin C loss.
- As a general rule, place vegetables grown below the ground into cold water and bring to the boil; place those grown above the ground into already boiling water. Some chefs prefer to start all vegetables in boiling water to preserve the Vitamin C content.
- Once the water has come back to the boil, simmer gently. Excessive liquid movement will cause some vegetables to break up, and does not decrease cooking time. When boiling vegetables to be glazed, the temperature should be raised towards the end of cooking time (see *Glazing,* page 322).
- Some types of vegetables need particular consideration. Tubers (e.g. potatoes) tend to break up easily and need careful handling. Do not peel any potatoes that are to be cooked further (e.g. sauté) or eaten cold (e.g. potato salad), as this will allow them to retain vitamins and maximum flesh. Take care not to overcook leaf vegetables or fruits. Green vegetables should be cooked quickly to retain colour and vitamins. White vegetables (e.g. artichoke bottoms) may need lemon juice added to the cooking liquor to avoid discolouration. Flower vegetables need attention to ensure that the stem is cooked but the flower is not overcooked; split the stalk before cooking if necessary.
- While cooking, skim off any impurities that rise to the surface to prevent them from discolouring the cooking liquid and affecting the flavour. Cook the vegetables until they are just firm to the bite, then drain well and serve immediately with butter or garnish as required.
- If vegetables are to be kept back after cooking, refresh them under cold running water, drain and keep cool until required. Reheat by microwaving, or by placing them into a colander, immersing it into boiling water for a few minutes until the vegetables are heated through, then drain and serve. Alternatively, keep the vegetables warm after cooking by placing them in a bain-marie or on a hotplate.

Recipe examples: Buttered carrots (*Carottes au beurre*), Leaf spinach (*Epinards en branches*).

Blanching
Vegetables may be part-boiled by a process known as *blanching.* The prepared vegetables are placed into boiling water for a very short time, refreshed under cold running water then refrigerated until required. The process can be used for a variety of reasons: to partly cook vegetables

that are to be completed by braising or roasting; to soften vegetables that are to be used for stuffings; to prepare vegetables for freezing; to help in removing the skin from vegetables (e.g. tomatoes); or to prevent discolouration.

To do

- Watch your supervisor preparing and baking potatoes. How do they check that they are fully cooked?
- Watch your supervisor preparing vegetables for grilling. Notice how much oil is used and the temperature setting of the grill or salamander.
- Find out what types of pans your establishment uses for boiling vegetables. Are different types of pan used for different vegetables?
- Find out how your establishment reheats vegetables that need to be held back for service after cooking.

Braising, stewing or poaching

These are all wet processes, where food is partially covered by water and cooked either in the oven or on the stove.

Braising

Here food is cooked in the oven, partially covered with liquid kept close to boiling point, and covered with a lid (see pages 11–13). The cooking liquor is often used after braising as a base for a sauce, and can include stock, wine, jus-lié, tomato purée, seasoning and herbs. Vegetables can be braised to make a complete vegetable dish (e.g Braised lettuce, cabbage or leek), or may be used as a base for a braised meat or poultry dish.

To braise a vegetable: roughly chop a mixture of vegetables and herbs (e.g. onions, carrots, celery, leeks, thyme, bay leaf) to form a mirepoix and place in the cooking pan. Do not use the main vegetable to be braised as part of the mirepoix. Blanch the main vegetable and place it on top of the mirepoix. Add stock to cover two thirds of the vegetable, and any salt and or bacon fat. Cover the pan with greaseproof paper and a lid, and place in a moderate oven (180 °C/360 °F) until cooked. Baste as necessary during cooking. Serve immediately when cooked.

Recipe examples: Braised stuffed cabbage (*Chou farci braisé*), Braised celery (*Céleri braisé*).

Stewing

Here pieces of food are simmered in a small amount of cooking liquor which forms part of the finished dish (see pages 9–11) and cooked either on top of the stove or in the oven. Vegetables are always used in meat, poultry and game stews, to enrich them (e.g. carrots, celery, onion, leeks) but may also be stewed to form complete vegetable dishes, e.g. ratatouille, vegetable curry.

Recipe examples: Marrow provencale (*Courge provençale*), Peas French style (*Petits pois à la française*), Ratatouille.

Poaching

Here food is covered either partially or wholly by liquid, and cooked at a temperature below boiling to prevent the food from breaking up. It differs from stewing or braising in that it cooks food quickly but at a lower temperature. It is unsuitable for most vegetables, as it is too slow and cooks at too low a temperature. However, some vegetables may be used in poaching recipes to enhance the flavour of the dish; generally mushrooms, onions, shallots and tomatoes.

Steaming

This is a wet cooking process, where food is cooked by the heat from steam (either through convection or condensation: see pages 9–10). Almost all vegetables that are suitable for boiling can be steamed successfully. The low water absorption rate while steaming makes it a very good method for cooking potatoes that are to be sautéed.

Steaming is a particularly good method for cooking vegetables for several reasons:

- they are less likely to break up than when being boiled, because there is little movement
- they lose fewer vitamins (especially the water-soluble ones, such as Vitamin C)
- they can be cooked very fast (especially by high-pressure steaming) even in large quantities, so can be cooked immediately before service.

The only disadvantage is that acids from the vegetables are retained (which are lost in boiling) which can adversely affect the colour of green vegetables. You also need to be very aware of timing when steaming vegetables as they can easily over-cook. Always use even-sized pieces to ensure that they all take the same amount of time to cook.

Recipe examples: Steamed potatoes (*Pommes vapeur*), Steamed green beans.

Deep-frying

This is a dry cooking process, where food is cooked by being totally immersed in hot (165–190 °C/320-375 °F) oil or fat (see pages 19–21). It is a fast method of cooking, as the entire surface of the food is kept in constant contact with the hot fat. Most vegetables can be cooked using this method, though some may need to be part-cooked before deep-frying. Deep-fried food should never be kept covered; it will lose its crispness and become soggy. As a safety precaution, always dry vegetables thoroughly before deep-frying: the hot oil will react violently to any water left on vegetables.

Potatoes

Potatoes are often cooked by deep-frying, especially the less floury varieties. They can be cut into various shapes and sizes (see the illustration on page 321), washed well (to remove starch), dried and placed raw into the hot fat. They may be fried twice, as they do not stay crisp after frying. The first frying or blanching is carried out at a low temperature (160 °C/325 °F) to soften the potatoes, after which they can be drained and kept cool until required. They are then fried again at a high temperature (180 °C/360 °F) until crisp and golden brown, drained and served. Allow frozen potatoes to defrost slightly, to a temperature just below freezing before deep-frying (check any instructions on the packaging).

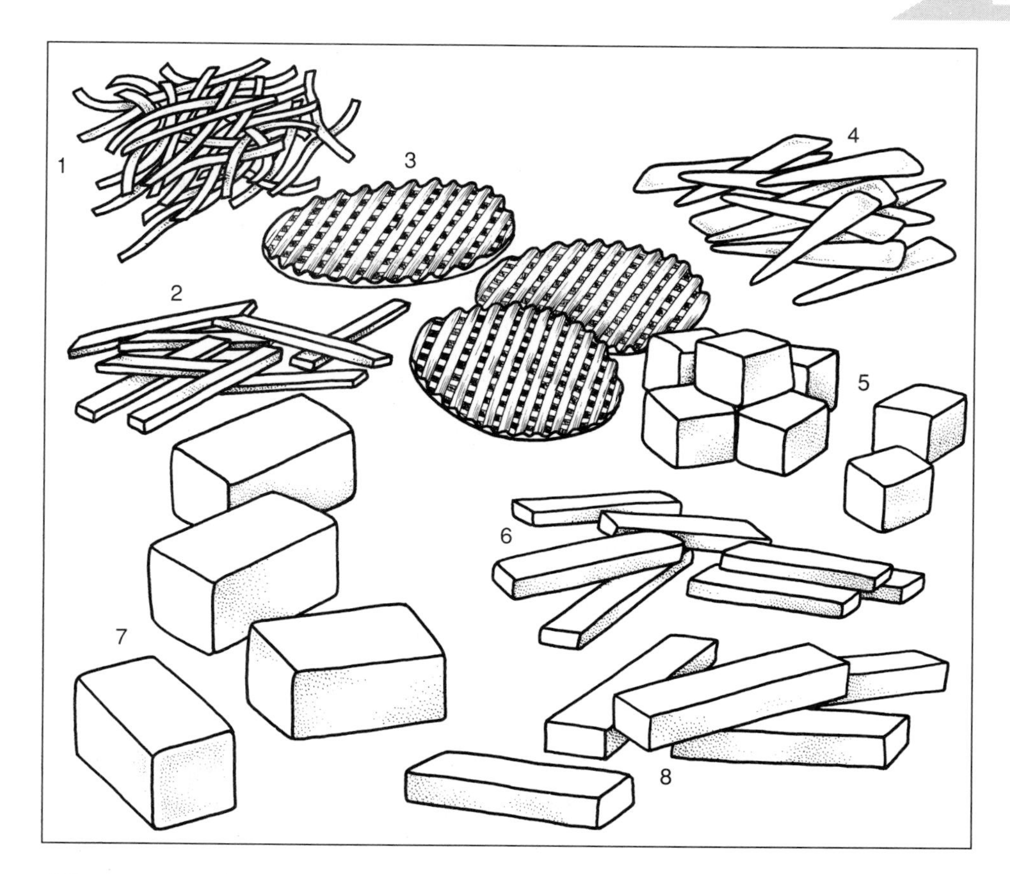

Potato cuts suitable for deep-frying:
1 straw
2 matchstick
3 wafer
4 woodchip
5 cubes
6 thick matchsticks
7 pont neuf
8 French fries

Other vegetables

Most vegetables are parboiled or steamed first, passed through flour to give an adhesive quality, dipped in batter and then deep-fried. Do not batter the vegetables until just before frying, and make sure the batter is thick enough to avoid dripping but thin enough to cover the vegetable. Some vegetables (e.g. onions, aubergine) do not need pre-cooking.

Considerations

- Cooking time depends on the vegetable type, its size, shape and texture. It also depends on the temperature when vegetables are immersed, how long it takes the oil to regain the correct temperature, and the amount of oil relative to the vegetables being fried.
- Do not deep-fry too many vegetables together: this reduces the oil temperature and increases fat absorption.
- Shake the frying basket during cooking to prevent the vegetables from sticking together.
- Always drain the vegetables well after cooking. Do not cover them.

Recipe examples: Fried aubergines (*Aubergines frites*), French fried onions (*Oignons frits à la française*).

Browning

Browning occurs when heat is directly applied to food via the surface of a pan, to colour the food surface (i.e. the outside of food only). Food is usually browned by shallow-frying, or occasionally by flash-roasting (large items only). It is also known as *searing,* and it used to be thought that the process sealed the food, though this is no longer considered to be true. The process enhances the flavours of foods, and is generally used on food that will afterwards be braised or used as a garnish for another dish.

To do

- Look at today's menu. Which processes will be used in preparing each dish?
- Watch some vegetables being shallow-fried and deep-fried. Notice how quickly they cook, and any differences in appearance when cooked.
- Check the packaging on at least four kinds of frozen vegetables. How are the instructions different on each one?
- Find out which vegetables your establishment cooks by steaming rather than boiling. Why?

Finishing vegetables

Glazing

Glazing is a process often used for finishing boiled vegetables to enhance flavour and give the finished vegetable dishes a shiny appearance. It is often used for root vegetables.

Method

1. Boil the turned or shaped vegetable in water to which you have added butter, sugar and salt (optional).
2. Allow the water to evaporate during cooking. Simmer the vegetable until nearly cooked, then increase the temperature in the final stages to encourage evaporation.
3. Toss the vegetables in the remaining syrup before serving immediately.

Recipe examples: Glazed carrots (*Carrottes Glacées*), Small white glazed onions (*Petits oignons glacés à blanc*).

Coating

Many boiled vegetables can be served with a coating of sauce to form finished dishes such as Cauliflower Mornay (*Chou-fleur Mornay*) or Braised leeks (*Poireaux braises*). The sauce is poured over the cooked vegetable to coat it before service. Timing is important: coating too long before service, or allowing the vegetables to cool and then trying to reheat them in the sauce will unpleasantly affect the colour, taste and texture of the dish.

Vegetables are also often served with a butter coating: *au beurre*. Here the cooked vegetable is added to a hot sauteuse containing melted butter, and tossed until completely coated with the butter.

Recipe examples: Buttered Brussels sprouts (*Choux de Bruxelles au beurre*), Buttered cabbage (*Chou au beurre*).

Colouring

Vegetable dishes do not usually feature colourings, though vegetables themselves may be the sources of natural colourings. The green colour (chlorophyll) of some vegetables, such as spinach, can be used to obtain a green colouring, while turmeric roots can be used to obtain a deep yellow colour.

In the past, sodium bicarbonate was occasionally added to green vegetables that had lost their colour as a result of over-boiling. This is no longer practised as it results in the destruction of Vitamin C and thiamine, and softens the texture. To prevent loss of colour as far as possible, renew the cooking water for each batch of vegetables, and return the water to boiling point as soon as possible (the acids causing discolouration stop being released from the vegetables at boiling point).

Some pale coloured vegetables, e.g. potato, parsnip, onion and chicory may take on a yellow colour during cooking in water. You can prevent this by adding a few drops of lemon juice to the cooking water. Note that red cabbage reacts with certain metals, especially aluminium, to produce bluish tints, so check the metal type of the pan you intend to use.

Using hot sauces

Hot sauces may be added to a vegetable at the final stage of cooking (e.g. dishes finished *à la crème*),or used either to coat or accompany many vegetable dishes. Hot sauces commonly used with vegetable dishes include: Sauce Mornay, Parsley sauce (*Sauce persil*), Cream sauce (*Sauce crème*), Jus lié, Melted butter (*Beurre clarifié*), Hollandaise sauce (*Sauce hollandaise*).

Using cold sauces

Cold sauces are usually served with cold or salad vegetables and include mayonnaise, vinaigrettes, etc. (see Unit 2D13: *Preparing food for cold presentation*).

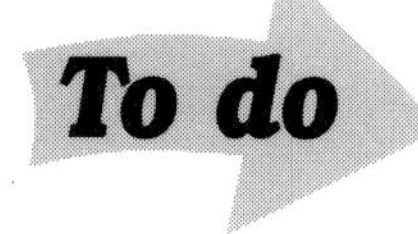

- Find out whether you have both low and high pressure steamers in your kitchen. Watch them being used. Which is more commonly used for cooking vegetables?
- Check today's menu to see what kinds of finishes are used on the vegetable dishes. Check that you are familiar with their preparation.
- Watch the preparation of two types of sauces to be used in vegetable dishes. When and how are they added to the vegetables?
- Check the colour of at least three types of vegetables before and after cooking. How has the colour changed, if at all?

Garnishing

Garnish is used to decorate a dish, and is always edible. Vegetable dishes may be accompanied by a hot garnish (such as cooked tomato concassée) or a simple sprinkling of herbs (parsley, chives, chervil, etc.). Certain styles of vegetable presentation require particular garnishes. The most common ones are listed below:

à la menthe — fresh mint leaves are added during cooking, and the finished dish is garnished with blanched mint leaves

amandine — almonds sautéed in butter are sprinkled over the vegetables and topped with chopped parsley

aux fines herbes — the cooked vegetables are brushed with butter and sprinkled with fine herbs (parsley, chives, chervil, etc.)

milanaise	the cooked vegetables are sprinkled with Parmesan cheese, gratinated under a salamander and finished with beurre noisette
persillés	the cooked vegetables are brushed with melted butter and sprinkled with chopped parsley
polonaise	the cooked vegetables are sprinkled with fried breadcrumbs, sieved hard-boiled egg, chopped parsley and finally *beurre noisette.*

Piping

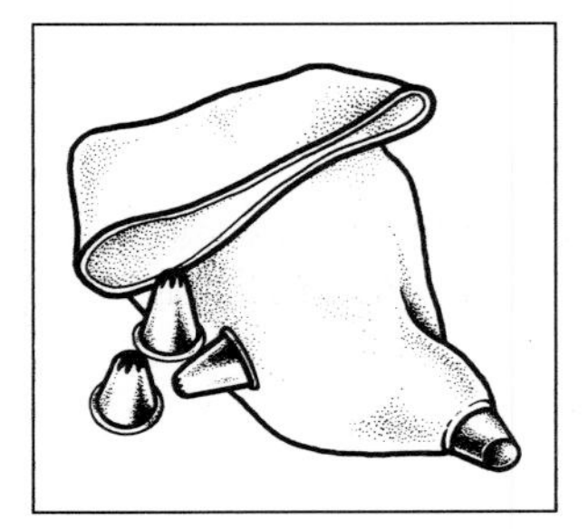

A Savoy bag (used for piping potato dishes)

Piping is one way of shaping vegetable mixtures. It is not commonly used in the preparation of vegetable dishes, with the notable exception of Duchesse and Marquis potatoes.

Duchesse potatoes (*Pommes Duchesse*) may be served as a potato dish or used in the preparation of many other dishes, e.g. Croquette potatoes (*Pommes croquettes*). Even-sized pieces of potato are boiled until tender, drained, passed through a sieve and mixed with egg yolk. The mixture is then placed into a piping bag with a large star-shaped tube, and piped out on to a greased tray in spiral shapes 4 cm (1 inch) high and 4 cm (1 inch) diameter. They are then reheated in a hot oven, brushed with eggwash and returned to the oven to develop a golden brown colour.

When stuffing small vegetables, e.g. tomatoes, the stuffing may be placed into a piping bag with a large star tube and piped into the tomato shells.

What have you learned?

1. Which process suits almost all types of vegetables? Why?
2. In what ways can you ensure that the least amount of Vitamin C is lost when cooking vegetables?
3. Why are some vegetables parboiled before deep-frying?
4. Name two ways of finishing vegetables. What characteristics does the finished dish have?

ELEMENT 3: Preparing and cooking rice dishes

Type and quality of rice

Long grain

This has had the hull, bran and most of the germ removed, so is less fibrous than brown rice. It is a narrow, pointed grain, best suited for savoury dishes and plain boiled rice because of its firm structure, which helps to keep the rice grains separate. It tends to be dry, flaky and separates easily when cooked.

Basmati rice

This type of rice is grown in the foothills of the Himalayas. It is a narrow form of long grain rice which needs to be soaked before cooking. Its distinctive flavour makes it particularly suitable for serving with Indian dishes.

Short grain or Italian rice

This is a short, rounded grain, traditionally used for milk puddings and sweet dishes because of its soft texture, but as more variations become available, this distinction is no longer so relevant. Short grain rice may be used when cooking Risotto (see page 327).

Brown rice

This form of rice has had only the hull (outer covering) removed and contains a great deal of fibre. It may take longer to cook than long grain rice, depending on the variety used.

Essential knowledge

Rice may become contaminated through:
- use of unhygienic equipment, utensils or preparation areas
- inadequate attention to personal hygiene (where bacteria passes from unclean hands, open cuts or sores, the nose or mouth to the rice)
- incorrect storage temperatures (cooked rice should always be stored below 5 °C/41 °F)
- incorrect waste disposal procedures
- being left uncovered at any stage
- excessive handling.

Preparing rice dishes

Health, safety and hygiene

Note all points given in Units G1 and G2 of the Core Units book. When cooking rice on top of the stove, make sure that no pan handles are sticking out beyond the edge of the stove. Act carefully when carrying pots of boiling liquid and wipe up any spilt liquid immediately. Also make sure that you are fully aware of what to do in the case of an emergency, particularly in the case of burns or cuts.

Planning your time

Be aware of what dishes you will be preparing, so that you know what equipment you will need to use, which dishes need to be prepared first, cooking times and so on. Preparing a production time plan will help you to think about priorities. When cooking rice, check which type of rice you will be using; remember that brown rice often takes longer to cook than long grain white varieties.

Preparing rice for cooking

Thoroughly wash all rice in cold water before cooking. Place the required amount of rice into a large bowl then fill the bowl with cold water. Using a large spoon, stir the rice, then carefully pour out the cloudy water. Repeat the process until the water remains clear after stirring.

Essential knowledge

It is important to keep preparation and storage areas and equipment hygienic in order to:

- prevent the transfer of food poisoning bacteria to the rice
- prevent pest infestation in storage areas
- ensure that standards of cleanliness are maintained
- comply with statutory health and safety regulations.

Cooking rice

Always cook rice in a pan large enough to allow room for stirring during cooking. If boiled rice is sticky when cooked, rinse it with very hot water then drain thoroughly. Note that the absorption rates of rice vary, and length and speed of cooking time will also dictate proportions of stock to rice.

Boiled rice

For boiling rice, use lightly salted water, and measure by volume, using 10 parts of water to 1 part of rice. For 4 portions, use 200 g (8 oz) long-grain rice (brown or white).

Method

1. Using a large saucepan, bring 2 litres (4 pt) water to the boil (add salt if required), then sprinkle the washed rice into the water.
2. Stir while the water comes back to the boil.
3. Simmer the rice gently for 12–15 minutes (brown rice may take longer), stirring occasionally to prevent the grains from sticking together. When stirring, take care not to damage the rice grains.
4. If not required for immediate service, refresh under cold running water and store in a chill in cold water.
5. When required, drain and reheat in water that is close to boiling point for 1 minute. Drain, place on a tray or in a colander or sieve, cover with a clean cloth and place in a moderate oven to dry slightly. Serve separately in a vegetable dish.

Pilau/Pilaf

This is a braised rice dish, cooked mainly in the oven. Use long-grain rice, as short-grain is liable to lose its shape. As a general rule, measure the rice by volume, and use use 2 parts of stock to 1 part of rice. For 4 portions, use 50 g (2 oz) long grain rice.

Method

1. Melt 25 g (1 oz) butter in a sauteuse, add 25 g (1 oz) chopped onion or shallot and cook gently until translucent, i.e. for approximately 3 minutes.
2. Add the rice and continue to cook for a further 3 minutes. Stir frequently to avoid burning.
3. Add 90 ml (4 fl oz) white stock, season and bring to the boil.
4. Cover with buttered greaseproof paper and a lid, and then place in a hot oven for approximately 15 minutes (until the rice is fully cooked: i.e. once all stock has been absorbed).
5. When cooked, remove from the oven and loosen the grains using a fork; they should be separate and fluffy.
6. Mix in 25 g (1 oz) butter, check the seasoning and serve.

Risotto

This rice dish is cooked (stewed) on the top of the stove, and results in a more moist dish than Pilau rice (above), using more stock. For 4 portions, use 100 g (4 oz) short-grain or brown rice.

Method

1. Melt 50 g (2 oz) butter in a sauteuse, add 25 g (1 oz) chopped onion or shallot and cook gently until translucent, i.e. for approximately 3 minutes.
2. Add the rice and continue to cook for a further 3 minutes. Stir frequently to avoid burning.
3. Add 250 ml ($^1/_2$ pt) white stock, season and allow to simmer. Stir frequently adding more stock if necessary until the rice is cooked and all stock has been absorbed.
4. Mix in 25 g (1 oz) Parmesan cheese using a fork, check seasoning and serve immediately.

Additional garnishes may be added to risottos either while still raw (at the start of the cooking process) or cooked (at varying stages of the cooking process); e.g. peas, beans, saffron, truffle, cooked meats, etc.

Mixed fried rice/stir-fried

A fried rice dish consists of a combination of pre-cooked rice and other ingredients (usually cooked meat, vegetables and eggs) where cooking is completed by shallow-frying or stir-frying using a frying pan or wok.

To achieve a good end result, ensure that the boiled rice is cool or, preferably, cold. The cooling process allows much of the moisture in the rice to evaporate, leaving drier grains for the oil to coat more effectively, so preventing sticking.

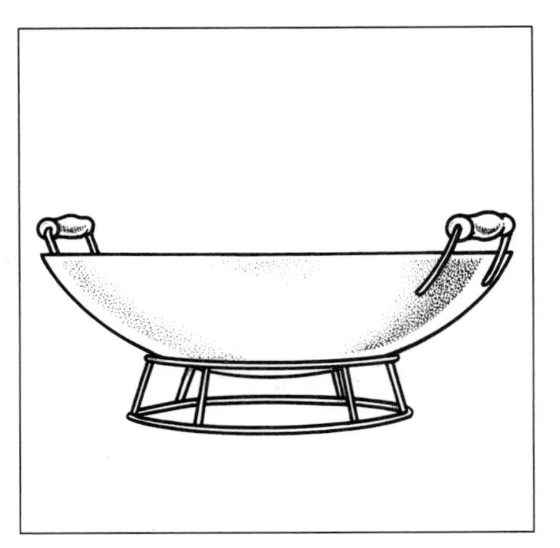

A wok

Vegetables suitable for stir-frying

Beansprouts
Bamboo shoots
Red/green/yellow peppers
Onions (especially spring onions)
Celery
Whole baby corn
Water chestnuts
Chinese greens
Cucumber
Carrots
Broccoli
Mange-tout
Spinach
Garlic

Method

1. Prepare and cook any meat or vegetables before beginning the frying process.
2. Heat the frying pan or wok until the pan is hot, then add the oil and allow it to heat until nearly smoking. (If the oil is not hot enough the rice will become saturated with oil and taste greasy.)
3. Add the cold rice and stir fry for 1 minute, then add any meat or vegetables. Continue stir frying for a further 5 minutes keeping the temperature high.

4 Add the beaten egg (and any delicate vegetables such as bean-sprouts) and fry for another 2 minutes. Do not cook the beaten egg prior to this stage, or it will become dry and hard.
5 Turn the mixture onto a service dish and serve immediately.

Steamed rice

Place the rice into a saucepan and add water until the water level is approximately 2.5 cm (1 in) above the top of the rice. Boil the rice, cooking it at a high heat until most of the water has evaporated. Then turn the heat down to a very low setting, cover the pan and allow the rice to complete cooking in the trapped steam.

Storing finished rice dishes

Cooked rice must always be stored at a temperature below 5 °C (41 °F) to prevent the growth of micro-organisms, particularly bacillus cereus. Always cover the rice and store it away from potential contaminants (such as raw foods).

What have you learned?

1 What are the main contamination threats when cooking and storing rice?
2 What are the four most commonly used types of rice?
3 How should you keep back cooked rice if not required for immediate service?
4 Why should you always wash rice before cooking it?
5 What are the main differences between pilau rice and risotto?

Extend your knowledge

1 Find out about vegetables not listed in the table on pages 307–9. What types of vegetable are they? How might you prepare and cook them?
2 What other ways of finishing vegetables are there? What vegetables are these methods suitable for, and how would you carry them out?
3 Find out what kinds of potatoes are available. Are they floury or waxy? What cooking methods would they suit best?
4 What other types of rice are available? What kinds of cooking processes and dishes might they be suitable for?

Glossary

accompaniments: items served with a particular dish, e.g. mint sauce with roast lamb

al dente: term used to describe lightly cooked pasta and vegetables, cooked firm to the bite

aspic: clear savoury jelly

au gratin: sprinkled with cheese or breadcrumbs then browned under a grill

bake blind: to cook a pastry case without a filling (beans are used to keep the shape)

bard: to cover the surface of meat, poultry or game with slices of fat to prevent the flesh drying out during cooking

barquette: boat-shaped short pastry case

baste: to coat a joint or bird with the roasting fat during cooking

bat out: to flatten out pieces of flesh with a cutlet bat (usually between polythene)

béchamel: white, roux-based sauce

beurre fondu: melted butter

beurre manié: mixture of butter and flour used to thicken sauces, soups, stews, etc.

blanch: to partly cook food by plunging into boiling water

blanquette: white stew cooked in stock from which the sauce is made

bouchée: bite-sized puff pastry case

bouillon: unclarified stock

bouquet garni: herbs bundled together in leek, celery or muslin; usually parsley stalks, thyme and bay leaf

browning: developing colour on the surface of food by exposing to high heat

canapé: hot or cold savoury consisting of pieces of food served on a base of bread, toast or biscuit

cartouche: name given to buttered greaseproof paper

chinois: fine meshed conical strainer

clarify: to make clear, e.g. consommés

coagulate: setting of a protein (e.g. egg)

contrefilet: piece of beef from a boned sirloin

court bouillon: liquor for cooking fish

crêpe: pancake

cross-contamination: transfer of bacteria from an infected source (e.g. raw food, dirty utensils) to previously uninfected food

croûte: toasted bread cut into required shape

croûtons: dice of crisp fried bread served with soup

darne: fish steak cut from a round fish, e.g. salmon

deglazing: the process of removing any caramelised meat juices from the bottom of the roasting tin

degraisser: skimming fat off a liquid

demi-glace: refined brown sauce made by reducing down an equal quantity of rich brown stock and espagnole

dice: to cut into cubes

disgorge: to soak meat in water in order to remove any visible blood

dishpaper: lining paper without perforations used for service dishes

docker: piece of equipment with spikes used for stabbing pastry before baking

duxelle: very finely chopped mushrooms cooked with shallots in a little oil or butter

eggwash: eggs whisked with a little water. Often brushed over food before baking (to develop colour)

emulsifier: agent used to preserve the texture of emulsions; e.g. egg yolk, gum arabic

emulsion: mixture of two liquids which would not normally mix together (e.g. oil and vinegar) and will only remain smooth and stable if bound with an emulsifier

entrecôte: steak cut from the contrefilet of beef

enzyme: special protein which breaks down other proteins

escalope: thin and flattened slice of meat, poultry or game

espagnole: basic brown sauce

estouffade: enriched brown meat stock

farce: forcemeat or stuffing

fermentation: stage in dough preparation when yeast produces carbon dioxide

filleting: removing the bones from fish to produce fillets

fines herbes: mixed herbs, eg. parsley, chives, chervil

flûtes: small circles of French bread served with soup

folding: blending a mixture by gently turning one part over the other using a spatula or spoon

fricassée: white stew made with fish, meat, poultry or vegetables cooked in a sauce

garnish: edible decoration used to enhance the appearance of a dish

gelatine: gum obtained from animal flesh used for setting jellies and bavarois. Not suitable for vegetarian dishes

glazing: giving food a glossy surface by coating with glaze, eggwash, etc.

gluten: vegetable protein found in wheat, rye, oats and barley

gnocchi: small savoury dumplings made with potato, choux paste or semolina

high risk foods: foods most at risk from bacterial infection

hors d'oeuvre: first dish to be served at a meal; hot or cold appetisers designed to stimulate the appetite

infra-red: type of light which is shone onto food to make it very hot

jardinière: vegetables cut into stick or baton shapes

julienne: thin strips

jus lié: roast gravy thickened with arrowroot

jus rôti: juice of the roast (roast gravy)

kneading: applying a gentle pressing and stretching action with the hand to dough, aerating the dough and making it smoother

knocking back: kneading dough to temporarily stop fermentation; the process is carried out twice during the preparation of the dough

lard: to insert strips of pork or bacon fat into raw flesh

liaison: mixture of cream and egg yolks, used to thicken sauces, soups or stews

mandoline: cutting implement used for slicing vegetables

marbling: fat found within the muscles of meat; desirable in beef as it helps to tenderise the cooking meat

marinade: liquor used to increase tenderness and add flavour to meat, game or poultry

mirepoix: roughly cut vegetables used for flavouring

noisette: cut of meat from a boned loin of lamb, mutton or venison

onion clouté: onion with a bay leaf studded by cloves; used for infusing flavour

pané: coated in breadcrumbs before cooking

pathogen: bacteria producing disease

paysanne: root vegetables cut into triangular, square or circle shapes

piquant: sharply flavoured

plat à sauter: shallow pan with straight sides used for frying

proving: placing dough in a warm environment to encourage fermentation, causing an increase in volume

pulses: edible seeds from pods of the legume family of plants

purée:
1 to use a food processor to very finely chop food into a paste
2 to rub cooked food through a sieve to form a paste
3 smooth mixture of cooked food obtained by passing the food through a sieve or liquidiser

quenelles: minced fish or meat mixed with cream and egg white, shaped by two spoons and lightly poached

reduction: concentration of flavour by evaporation

regeneration: reheating of chilled, frozen or processed food

rissoles: fried balls or cakes of prepared food

roux: mixture of fat and flour used to thicken sauces, soups, stews, etc.

rubber: dry cloth used for handling hot utensils

sabayon: light and aerated mixture of egg yolks and a little water cooked until it forms stable peaks

salamander: grill where food is placed under the heat source

satellite kitchen: kitchen where previously prepared or processed food is heated for service

sauteuse: shallow saucepan with sloping sides usually made from copper and lined with tin

score: to mark or cut the outer surface of food lightly before roasting

sear: to shallow-fry or oven-cook the surface of food until brown to develop colour and flavour

silicone paper: special type of paper which can be used when baking; has non-stick properties

soft flour: flour with a low protein content and weak glutens

soufflé: very light sweet or savoury dish served hot or cold

strong flour: flour with a high protein content and strong glutens

tartlette: round, short pastry case

tenderloin: alternative name for a fillet of beef

tomato concassée: chopped flesh of tomatoes with the skin and seeds removed

tronçon: fish steak cut from a flat fish, e.g. turbot

truss: to tie poultry and feathered game using a trussing needle and string before cooking

velouté: roux-based sauce made with white stock

vol-au-vent: puff pastry case

Index